Civil Aircraft Markings 1982

THIRTY-SECOND EDITION

Ala[illegible]**t**

LONDON

IAN ALLAN LTD

Introduction

The 'G' prefixed four letter registration system was adopted in 1919 after a short-lived spell with serial numbers beginning at K–100. Until 1928 the UK allocations were in the G–Exxx range, but as a result of further International agreements, this series was ended at G–EBZZ, the replacement becoming G–Axxx. From this point the registrations were issued in a reasonably orderly manner through to G–AZZZ. To avoid possible confusion with signal codes the G–AQxx sequence was omitted, while G–AUxx was reserved for Australian use originally. However, individual requests for marks in the latter batch have been granted by the Authorities. In July 1972 the next logical sequence was commenced at G–BAAA.

Unfortunately in recent years the strictly applied rules relating to aircraft registration have largely been relaxed, allowing widespread use of personalised marks upon application. Re-registrations have also become quite a common feature, a practice virtually unheard of in the past. Some aircraft have also been allowed to wear military markings without displaying their civil identity. In this volume the serial number actually carried in such cases is shown in parenthesis after the type's name. For example Gladiator G–AMRK flies as L8032 in RAF colours. As a further aid a military conversion list is provided.

Where re-registration has taken place at some time, the previous UK civil identity appears in parenthesis after the owner's/operator's name. An example of this practice is One-Eleven G–BBMG of British Airways which originally carried G–AWEJ.

Preserved aircraft which either fly infrequently or are purely static exhibits are shown with an asterisk in front of the type.

Any new registrations issued by the Civil Aviation Authority after this publication went to press will inevitably not be included until the next edition. To aid the recording of later marks logged during the year, grids have been provided at the end of the book.

The two-letter codes used by airlines to prefix flight numbers in timetables, airport movements boards, etc. are included for those carriers appearing in the book. Radio frequencies for the larger airfields/airports are also listed.

Acknowledgements

Once again thanks are extended to the Registration Department of the Civil Aviation Authority for continuing to allow regular researches into their files. The useful suggestions and information supplied by M. Drake, H. W. Gandy, H. W. Kohne, D. Lewis, P. R. March, B. R. Robinson, T. Patten, Stansted Aviation Society and the PR staffs of numerous airlines was gratefully received, while the valuable assistance of A. S. Wright and C. P. Wright must also be recorded.

AJW

Cover: Douglas DC-10-30 of Alitalia./*McDonnell Douglas*

This edition published 1982

ISBN 0 7110 1197 4

Published by Ian Allan Ltd, Shepperton, Surrey, and printed in the United Kingdom by Jarrold and Sons Ltd, Norwich

International Civil Aircraft Markings

A2–	Botswana
A6–	United Arab Emirates
A7–	Qatar
A9C–	Bahrain
A40–	Oman
AP–	Pakistan
B–	China/Taiwan
C–	Canada
C2–	Nauru
C5–	Gambia
C6–	Bahamas
C9–	Mozambique
CC–	Chile
CCCP–*	Soviet Union
CN–	Morocco
CP–	Bolivia
CR–	Portuguese Overseas Provinces
CS–	Portugal
CU–	Cuba
CX–	Uruguay
D–	German Federal Republic (West)
D2–	Angola
D6	Comores Islands
DDR–	German Democratic Republic (East)
DQ–	Fiji
EC–	Spain
EI, EJ–	Republic of Ireland
EL–	Liberia
EP–	Iran
ET–	Ethiopia
F–	France, Colonies and Protectorates
G–	United Kingdom
H4–	Solomon Islands
HA–	Hungarian People's Republic
HB–	Switzerland and Liechtenstein
HC–	Ecuador
HH–	Haiti
HK–	Colombia
HL–	Korea (South)
HP–	Panama
HR–	Honduras
HS–	Thailand
HZ–	Saudi Arabia
I–	Italy
J2–	Djibouti
J5–	Guinea Bissau
J6–	St. Lucia
J7	Dominican Republic
JA–	Japan
JY–	Jordan
LN–	Norway
LQ–, LV–	Argentine Republic
LX–	Luxemburg
LZ–	Bulgaria
N–	United States of America
OB–	Peru
OD–	Lebanon
OE–	Austria
OH–	Finland
OK–	Czechoslovakia
OO–	Belgium
OY–	Denmark
P–	Korea (North)
P2–	Papua New Guinea
PH–	Netherlands
PJ–	Netherlands Antilles
PK–	Indonesia and West Irian
PP–, PT–	Brazil
PZ–	Surinam
RDPL–	Laos
RP–	Philippine Republic
S2–	Bangladesh
S7–	Seychelles
S9–	São Tomé
SE–	Sweden
SP–	Poland
ST–	Sudan
SU–	Egypt
SX–	Greece
TC–	Turkey
TF–	Iceland
TG–	Guatemala
TI–	Costa Rica
TJ–	United Republic of Cameroon
TL–	Central African Republic
TN–	Republic of Congo (Brazzaville)
TR–	Gabon
TS–	Tunisia
TT–	Chad
TU–	Ivory Coast
TY–	Benin
TZ–	Mali
VH–	Australia
VP–F	Falkland Islands
VP–H	Belize
VP–LAA/ LJZ	Antigua
VP–LKA/ LLZ	St. Kitts-Nevis
VP–LMA/ LUZ	Montserrat
VP–LVA/ LZZ	Virgin Islands
VP–P	Kiribati
VP–V	St. Vincent
VP–W, VP–Y	Zimbabwe/Rhodesia
VQ–G	Grenada
VQ–H	St. Helena
VR–B	Bermuda
VR–C	Cayman Islands
VR–G	Gibraltar (not used: present Gibraltar Airways aircraft registered G–)
VR–H	Hong Kong
VR–U	Brunei
VT–	India
XA–, XB–, XC–	Mexico
XT–	Upper Volta
XU–	Kampuchea (formerly Khmer Republic)
XV–	Vietnam
XY–, XZ–	Burma
YA–	Afghanistan
YI–	Iraq
YK–	Syria
YN–	Nicaragua
YR–	Romania
YS–	El Salvador
YU–	Yugoslavia
YV–	Venezuela
ZA–	Albania
ZK–, ZL–, ZM–	New Zealand
ZP–	Paraguay
ZS–, ZT–, ZU–	South Africa
3A–	Monaco
3B–	Mauritius
3C–	Equatorial Guinea
3D–	Swaziland
3X–	Guinea
4R–	Sri Lanka
4W–	Yemen Arab Republic
4X–	Israel

* Cyrillic letters for SSSR.

5A–	Libya
5B–	Cyprus
5C–	Equatorial Guinea
5H–	Tanzania
5N–	Nigeria
5R–	Malagasy Republic (Madagascar)
5T–	Mauritania
5U–	Niger
5V–	Togo
5W–	Western Samoa (Polynesia)
5X–	Uganda
5Y–	Kenya
6O–	Somalia
6V–, 6W–	Senegal
6Y–	Jamaica
7O–	Democratic Yemen
7P–	Lesotho
7Q–	Malawi
7T–	Algeria
8P–	Barbados
8Q–	Maldives
8R–	Guyana
9G–	Ghana
9H–	Malta
9J	Zambia
9K–	Kuwait
9L–	Sierra Leone
9M–	Malaysia
9N–	Nepal
9Q–	Zaïre
9U–	Burundi
9V–	Singapore
9XR–	Ruanda
9Y–	Trinidad and Tobago

Aircraft Type Designations

(e.g. PA–28 Piper Type 28)

A.	Beagle, Auster
AA–	American Aviation, Grumman American
AB	Agusta-Bell
AS	Aerospatiale
A.S.	Airspeed
A.W.	Armstrong Whitworth
B.	Blackburn, Bristol, Boeing, Beagle
BAC	British Aircraft Corporation
BAe	British Aerospace
BN	Britten-Norman
Bo	Bolkow
Bu	Bucker
C.H.	Chrislea
CLA	Comper
CP.	Piel
D	Druine
DC–	Douglas Commercial
D.H.	de Havilland
D.H.C.	de Havilland Canada
DR.	Jodel (Robin-built)
EP	Edgar Percival
F.	Fairchild, Fokker
G.	Grumman
GA	Gulfstream American
G.A.L.	General Aircraft
G.C.	Globe
GY	Gardan
H	Helio
HM.	Henri Mignet
HP.	Handley Page
HR.	Robin
H.S.	Hawker Siddeley
IL	Ilyushin
J.	Auster
L.	Lockheed
L.A.	Luton
M.	Miles, Mooney
MBB	Messerschmitt-Bölkow-Blohm
M.S.	Morane-Saulnier
P.	Hunting (formerly Percival), Piaggio
PA–	Piper
PC.	Pilatus
R.	Rockwell
S.	Short, Sikorsky
SA., SE, SO.	Sud-Aviation, Aérospatiale, Scottish Aviation
S.R.	Saunders-Roe, Stinson
ST	SOCATA
T.	Tipsy
Tu	Tupolev
UH.	United Helicopters (Hiller)
V.	Vickers- Armstrongs, BAC
V.S.	Vickers-Supermarine
W. S.	Westland
Z.	Zlin

BRITISH CIVIL AIRCRAFT REGISTRATIONS

Reg.	Type	Owner or Operator	Notes
G-EACN	*BAT BK23 Bantam (K123)	Shuttleworth Trust	
G-EAVX	Sopwith Pup (B1807)	K. A. M. Baker	
G-EBHX	*D.H.53 Humming Bird	Shuttleworth Trust	
G-EBIA	*S.E.5A (F904)	Shuttleworth Trust	
G-EBIB	*S.E.5A (F939)	Science Museum	
G-EBIC	*S.E.5A (F938)	RAF Museum	
G-EBIR	*D.H.51	Shuttleworth Trust *Miss Kenya*	
G-EBJG	*Parnall Pixie III	Midland Aircraft Preservation Soc.	
G-EBJO	*ANEC II	Shuttleworth Trust	
G-EBKN	*Avro 504K (E449)	RAF Museum	
G-EBKY	*Sopwith Pup (N5180)	Shuttleworth Trust	
G-EBLV	D.H.60 Cirrus Moth	British Aerospace	
G-EBMB	*Hawker Cygnet I	RAF Museum	
G-EBNV	*English Electric Wren	Shuttleworth Trust	
G-EBQP	*D.H.53 Humming Bird	Russavia Collection	
G-EBWD	D.H.60X Hermes Moth	Shuttleworth Trust	
G-EBYY	*Cierva C.8L	Musée de l'Air, Paris	
G-EBZM	*Avro 594 Avian IIIA	Torbay Aircraft Museum	
G-AAAH	*D.H.60G Gipsy Moth	Science Museum	
G-AACN	*H.P.39 Gugnunc	Science Museum, Wroughton	
G-AAHW	Klemm L.25-1A	Anvil Aviation Ltd.	
G-AAIN	*Parnall Elf II	Shuttleworth Trust	
G-AAMY	D.H.60M Moth	C. C. & Mrs. J. M. Lovell	
G-AANG	*Blériot Monoplane	Shuttleworth Trust	
G-AANH	*Deperdussin Monoplane	Shuttleworth Trust	
G-AANI	*Blackburn Monoplane	Shuttleworth Trust	
G-AANJ	*L.V.G.-C VI (7198/19)	Shuttleworth Trust	
G-AAPZ	*Desoutter I (mod.)	Shuttleworth Trust	
G-AARO	Arrow A2-60 Sport	P. A. Mann	
G-AAUP	Klemm L.25-1A	R. S. Russell	
G-AAWO	D.H.60G Gipsy Moth	J. F. W. Reid	
G-AAYX	*Southern Martlet	Shuttleworth Trust	
G-AAZP	D.H.80A Puss Moth	G. R. Mollison	
G-ABAA	*Avro 504K (H2311)	RAF Museum	
G-ABAG	D.H.60G Moth	Shuttleworth Trust	
G-ABDW	D.H.80A Puss Moth	Museum of Flight, E. Fortune	
G-ABEE	*Avro 594 Avian IVM (Sports)	Aeroplane Collection Ltd.	
G-ABEV	D.H.60G Moth	R. I. & Mrs. J. O. Souch	
G-ABEW	D.H.60G-III Moth Major	B. P. Gardner	
G-ABLM	*Cierva C.24	Mosquito Aircraft Museum	
G-ABLS	D.H.80A Puss Moth	R. C. F. Bailey	
G-ABMR	*Hart 2 (J9941)	RAF Museum	
G-ABNT	Civilian Coupe	Shipping & Airlines Ltd.	
G-ABNX	Redwing 2	J. Pothecary	
G-ABOI	Wheeler Slymph	Midland Air Museum	
G-ABTC	CLA.7 Swift	P. Channon	
G-ABUS	CLA.7 Swift	R. C. F. Bailey	
G-ABUU	CLA.7 Swift	J. Pothecary	
G-ABVE	Arrow Active 2	J. D. Penrose	
G-ABWP	Spartan Arrow	R. E. Blain	
G-ABXL	*Granger Archaeopteryx	Shuttleworth Trust	
G-ABYA	D.H.60G Gipsy Moth	Dr. I. D. C. Hay & J. F. Moore	
G-ABZB	D.H.60G-III Moth	R. E. & B. A. Ogden	
G-ACCB	*D.H.83 Fox Moth	Midland Aircraft Preservation Soc.	
G-ACDC	D.H.82A Tiger Moth	Tiger Club Ltd.	
G-ACDJ	D.H.82A Tiger Moth	F. J. Terry	
G-ACEJ	D.H.83 Fox Moth	A. Haig-Thomas	
G-ACGT	*Avro 594 Avian IIIA	K. Smith	
G-ACIT	D.H.84 Dragon	J. Beaty (Historic Aircraft Museum)	
G-ACLL	D.H.85 Leopard Moth	H.L.S. Developments Ltd.	
G-ACMA	D.H.85 Leopard Moth	S. J. Filhol	
G-ACMN	D.H.85 Leopard Moth	H. D. Labouchere	
G-ACNB	*Avro 504K	Shuttleworth Trust	
G-ACSP	*D.H.88 Comet	Veteran & Vintage Aircraft (Engineering) Ltd.	

Notes	Reg.	Type	Owner or Operator
	G–ACSS	*D.H.88 Comet	Shuttleworth Trust
	G–ACTF	CLA.7 Swift	A. J. Chalkley
	G–ACUS	D.H.85 Leopard Moth	C. W. Annis
	G–ACUU	*Cierva C.30A	G. S. Baker (Skyfame Collection)
	G–ACUX	*S.16 Scion	Ulster Folk & Transport Museum
	G–ACVA	*Kay Gyroplane	Glasgow Museum of Transport
	G–ACWP	*Cierva C.30A (AP507)	Science Museum
	G–ACXE	B.K. L-25C Swallow	D. G. Ellis
	G–ADAH	*D.H.89A Dragon Rapide	Museum of Flight, E. Fortune
	G–ADEV	*Avro 504K (E3404)	Shuttleworth Trust
	G–ADFV	*Blackburn B-2	Humberside Aircraft Preservation Soc.
	G–ADGP	M.2L Hawk Speed Six	A. M. Stow
	G–ADGT	D.H.82A Tiger Moth	D. R. & Mrs. M. Wood
	G–ADGV	D.H.82A Tiger Moth	Air Cdre. A. H. Wheeler
	G–ADIA	D.H.82A Tiger Moth	J. Beaty
	G–ADJJ	D.H.82A Tiger Moth	J. M. Preston
	G–ADKC	D.H.87B Hornet Moth	E. J. Roe
	G–ADKK	D.H.87B Hornet Moth	C. W. Annis
	G–ADKM	D.H.87B Hornet Moth	F. R. E. Hayter
	G–ADLY	D.H.87B Hornet Moth	A. Wood
	G–ADMT	D.H.87B Hornet Moth	Scottish Aircraft Collection, Perth
	G–ADMW	*M.2H Hawk Major (DG590)	RAF Museum
	G–ADND	D.H.87B Hornet Moth	Shuttleworth Trust
	G–ADNE	D.H.87B Hornet Moth	P. A. Mann
	G–ADNZ	*D.H.82A Tiger Moth	R. W. & Mrs. S. Pullan
	G–ADOT	*D.H.87B Hornet Moth	Mosquito Aircraft Museum
	G–ADPJ	B.A.C. Drone	G. A. R. Eastell
	G–ADPR	*P.3 Gull	Shuttleworth Trust *Jean*
	G–ADPS	Swallow 2	Strathallan Aircraft Collection
	G–ADRA	Pietenpol Aircamper	A. J. Mason & R. J. Barrett
	G–ADRC	K. & S. Jungster J-1	J. J. Penney & L. R. Williams
	G–ADRY	*Pou-du-Ciel (Replica) (BAPC 29)	P. Roberts
	G–ADUR	D.H.87B Hornet Moth	R. C. Lenton
	G–ADXS	*Pou-du-Ciel	Historic Aircraft Museum, Southend
	G–ADXT	D.H.82A Tiger Moth	J. & J. M. Pothecary
	G–ADYS	Aeronca C.3	J. Willmot
	G–AEBB	*Pou-du-Ciel	Shuttleworth Trust
	G–AEBJ	Blackburn B-2	British Aerospace
	G–AEDB	Kronfield Drone	M. C. Russell
	G–AEDU	D.H.90 Dragonfly	A. Haig-Thomas & M. Barraclough
	G–AEEG	M.3A Falcon	Shipping & Airlines Ltd.
	G–AEEH	*Pou-du-Ciel	RAF St. Athan
	G–AEFT	Aeronca C-3	C. E. Humphreys & ptnrs.
	G–AEGV	*HM.14 Pou-du-Ciel	Midland Aircraft Preservation Soc.
	G–AEHM	*Pou-du-Ciel	Science Museum
	G–AEKR	*Flying Flea (Replica) (BAPC 121)	Nostell Aviation Museum
	G–AEKV	Kronfield Drone	Wg. Cdr. J. E. McDonald
	G–AELO	D.H.87B Hornet Moth	S. N. Bostock
	G–AEML	D.H.89 Dragon Rapide	J. P. Filhol Ltd.
	G–AENP	*Hawker Hind	Shuttleworth Trust
	G–AEOA	D.H.80A Puss Moth	A. Haig-Thomas (Shuttleworth)
	G–AEOE	*HM.14 Pou-du-Ciel (BAPC 22)	Newark Air Museum
	G–AEOH	*HM.14 Pou-du-Ciel	Midland Air Museum
	G–AEPH	*Bristol F.2B (D8096)	Shuttleworth Trust
	G–AERD	P.3 Gull Six	C. C. & Mrs. J. M. Lovell
	G–AERV	M.11A Whitney Straight	Ulster Folk & Transport Museum
	G–AESE	D.H.87B Hornet Moth	H. J. Shaw
	G–AETA	*Caudron G.3 (3066)	RAF Museum
	G–AEUJ	M.11A Whitney Straight	R. E. Mitchell
	G–AEVS	Aeronca 100	R. & M. Nerou
	G–AEVZ	*Swallow 2	Museum of Flight, E. Fortune
	G–AEXD	Aeronca 100	Mrs. M. A. & R. W. Mills
	G–AEXF	P.6 Mew Gull	M. C. Barraclough & T. M. Storey
	G–AEXT	Dart Kitten II	R. E. Nerou
	G–AEXZ	Piper J-2 Cub	Mrs. M. & J. R. Dowson
	G–AEYY	*Martin Monoplane	Martin Monoplane Syndicate

Reg.	Type	Owner or Operator	Notes
G–AFBS	*M14A Hawk Trainer	Skyfame Collection (G–AKKU)	
G–AFCL	B. A. Swallow 2	A. M. Dowson	
G–AFDX	*Hanriot HD.1 (75)	RAF Museum	
G–AFFD	*Percival Q-6	K. Gomm	
G–AFFI	*Pou-du-Ciel (BAPC 76)	Nostell Aviation Museum	
G–AFGC	B. A. Swallow 2	H. Plain	
G–AFGD	B. A. Swallow 2	A. T. Williams & ptnrs.	
G–AFGE	B. A. Swallow 2	Donald G. Ellis	
G–AFGH	Chilton D.W.I.	M. L. & G. L. Joseph	
G–AFGI	*Chilton D.W.I.	J. E. McDonald	
G–AFHA	*Mosscraft M.A.I.	C. V. Butler	
G–AFIN	*Chrislea Airguard	Aeroplane Collection Ltd.	
G–AFIU	*Luton Minor (Parker C.A.4)	S. P. Connatty & I. V. Jones (N.A.P.S.)	
G–AFJA	Watkinson Dingbat	K. Woolley	
G–AFJB	Foster-Wickner G.M.I. Wicko	K. Woolley	
G–AFJR	Tipsy Trainer I	M. E. Vaisey	
G–AFJV	Mosscraft MA.2	C. V. Butler	
G–AFLW	M.17 Monarch	N. Jensen	
G–AFNG	D.H.94 Moth Minor	A. Haig-Thomas (Shuttleworth)	
G–AFNI	D.H.94 Moth Minor	B. M. Welford	
G–AFOB	D.H.94 Moth Minor	R. E. Ogden	
G–AFOH	Cessna 145	R. I. & Mrs. J. O. Souch	
G–AFOJ	D.H.94 Moth Minor	R. M. Long	
G–AFPN	D.H.94 Moth Minor	—	
G–AFRV	Tipsy Trainer I	Capt. R. C. F. Bailey	
G–AFSC	Tipsy Trainer I	G. P. Hermer & ptnrs.	
G–AFSV	Chilton D.W.IA	R. Nerou	
G–AFTA	*Hawker Tomtit (K1786)	Shuttleworth Trust	
G–AFTN	*Taylorcraft Plus C2	Leicestershire County Council Museums	
G–AFVE	D.H.82 Tiger Moth	Swinstead Aviation Ltd.	
G–AFVN	Tipsy Trainer I	W. Callow & ptnrs.	
G–AFWI	D.H.82A Tiger Moth (BB814)	Portsmouth Naval Gliding Club	
G–AFWT	Tipsy Trainer I	H. Scanlon	
G–AFYD	Luscombe 8E Silvaire	J. D. Iliffe	
G–AFYO	Stinson H.W.75	B. A. Dunlop & P. Elliott	
G–AFZE	Heath Parasol	K. C. D. St. Cyrien	
G–AGBN	G.A.L.42 Cygnet 2	Museum of Flight, E. Fortune	
G–AGJG	D.H.89A Dragon Rapide	Aerial Enterprises Ltd.	
G–AGLK	Auster 5D	W. C. E. Tazewell	
G–AGNV	*Avro 685 York I (MW100)	Aerospace Museum, Cosford	
G–AGOH	J/I Autocrat	Leicester Museum of Technology	
G–AGOS	R.S.3 Desford I (VZ728)	Scottish Aircraft Collection, Perth	
G–AGOY	M.48 Messenger 3	T. Clark	
G–AGPG	*Avro 19 Srs. I	Historic Aircraft Museum, Southend	
G–AGRU	*V.498 Viking IA	Aerospace Museum, Cosford	
G–AGSH	D.H.89A Dragon Rapide 6	Pioneer Aviation Ltd.	
G–AGTM	D.H.89A Dragon Rapide 6 (NF875)	Airborne Taxi Services Ltd.	
G–AGTO	J/I Autocrat	M. J. Barnett & D. J. T. Miller	
G–AGTT	J/I Autocrat	D. W. Prince & R. K. J. Hadlow	
G–AGVG	J/I Autocrat	T. F. McDonald	
G–AGVN	J/I Autocrat	P. J. Elliott	
G–AGVV	Piper L-4H Cub	A. R. W. Taylor & D. Lofts	
G–AGWE	Avro 19 Srs. 2	—	
G–AGXN	J/IN Alpha	P. A. Davey	
G–AGXT	*J/IN Alpha	The Aeroplane Collection	
G–AGXU	J/IN Alpha	Mrs. J. Lewis	
G–AGXV	J/I Autocrat	P. G. & P. G. Morton	
G–AGYD	J/IN Alpha	F. K. Smith & P. Herring	
G–AGYH	J/IN Alpha	G. E. Twyman & P. J. Rae	
G–AGYK	J/I Autocrat	R. W. & Mrs. N. M. Biggs	
G–AGYT	J/IN Alpha	J. C. Braybow & I. R. F. Hammond	
G–AGYU	D.H.82A Tiger Moth (DE208)	A. Wood	
G–AHAL	J/IN Alpha	Skegness Air Taxi Services Ltd.	
G–AHAM	J/I Autocrat	D. W. Philp	
G–AHAU	J/I Autocrat	B. J. W. Foley	
G–AHAV	J/I Autocrat	R. H. G. Kingsmill	
G–AHAY	J/I Autocrat	Aviation Advisory Services Ltd.	
G–AHBL	D.H.87B Hornet Moth	Dr. Ursula H. Hamilton	
G–AHBM	D.H.87B Hornet Moth	P. Franklin	

Notes	Reg.	Type	Owner or Operator
	G–AHCK	J/1N Alpha	P. A. Woodman
	G–AHCL	J/1N Alpha	R. Willis & ptnrs.
	G–AHCN	J/1N Alpha	FTS Flying Group
	G–AHCR	Gould-Taylorcraft Plus D Special	D. E. H. Balmford & D. R. Shepherd
	G–AHED	*D.H.89A Dragon Rapide (RL962)	RAF Museum (Cardington)
	G–AHGD	D.H.89A Dragon Rapide	M. R. L. Astor *Women of the Empire*
	G–AHGW	Taylorcraft Plus D	C. V. Butler
	G–AHGZ	Taylorcraft Plus D	S. J. Ball
	G–AHHH	J/1 Autocrat	D. Bridge
	G–AHHK	J/1 Autocrat	W. J. Ogle
	G–AHHN	J/1 Autocrat	KK Aviation
	G–AHHT	J/1N Alpha	P. M. A. Parrett
	G–AHIC	*Avro 19 Srs. 2	Strathallan Aircraft Collection
	G–AHIZ	D.H.82A Tiger Moth	C.F.G. Flying Ltd.
	G–AHKX	Avro 19 Srs. 2	BAe (Manchester) Anson Preservation Group
	G–AHKY	*Miles M.18 Series 2	Scottish Aircraft Collection, Perth
	G–AHLI	Auster 3	G. A. Leathers
	G–AHLK	Auster 3	A. A. Bucknell
	G–AHLT	D.H.82A Tiger Moth	R. C. F. Bailey
	G–AHMJ	*Cierva C.30A (K4235)	Shuttleworth Trust
	G–AHMN	D.H.82A Tiger Moth (N6985)	George House (Holdings) Ltd.
	G–AHNG	Taylorcraft Plus D	D. J. Pratt
	G–AHRI	*D.H.104 Dove 1	Lincolnshire Aviation Museum
	G–AHSA	*Avro 621 Tutor (K3215)	Shuttleworth Trust
	G–AHSD	Taylorcraft Plus D	A. Tucker
	G–AHSO	J/1N Alpha	Skegness Air Taxi Services Ltd.
	G–AHSP	J/1 Autocrat	F. A. L. Castleden & ptnrs.
	G–AHSS	J/1N Alpha	D. B. Connelly
	G–AHST	J/1N Alpha	P. H. Lewis
	G–AHSW	J/1 Autocrat	K. W. Brown
	G–AHTE	*P.44 Proctor V	S. Wales Historical Aircraft Preservation Soc.
	G–AHTW	*A.S.40 Oxford (V3388)	Skyfame Collection
	G–AHUI	*M.38 Messenger 2A	Humberside Aircraft Preservation Soc.
	G–AHUJ	M.14A Hawk Trainer 3 (R1914)	Vintage Aircraft Team
	G–AHUV	D.H.82A Tiger Moth	W. G. Gordon
	G–AHVV	D.H.82A Tiger Moth	R. Jones
	G–AHWJ	Taylorcraft Plus D	A. Tucker
	G–AHXE	Taylorcraft Plus D (LB312)	Mrs. J. M. C. Pothecary
	G–AIBE	*Fulmar II (N1854)	F.A.A. Museum
	G–AIBH	J/1N Alpha	Skegness Air Taxi Services Ltd.
	G–AIBM	J/1 Autocrat	J. L. Goodley
	G–AIBW	J/1N Alpha	J. L. Way
	G–AIBX	J/1 Autocrat	Wasp Flying Group
	G–AIBY	J/1 Autocrat	D. Morris
	G–AIDL	D.H.89A Dragon Rapide 6	Southern Joyrides Ltd.
	G–AIDN	V.S.502 Spitfire T.8	G. F. Miller
	G–AIDS	D.H.82A Tiger Moth	BSP Electric & Maintenance Co. Ltd.
	G–AIEK	M.38 Messenger 2A (RG333)	J. Buckingham
	G–AIFZ	J/1N Alpha	C. P. Humphries
	G–AIGD	J/1 Autocrat	A. G. Batchelor
	G–AIGF	J/1N Alpha	A. R. C. Mathie
	G–AIGM	J/1N Alpha	S. T. Raby
	G–AIGR	J/1N Alpha	Sywell Aviation Services Ltd.
	G–AIGU	J/1N Alpha	T. Pate
	G–AIIH	Piper L-4H Cub	J. W. Benson
	G–AIIZ	D.H.82A Tiger Moth	D. E. Baker
	G–AIJI	*J/1N Alpha	Humberside Aircraft Preservation Soc.
	G–AIJM	Auster J/4	A. B. Graham
	G–AIJR	Auster J/4	B. A. Harris
	G–AIJT	Auster J/4	Aberdeen Flying Group
	G–AILL	M. 38 Messenger 2A	H. Best-Devereux
	G–AIPR	Auster J/4	R. W. & M. A. Mills
	G–AIPV	J/1 Autocrat	J. Linegar
	G–AIPW	J/1 Autocrat	J. Buckingham
	G–AIRC	J/1 Autocrat	A. G. Martlew
	G–AIRI	D.H.82A Tiger Moth	E. R. Goodwin

Reg.	*Type*	*Owner or Operator*	*Notes*
G–AIRK	D.H.82A Tiger Moth	R. C. Teverson & ptnrs.	
G–AISA	Tipsy B Srs. I	B. T. Morgan & A. Liddiard	
G–AISB	Tipsy B Srs. I	P. E. Barker	
G–AISC	Tipsy B Srs. I	Wagtail Flying Group	
G–AIST	V.S.300 Spitfire I (AR213)	The Hon. P. Lindsay	
G–AISU	V.S. Spitfire 5B (AB910)	RAF Battle of Britain Historic Aircraft Flight	
G–AITB	*A.S.10 Oxford (MP425)	RAF Museum	
G–AITF	*A.S.40 Oxford	RAF Museum, Henlow	
G–AIUA	M.14A Hawk Trainer 3	Shuttleworth Trust	
G–AIUL	D.H.89A Dragon Rapide 6	A. F. Ward	
G–AIVW	D.H.82A Tiger Moth Seaplane	Tiger Club Ltd.	
G–AIWA	P.28B Proctor I (R7524)	N. C. Jensen	
G–AIXA	Taylorcraft Plus D	P. J. Anderson	
G–AIXD	D.H.82A Tiger Moth	D. L. Lloyd	
G–AIXN	Benes-Mraz M.Ic Sokol	J. F. Evetts & D. Patel	
G–AIYR	D.H.89A Dragon Rapide	C. D. Cyster & ptnrs.	
G–AIYS	D.H.85 Leopard Moth	M. V. Gauntlett	
G–AIZE	*F.24W Argus 2	RAF Museum	
G–AIZG	*V.S. Walrus (L2301)	Historic Aircraft Preservation Soc./ Royal Navy	
G–AIZU	J/I Autocrat	T. A. Collins & A. H. R. Stansfield	
G–AIZY	J/I Autocrat	B. J. Richards	
G–AIZZ	J/I Autocrat	Peter Long International Ltd.	
G–AJAB	J/IN Alpha	A. E. Poulson	
G–AJAC	J/IN Alpha	R. C. Hibberd	
G–AJAE	J/IN Alpha	M. G. Stops	
G–AJAJ	J/IN Alpha	A. R. Milne & T. F. L. Bessant	
G–AJAM	J/2 Arrow	D. A. Porter	
G–AJAS	J/IN Alpha	C. J. Baker	
G–AJCP	D.31 Turbulent	H. J. Shaw	
G–AJDW	J/I Autocrat	D. R. Hunt	
G–AJEB	*J/IN Alpha	Aeroplane Collection Ltd.	
G–AJEE	J/I Autocrat	F. A. Arnold	
G–AJEH	J/IN Alpha	Astral Surveys Ltd.	
G–AJEI	J/IN Alpha	Skegness Air Taxi Services Ltd.	
G–AJEM	J/I Autocrat	H. A. Nind	
G–AJGJ	Auster 5	J. J. McLaughlin & ptnrs.	
G–AJHO	D.H.89A Dragon Rapide	East Anglian Aviation Soc. Ltd.	
G–AJHS	D.H.82A Tiger Moth	W. G. Fisher	
G–AJHU	D.H.82A Tiger Moth	F. P. Le Coyte	
G–AJID	J/I Autocrat	L. W. Usherwood	
G–AJIH	J/I Autocrat	D. F. Campbell & ptnrs.	
G–AJIS	J/IN Alpha	A. Tucker	
G–AJIU	J/I Autocrat	P. D. Webb	
G–AJIW	J/IN Alpha	B. M. Baker (Hatfield) Ltd.	
G–AJJP	*Jet Gyrodyne (XJ389)	Aerospace Museum, Cosford	
G–AJKK	M.38 Messenger 2A	A. M. Lambourne & T. C. Eaves	
G–AJOA	D.H.82A Tiger Moth (T5424)	F. P. Le Coyte	
G–AJOC	M.38 Messenger 2A	Ulster Folk & Transport Museum	
G–AJOE	M.38 Messenger 2A (RH378)	Aircraft Mart	
G–AJOV	*Sikorsky S-51	Aerospace Museum, Cosford	
G–AJOZ	*F.24W Argus 2	Lincolnshire Aviation Museum	
G–AJPI	F.24W-41a Argus 2 (EV851)	Leisure Sport Ltd.	
G–AJPZ	J/I Autocrat	K. Pyle	
G–AJRB	J/I Autocrat	T. Booth	
G–AJRC	J/I Autocrat	S. W. Watkins & ptnrs.	
G–AJRE	J/I Autocrat	D. J. Ronayne	
G–AJRH	J/IN Alpha	N. H. Ponsford	
G–AJRS	*M.14A Hawk Trainer 3 (P6382)	Shuttleworth Trust	
G–AJTH	M.65 Gemini IA	K. A. Learmonth	
G–AJUD	J/I Autocrat	J. A. Hughes & K. M. Bowen	
G–AJUE	J/I Autocrat	M. A. G. Westman	
G–AJUL	J/IN Alpha	M. J. Crees	
G–AJVE	D.H.82A Tiger Moth	M. J. Abbott & I. J. Jones	
G–AJVH	Swordfish (LS326)	F.A.A. Museum	
G–AJVT	Auster 5	I. N. M. Cameron	
G–AJWB	M.38 Messenger 2A	W. F. & M. R. Higgins	
G–AJXC	Auster 5	J. E. Graves	
G–AJXV	Auster 4 (NJ695)	P. C. J. Farries	
G–AJYO	J/5B Autocar	A. H. Chaplin & ptnrs.	

Notes	Reg.	Type	Owner or Operator
	G-AKAA	Piper L-4H Cub	J. W. Benson
	G-AKAT	*M.14A Magister (T9738)	Newark Air Museum
	G-AKBM	*M.38 Messenger 2A	Bristol Plane Preservation Unit
	G-AKBO	M.38 Messenger 2A	J. R. A. Ramshaw
	G-AKDN	D.H.C. 1A Chipmunk 10	K. R. Nunn
	G-AKEL	M.65 Gemini 1A	J. M. Bisco
	G-AKER	*M.65 Gemini 1A	Vintage Aircraft Team
	G-AKEZ	*M.38 Messenger 2A (RG333)	Wales Aircraft Museum
	G-AKGD	*M.65 Gemini 1A	D. G. Addicott
	G-AKGE	M.65 Gemini 3C	R. E. Winn
	G-AKHP	M.65 Gemini 1A	Shipping & Airlines Ltd.
	G-AKHW	M.65 Gemini 1A	Bee Gee Aviation
	G-AKHZ	*M.65 Gemini 7	Vintage Aircraft Team
	G-AKIF	D.H.89A Dragon Rapide	Airborne Taxi Services Ltd.
	G-AKIN	M.38 Messenger 2A	A. J. Spiller
	G-AKJU	J/1N Alpha	Arrow Air Services (Engineering) Ltd.
	G-AKKB	M.65 Gemini 1A	S.A.C. Bristol Ltd.
	G-AKKH	M.65 Gemini 1A	M. C. Russell
	G-AKKR	*M.14A Magister (T9707)	RAF Museum
	G-AKOE	D.H.89A Dragon Rapide 4	J. E. Pierce
	G-AKOW	Auster 5	A. M. L. McLean
	G-AKPF	*M.14A Hawk Trainer (N3788)	L. N. D. Taylor
	G-AKPI	Auster 5 (NJ703)	B. H. Hargrave
	G-AKSZ	Auster 5	A. R. C. Mathie
	G-AKUW	C.H.3 Super Ace	C. V. Butler
	G-AKVF	C.H.3 Super Ace	P. V. B. Longthorp
	G-AKVZ	M.38 Messenger 4B	Shipping & Airlines Ltd.
	G-AKWS	Auster 5	J. E. Homewood
	G-AKWT	*Auster 5 (MT360)	Humberside Aircraft Preservation Soc.
	G-AKXP	Auster 5	F. E. Telling
	G-AKXS	D.H.82A Tiger Moth	P. A. Colman
	G-AKZN	*P.30 Proctor 2E (Z7197)	RAF Museum
	G-ALAH	*M.38 Messenger 4A (RH377)	Aeroplane Collection Ltd.
	G-ALAX	*D.H.89A Dragon Rapide	Durney Aeronautical Collection
	G-ALBJ	Auster 5	R. H. Elkington
	G-ALBK	Auster 5	S. J. Wright & Co. (Farmers) Ltd.
	G-ALBN	*Bristol 173 (XF785)	RAF Museum
	G-ALCK	*Proctor (LZ766)	Skyfame Collection
	G-ALCS	M.65 Gemini 3C	R. E. Winn
	G-ALCU	*D.H.104 Dove 2	Midland Air Museum
	G-ALEH	PA-17 Vagabond	J. O. Souch
	G-ALFA	Auster 5	Alpha Flying Group
	G-ALFT	*D.H.104 Dove 6	Torbay Aircraft Museum
	G-ALFU	*D.H.104 Dove 6	Imperial War Museum
	G-ALGT	V.S.379 Spitfire 14 (RM619)	Rolls-Royce Ltd.
	G-ALIW	D.H.82A Tiger Moth	D. I. M. Geddes & F. Curry
	G-ALJF	P.34A Proctor 3	J. F. Moore
	G-ALNA	D.H.82A Tiger Moth	Wessex Flying Group
	G-ALND	D.H.82A Tiger Moth (N9191)	Arrow Air Services (Engineering) Ltd.
	G-ALSP	*Bristol 171 (WV783)	RAF Museum
	G-ALSS	*Bristol 171 (WA576)	—
	G-ALST	*Bristol 171 (WA577)	N.E. Aircraft Museum
	G-ALSW	*Bristol 171 (WT933)	Newark Air Museum
	G-ALSX	*Bristol 171 (G-48-1)	Rotorcraft Museum (Duxford)
	G-ALTW	D.H.82A Tiger Moth	C. J. Musk
	G-ALUC	D.H.82A Tiger Moth	D. R. & Mollie Wood
	G-ALWB	D.H.C1. Chipmunk 22A	K. C. Keegan & ptnrs.
	G-ALWC	Dakota 4	Clyde Surveys Ltd.
	G-ALWF	*V.701 Viscount	Viscount Preservation Trust
	G-ALWW	D.H.82A Tiger Moth	F. W. Fay & ptnrs.
	G-ALXT	*D.H.89A Dragon Rapide	Science Museum, Wroughton
	G-ALXZ	Auster 5-150	T. Bessart & R. Witheridge
	G-ALYB	Auster 5	G. F. Kilsby
	G-ALYG	Auster 5D	A. L. Young
	G-ALZE	*BN-1F	Aerospace Museum, Cosford
	G-ALZO	*A.S.57 Ambassador	Dan-Air Preservation Group
	G-AMAU	Hurricane 2C (PZ865)	RAF Battle of Britain Historic Aircraft Flight

Reg.	Type	Owner or Operator	Notes
G–AMAW	Luton L.A.4 Minor	J. R. Coates	
G–AMCA	Dakota 4	Air Atlantique Ltd.	
G–AMDA	*Avro 652A Anson I (N4877)	Skyfame Collection	
G–AMDN	Hiller UH-12A	Bristow Helicopters Ltd.	
G–AMHJ	Dakota 6	Air Atlantique Ltd.	
G–AMIU	D.H.82A Tiger Moth	R. & Mrs. J. L. Jones	
G–AMKU	J/1B Aiglet	Southdown Flying Group	
G–AMLZ	P.50 Prince 6E	J. F. Coggins	
G–AMMS	J/5K Aiglet Trainer.	G. White & Sons	
G–AMOG	*V.701 Viscount	Aerospace Museum, Cosford	
G–AMPO	Dakota 4	Air Atlantique Ltd.	
G–AMPP	*Dakota 3 (G–AMSU)	Dan-Air Preservation Group	
G–AMPW	J/5B Autocar	J. V. Inglis	
G–AMPY	Dakota 4	Air Atlantique Ltd.	
G–AMRA	Dakota 6	Air Atlantique Ltd.	
G–AMRF	J/5F Aiglet Trainer	A. I. Topps	
G–AMRK	G.37 Gladiator (L8032)	Shuttleworth Trust	
G–AMSZ	Auster 5	G. & D. Knight & D. G. Pridham	
G–AMTA	J/5F Aiglet Trainer	H. J. Jauncey	
G–AMTM	J/I Autocrat	R. Stobo & D. Clewley	
G–AMUF	D.H.C.I Chipmunk 21	Chipmunk Flying Group	
G–AMUH	D.H.C.I Chipmunk 21	D. H. Whalley	
G–AMVD	Auster 5	C. G. Clarke	
G–AMVP	Tipsy Junior	A. R. Wershat	
G–AMXA	*D.H.106 Comet C.2 (XK655)	Strathallan Aircraft Collection	
G–AMYD	J/5L Aiglet Trainer	G. H. Maskell	
G–AMYJ	Dakota 6	Air Atlantique Ltd.	
G–AMZI	J/5F Aiglet Trainer	J. F. Moore	
G–AMZT	J/5F Aiglet Trainer	D. Hyde & J. W. Saull	
G–AMZU	J/5F Aiglet Trainer	R. N. Goode & ptnrs.	
G–ANAF	Dakota 4	Air Atlantique Ltd.	
G–ANAP	*D.H.104 Dove 6	Brunel Technical College, Lulsgate	
G–ANCF	B.175 Britannia 308F	Invicta International Airlines Ltd.	
G–ANCS	D.H.82A Tiger Moth (R4907)	R. H. Reeves	
G–ANCX	D.H.82A Tiger Moth	D. R. Wood	
G–ANDE	D.H.82A Tiger Moth	Stapleford Tiger Group	
G–ANDM	D.H.82A Tiger Moth	J. Green	
G–ANDP	D.H.82A Tiger Moth	A. H. Diver	
G–ANEF	D.H.82A Tiger Moth (T5493)	RAF College Flying Club Co. Ltd.	
G–ANEL	D.H.82A Tiger Moth (N9238)	W. P. Maynall	
G–ANEW	D.H.82A Tiger Moth	A. L. Young	
G–ANEZ	D.H.82A Tiger Moth	J. W. Benson	
G–ANFC	*D.H.82A Tiger Moth	Mosquito Aircraft Museum	
G–ANFH	*Westland S.55	British Rotorcraft Museum	
G–ANFI	D.H.82A Tiger Moth	K. L. Backus & D. S. Kirkham	
G–ANFM	D.H.82A Tiger Moth	S. A. Brook & ptnrs.	
G–ANFP	*D.H.82A Tiger Moth	Mosquito Aircraft Museum	
G–ANFV	D.H.82A Tiger Moth (DF155)	R. A. L. Falconer	
G–ANFW	D.H.82A Tiger Moth	G. M. Fraser	
G–ANHK	D.H.82A Tiger Moth	J. D. Iliffe	
G–ANHR	Auster 5	J. O. C. Warner	
G–ANHS	Auster 4	P. S. Sturdgess	
G–ANHX	Auster 5D	D. J. Baker	
G–ANHZ	Auster 5	J. H. D. Newman	
G–ANIE	Auster 5	H. O. Stibbard	
G–ANIJ	Auster 5D	Army Aviation Museum	
G–ANIS	Auster 5	J. Clarke-Cockburn	
G–ANJA	D.H.82A Tiger Moth	J. J. Young	
G–ANJD	D.H.82A Tiger Moth	H. J. Jauncey	
G–ANJK	D.H.82A Tiger Moth	Montgomery Ultra Light Flying Club	
G–ANJV	*Westland S.55 Srs. 3	British Rotorcraft Museum	
G–ANKK	D.H.82A Tiger Moth (T5854)	Birmingham Tiger Group	
G–ANKT	D.H.82A Tiger Moth (T6818)	The Shuttleworth Collection	
G–ANLS	D.H.82A Tiger Moth	P. A. Gliddon	
G–ANLW	W.B.I. Widgeon (MD497)	Helicopter Hire Ltd.	
G–ANMV	D.H.82A Tiger Moth (T7404)	George House (Holdings) Ltd.	
G–ANMZ	D.H.82A Tiger Moth	The Tiger Club	
G–ANNK	D.H.82A Tiger Moth	Mrs. P. J. Wilcox	
G–ANOD	D.H.82A Tiger Moth	D. R. & Mrs. M. Wood	
G–ANOH	D.H.82A Tiger Moth	S. J. Carr	
G–ANOK	*S-91C Safir	Museum of Flight, E. Fortune	
G–ANOM	D.H.82A Tiger Moth	P. A. Colman	

Notes	Reg.	Type	Owner or Operator
	G-ANON	D.H.82A Tiger Moth (T7909)	A. C. Mercer
	G-ANOO	D.H.82A Tiger Moth	T. J. Hartwell & ptnrs.
	G-ANOR	D.H.82A Tiger Moth	A. J. Cheshire
	G-ANOV	*D.H.104 Dove 6	Museum of Flight, E. Fortune
	G-ANPE	D.H.82A Tiger Moth	R. W. & P. R. Budge
	G-ANPK	D.H.82A Tiger Moth	J. W. Benson
	G-ANPP	P.34A Proctor 3	C. P. A. & Mrs. J. Jeffery
	G-ANRF	D.H.82A Tiger Moth	C. D. Cyster
	G-ANRN	D.H.82A Tiger Moth	J. J. V. Elwes
	G-ANRP	Auster 5 (TW439)	Warnham Air Museum
	G-ANRX	D.H.82A Tiger Moth	Mosquito Aircraft Museum
	G-ANSM	D.H.82A Tiger Moth	Torbay Aviation Ltd.
	G-ANSZ	D.H.114 Heron 1B/C	Hurst Rent-a-Car Ltd.
	G-ANTE	D.H.82A Tiger Moth	T. I. Sutton & B. J. Champion
	G-ANTK	*Avro 685 York	Dan Air Preservation Group
	G-ANTS	*D.H.82A Tiger Moth	Strathallan Aircraft Collection
	G-ANUO	D.H.114 Heron 2D	The GEC Co. Ltd. & National Nuclear Corporation Ltd.
	G-ANUW	D.H.104 Dove 6	Civil Aviation Authority
	G-ANVU	D.H.104 Dove 1B	T. D. Keegan
	G-ANWB	D.H.C.1 Chipmunk 21	G. Briggs
	G-ANWX	J/5L Aiglet Trainer	Newcastle & Tees-side Gliding Club
	G-ANXB	D.H.114 Heron 1B	Pan Universal Aircraft Services (CI) Ltd.
	G-ANXR	P.31C Proctor 4 (RM221)	J. R. Batt
	G-ANYP	*P.31C Proctor 4 (NP184)	Torbay Aircraft Museum
	G-ANZJ	*P.31C Proctor 4 (NP303)	Historic Aircraft Museum, Southend
	G-ANZR	D.H.82A Tiger Moth	D. R. & Mrs. M. Wood
	G-ANZU	D.H.82A Tiger Moth	P. A. Jackson
	G-ANZZ	D.H.82A Tiger Moth	Tiger Club Ltd.
	G-AOAA	D.H.82A Tiger Moth	Tiger Club Ltd.
	G-AOAR	*P.31C Proctor 4 (NP181)	Historic Aircraft Preservation Soc.
	G-AOBH	D.H.82A Tiger Moth (T7997)	I. B. & D. W. Grace
	G-AOBO	D.H.82A Tiger Moth	T. J. Bolt & J. N. Moore
	G-AOBU	*P.84 Jet Provost	Shuttleworth Trust
	G-AOBV	J/5P Autocar	P. E. Champney
	G-AOBX	D.H.82A Tiger Moth (T7187)	Vintage Displays & Training Services Ltd.
	G-AOCR	Auster 5D	C. E. Tyers
	G-AOCU	Auster 5	S. J. Ball
	G-AODA	Westland S.55 Srs.3	Bristow Helicopters Ltd.
	G-AODR	D.H.82A Tiger Moth	D. R. & Mrs. M. Wood
	G-AODT	D.H.82A Tiger Moth	N. A. Brett & A. H. Warminger
	G-AOEG	D.H.82A Tiger Moth	Truman Aviation Ltd.
	G-AOEH	Aeronca 7AC Champion	M. Weeks & ptnrs.
	G-AOEI	D.H.82A Tiger Moth	C.F.G. Flying Ltd.
	G-AOEL	D.H.82A Tiger Moth	Museum of Flight, E. Fortune
	G-AOES	D.H.82A Tiger Moth	Mrs. J. M. Goodger & G. A. Cordery
	G-AOET	D.H.82A Tiger Moth	Jack W. Benson
	G-AOFE	D.H.C.1 Chipmunk 22A	B. Webster
	G-AOFJ	Auster 5	Miss M. R. Innocent
	G-AOFM	J/5P Autocar	C. M. Barnes
	G-AOFS	J/5L Aiglet Trainer	G. W. Howard
	G-AOFW	*ATL.98 Carvair	British Air Ferries Ltd. *Big John*
	G-AOGA	M.75 Aries	R. E. Winn
	G-AOGE	P.34A Proctor 3	N. I. Dalziel
	G-AOGR	D.H.82A Tiger Moth	H. C. Adkins & L. G. Wright
	G-AOGV	J/5R Alpine	ABH Aviation
	G-AOHL	*V.802 Viscount	British Air Ferries (Cabin Trainer)
	G-AOHM	V.802 Viscount	British Air Ferries *Anne-Marie*
	G-AOHT	V.802 Viscount	Aqua-Avia Skybus (NZ)
	G-AOHV	V.802 Viscount	British Air Ferries *Winn*
	G-AOHZ	J/5P Autocar	M. R. Gibbons & G. W. Brown
	G-AOIL	D.H.82A Tiger Moth	The Shuttleworth Collection
	G-AOIM	D.H.82A Tiger Moth	R. M. Wade & F. J. Terry
	G-AOIR	Thruxton Jackaroo	Stevenage Flying Club
	G-AOIS	D.H.82A Tiger Moth	V. B. & R. G. Wheele
	G-AOIY	J/5G Autocar	L. N. Garner
	G-AOJC	*V.802 Viscount	Wales Aircraft Museum
	G-AOJH	D.H.83C Fox Moth	J. S. Lewery
	G-AOJJ	D.H.82A Tiger Moth	E. Lay
	G-AOKH	P.40 Prentice 1	J. F. Moore

Reg.	Type	Owner or Operator	Notes
G-AOKL	P.40 Prentice I (VS610)	J. R. Batt	
G-AOKO	*P.40 Prentice I	J. F. Coggins	
G-AOLK	P.40 Prentice I	Hilton Aviation Ltd.	
G-AOLP	P.40 Prentice I	D. N. Downton & ptnrs.	
G-AOLU	*P.40 Prentice I (VS356)	Scottish Aircraft Collection, Perth	
G-AORL	D.H.C.I Chipmunk 22	D. Gardner	
G-AORR	D.H.C.I Chipmunk 22A	T. L. P. Delaney	
G-AORW	D.H.C.I Chipmunk 22A	A. R. Lockyer	
G-AOSK	D.H.C.I Chipmunk 22	J. G. Cullen	
G-AOSO	D.H.C.I Chipmunk 22	D. Blackburn	
G-AOSU	D.H.C.I Chipmunk 22 (Lycoming)	RAFGSA	
G-AOSY	D.H.C.I Chipmunk 22	J. A. W. Clowes	
G-AOSZ	D.H.C.I Chipmunk 22A	J. Pothecary & P. Villa	
G-AOTD	D.H.C.I Chipmunk 22	Shuttleworth Trust	
G-AOTF	D.H.C.I Chipmunk 23	RAFGSA	
G-AOTI	D.H.114 Heron 2D	Rolls-Royce Ltd.	
G-AOTK	D.53 Turbi	The T. K. Flying Group	
G-AOTR	D.H.C.I Chipmunk 22	London Gliding Club (Pty) Ltd.	
G-AOTY	D.H.C.I Chipmunk 22A	West London Aero Services Ltd.	
G-AOUJ	*Fairey Ultra-light	British Rotorcraft Museum	
G-AOUO	D.H.C.I Chipmunk 22 (Lycoming)	RAFGSA	
G-AOUP	D.H.C.I Chipmunk 22	Wessex Flying Group	
G-AOVF	B.175 Britannia 312F	Invicta International Airways Ltd.	
G-AOVT	*B.175 Britannia 312	Duxford Aviation Soc.	
G-AOVW	Auster 5	B. Marriott	
G-AOXG	D.H.82A Tiger Moth (XL717)	F.A.A. Museum	
G-AOXN	D.H.82A Tiger Moth	D. W. Mickleburgh	
G-AOYG	V.806 Viscount	British Airways	
G-AOYH	V.806 Viscount	British Airways	
G-AOYI	V.806 Viscount	British Air Ferries	
G-AOYJ	V.806 Viscount	British Air Ferries	
G-AOYL	V.806 Viscount	British Airways	
G-AOYM	V.806 Viscount	British Airways	
G-AOYN	V.806 Viscount	British Air Ferries *Diane*	
G-AOYO	V.806 Viscount	British Airways	
G-AOYP	V.806 Viscount	British Air Ferries	
G-AOYR	V.806 Viscount	British Airways	
G-AOYS	V.806 Viscount	British Air Ferries	
G-AOZB	D.H.82A Tiger Moth	Structure Flex Ltd.	
G-AOZH	D.H.82A Tiger Moth (K2572)	V. B. & R. G. Wheele	
G-AOZL	J/5Q Alpine	L. C. Cole	
G-AOZP	D.H.C.I Chipmunk 22	M. E. Darlington	
G-APAF	Auster 5	Globalpost Ltd.	
G-APAH	Auster 5	Executive Flying Services Ltd.	
G-APAM	Thruxton Jackaroo	R. P. Williams	
G-APAO	D.H.82A Tiger Moth	C. K. Irvine	
G-APAP	Thruxton Jackaroo	H. D. V. Leech	
G-APAS	*D.H.106 Comet 1XB	Aerospace Museum, Cosford	
G-APBD	PA-23 Apache 160	E. A. Clack & T. Pritchard	
G-APBE	Auster 5	G. W. Clarke	
G-APBI	D.H.82A Tiger Moth (EM903)	R. Devaney & ptnrs.	
G-APBO	D.53 Turbi	R. Johnson	
G-APBW	Auster 5	F. J. Wiseman	
G-APCB	J/5Q Alpine	G. Jones	
G-APCC	D.H.82A Tiger Moth	L. J. Rice	
G-APCU	D.H.82A Tiger Moth	K. C. K. Virtue	
G-APCY	J/1N Alpha	J. R. Pearson	
G-APDB	*D.H.106 Comet 4	Duxford Aviation Soc.	
G-APDT	*D.H.106 Comet 4	British Airways (used by Apprentices at Heathrow)	
G-APDV	Hiller UH-12C	S. E. Davidson	
G-APEG	V.953C Merchantman	Air Bridge Carriers Ltd.	
G-APEJ	V.953C Merchantman	Air Bridge Carriers Ltd.	
G-APEK	V.953C Merchantman	Air Bridge Carriers Ltd.	
G-APEP	V.953C Merchantman	Air Bridge Carriers Ltd.	
G-APES	V.953C Merchantman	Air Bridge Carriers Ltd.	
G-APET	V.953C Merchantman	Air Bridge Carriers Ltd.	
G-APEX	V.806 Viscount	British Air Ferries *Sun Seeker*	
G-APEY	V.806 Viscount	British Air Ferries	

Notes	Reg.	Type	Owner or Operator
	G–APFA	D.54 Turbi	A. Eastelow & F. J. Keitch
	G–APFJ	*Boeing 707-436	Aerospace Museum, Cosford
	G–APFU	D.H.82A Tiger Moth	M. R. Coward & D. M. White
	G–APGL	*D.H.82A Tiger Moth (NM140)	—
	G–APHV	Avro 19 Srs. 2 (VM360)	—
	G–APIE	Tipsy Belfair B	J. J. Penney & ptnrs.
	G–APIG	D.H.82A Tiger Moth	Evans Estates Ltd.
	G–APIH	D.H.82A Tiger Moth	A. J. Detheridge
	G–APIK	J/1N Alpha	T. D. Howe
	G–APIM	V.806 Viscount	British Airways
	G–APIT	P.40 Prentice 1	J. R. Batt
	G–APIY	*P.40 Prentice 1 (VR249)	Newark Air Museum
	G–APJB	P.40 Prentice 1	City Airways
	G–APJJ	*Fairey Ultra-light	Midland Aircraft Preservation Soc.
	G–APJN	Hiller UH-12B	Bristow Helicopters Ltd.
	G–APJO	D.H.82A Tiger Moth	D. R. & Mrs. M. Wood
	G–APJZ	J/1N Alpha	L. Goddard & E. Amey
	G–APKH	D.H.85 Leopard Moth	P. Franklin (G–ACGS)
	G–APKM	J/1N Alpha	K. C. Bachmann
	G–APKN	J/1N Alpha	Felthorpe Auster Group
	G–APKY	Hiller UH-12B	Shackleton Aviation Ltd.
	G–APLG	J/5L Aiglet Trainer	D. H. Furnell
	G–APLK	M-100 Student	Discovery (R & D) Ltd.
	G–APLO	D.H.C.1 Chipmunk 22A	Channel Islands Aero Holdings Ltd.
	G–APMH	J/1U Workmaster	R. E. Neal and S. R. Stevens
	G–APML	Dakota 6	Air Atlantique Ltd.
	G–APMM	D.H.82A Tiger Moth	R. K. J. Hadlow
	G–APMP	Hiller UH-12C	Morland Beazley Helicopters Ltd.
	G–APMR	Hiller UH-12C	Bristow Helicopters Ltd.
	G–APMS	Hiller UH-12C	Bristow Helicopters Ltd.
	G–APMX	D.H.82A Tiger Moth	K. B. Palmer
	G–APMY	PA-23 Apache 160	Mooney Aviation Ltd.
	G–APNJ	Cessna 310	Shackleton Aviation Ltd.
	G–APNR	Hiller UH-12C	Bristow Helicopters Ltd.
	G–APNS	Garland-Bianchi Linnet	Mrs. P. B. Voice
	G–APNT	Currie Wot	L. W. Richardson & ptnrs.
	G–APNZ	D.31 Turbulent	Tiger Club Ltd.
	G–APOA	J/1N Alpha	Bristow Helicopters Ltd.
	G–APOD	Tipsy Belfair	R. J. Miller
	G–APOI	Saro Skeeter Srs. 8	B. G. Heron
	G–APOL	D.36 Turbulent	A. R. Leith
	G–APPA	D.H.C.1 Chipmunk 22	D. H. Parkhouse & R. Turnor-Warwick
	G–APPL	P.40 Prentice 1	Miss S. J. Saggers
	G–APPM	D.H.C.1 Chipmunk 22	Andrew Edie Aviation
	G–APRF	Auster 5	P. Elliott & ptnrs.
	G–APRJ	*Avro 694 Lincoln B.2 (G-29-1)	Historic Aircraft Museum, Southend
	G–APRL	AW650 Argosy 101	Air Bridge Carriers Ltd.
	G–APRM	AW650 Argosy 102	Rolls-Royce Ltd.
	G–APRN	AW650 Argosy 101	Air Bridge Carriers Ltd.
	G–APRO	Auster 6A	A. H. Wheeler
	G–APRR	Super Aero 45	C. J. Reed
	G–APRT	Taylor JT-1 Monoplane	C. J. Chambers & ptnrs.
	G–APRU	M.S.760 Paris	Cranfield Institute of Technology
	G–APSH	Hiller UH-12B	Bristow Helicopters Ltd.
	G–APSO	D.H.104 Dove 5	Miss J. D. Baker
	G–APSZ	Cessna 172	M. J. Butler & ptnrs.
	G–APTH	Agusta-Bell 47J	W. R. Finance Ltd.
	G–APTM	Hiller UH-12B	Bristow Helicopters Ltd.
	G–APTP	PA-22 Tri-Pacer 150	J. R. Williams
	G–APTR	J/1N Alpha	C. J. & D. J. Baker
	G–APTS	D.H.C.1 Chipmunk 22A	B. R. Pickard
	G-APTU	Auster 5	P. Bowers
	G–APTW	*W.B.1 Widgeon	Cornwall Air Park
	G–APTY	Beech G.35 Bonanza	G. E. Brennand & J. M. Fish
	G–APTZ	D.31 Turbulent	G. Edmiston
	G–APUD	*Bensen B.7M (modified)	Aeroplane Collection Ltd.
	G–APUE	L-40 Meta Sokol	P. Phipps
	G–APUK	J/1 Autocrat	P. L. Morley
	G–APUP	Sopwith Pup (N5182)	D. W. Arnold
	G–APUR	PA-22 Tri-Pacer 160	G. A. Allen & ptnrs.
	G–APUW	Auster J-5V-160	Anglia Auster Syndicate

Reg.	Type	Owner or Operator	Notes
G–APUY	D.31 Turbulent	C. Jones & ptnrs.	
G–APUZ	PA-24 Comanche 250	P. N. Martin & ptnrs.	
G–APVG	J/5L Aiglet Trainer	CIBA-Geigy (UK) Ltd.	
G–APVN	D.31 Turbulent	R. Sherwin	
G–APVS	Cessna 170B	P. E. L. Lamyman	
G–APVU	L-40 Meta-Sokol	D. Kirk	
G–APVV	Mooney M-20A	Telecon Associates	
G–APVY	PA-25 Pawnee 150	A.D.S. (Aerial) Ltd.	
G–APVZ	D.31 Turbulent	A. F. Bullock	
G–APWA	HPR-7 Herald 100	British Air Ferries *Mandy Keegan*	
G–APWE	HPR-7 Herald 201	Air UK	
G–APWF	HPR-7 Herald 201	Air UK	
G–APWG	HPR-7 Herald 201	Air UK	
G–APWH	HPR-7 Herald 201	Air UK	
G–APWJ	HPR-7 Herald 201	Air UK	
G–APWR	PA-22 Tri-Pacer 160	Bencray Ltd.	
G–APWY	*Piaggio P. 166	Historic Aircraft Museum, Southend	
G–APWZ	EP.9 Prospector	Sussex Agricultural Services	
G–APXJ	PA-24 Comanche 250	J. W. Quine & G. F. Lindsay	
G–APXR	PA-22 Tri-Pacer 160	D. H. Rees & E. Hooton	
G–APXT	PA-22 Tri-Pacer 150	McKenzie & Tapp Ltd.	
G–APXU	PA-22 Tri-Pacer 125	I. V. & K. Fairhurst	
G–APXW	EP.9 Prospector	Sussex Agricultural Services	
G–APXX	*D.H.A.3 Drover 2	Historic Aircraft Museum, Southend	
G–APXY	Cessna 150	Merlin Flying Club Ltd.	
G–APYB	T.66 Nipper 2	B. O. Smith	
G–APYD	*D.H.106 Comet 4B	Science Museum Store, Wroughton	
G–APYG	D.H.C.1 Chipmunk 22	E. J. I. Musty & P. A. Colman	
G–APYI	PA-22 Tri-Pacer 135	Air Farm Ltd.	
G–APYN	PA-22 Tri-Pacer 160	W. D. Stephens	
G–APYT	7FC Tri-Traveller	C. H. Morris & R. W. Brown	
G–APYU	7FC Tri-Traveller	K. Collins	
G–APYW	PA-22 Tri-Pacer 150	B. A. Wallace	
G–APYX	PA-23 Aztec 250	Tamavia Ltd.	
G–APZE	PA-23 Apache 160	J. P. Dodd	
G–APZG	PA-24 Comanche 250	Steve Stephens Ltd.	
G–APZJ	PA-18 Super Cub 150	Southern Sailplanes	
G–APZK	PA-18 Super Cub 95	W. T. Knapton	
G–APZL	PA-22 Tri-Pacer 160	F. J. Holdcroft Ltd.	
G–APZR	Cessna 150	C. R. E. S. Taylor	
G–APZS	Cessna 175A	A. J. House	
G–APZU	D.H.104 Dove 6	Acraman Holdings Ltd.	
G–APZX	PA-22 Tri-Pacer 150	M. G. Montgomerie & ptnrs.	
G–ARAB	Cessna 150	A. H. Nicholas	
G–ARAI	PA-22 Tri-Pacer 160	J. E. Fox	
G–ARAJ	PA-22 Tri-Pacer 160	Wearside Flying Group	
G–ARAM	PA-18 Super Cub 150	E. Sussex Gliding Club Ltd.	
G–ARAN	PA-18 Super Cub 150	Yorkshire Gliding Club (Pty) Ltd.	
G–ARAO	PA-18 Super Cub 95	Alidair Ltd.	
G–ARAP	7EC Traveller	P. J. Heron	
G–ARAS	7FC Tri-Traveller	A. Bruniges	
G–ARAT	Cessna 180C	R. E. Styles & ptnrs.	
G–ARAU	Cessna 150	R. N. R. Bellamy	
G–ARAW	Cessna 182C Skylane	G. Grenall	
G–ARAX	PA-22 Tri-Pacer 150	Megacirc Ltd.	
G–ARAY	H.S.748 Srs. 2	Dan-Air Services Ltd.	
G–ARAZ	D.H.82A Tiger Moth (R4959)	M. V. Gauntlett	
G–ARBE	D.H.104 Dove 8	British Aerospace	
G–ARBG	T.66 Nipper 2	Felthorpe Tipsy Group	
G–ARBL	D.31 Turbulent	C. C. Taylor	
G–ARBN	PA-23 Apache 160	H. Norden & H. J. Liggins	
G–ARBO	PA-24 Comanche 250	R. K. Blewett	
G–ARBP	T.66 Nipper 2	A. Cambridge & D. B. Winstanley	
G–ARBS	PA-22 Tri-Pacer 160	Garb Enterprises	
G–ARBV	PA-22 Tri-Pacer 150	C. R. Turner	
G–ARBZ	D.31 Turbulent	D. G. H. Hilliard	
G–ARCC	PA-22 Tri-Pacer 150	Fainville Ltd.	
G–ARCF	PA-22 Tri-Pacer 150	A. L. Scadding	
G–ARCI	Cessna 310D	R. H. B. & M. C. le Brocq	
G–ARCL	Cessna 175A	C. E. Sharp Plant Hire and Sales Ltd.	
G–ARCS	Auster D6/180	E. A. Matty	

Notes	Reg.	Type	Owner or Operator
	G-ARCT	PA-18 Super Cub 95	M. Kirk
	G-ARCV	Cessna 175A	P. J. O. Hanlon
	G-ARCW	PA-23 Apache 160	E. M. Brain & R. Chew
	G-ARCX	*AW Meteor 14	Museum of Flight, E. Fortune
	G-ARCZ	D.31 Turbulent	Stapleford Turbulent Group
	G-ARDB	PA-24 Comanche 250	G. K. Hare
	G-ARDC	Cessna 210 Centurion	Northern Executive Aviation Ltd.
	G-ARDD	CP.301C1 Emeraude	A. Mackintosh
	G-ARDE	D.H.104 Dove 6	Hunting Surveys & Consultants Ltd.
	G-ARDG	EP.9 Prospector	Sussex Agricultural Services
	G-ARDJ	Auster D.6/180	J. D. H. Radford
	G-ARDO	Jodel D.112	P. J. H. McCraig
	G-ARDP	PA-22 Tri-Pacer 150	G. M. Jones
	G-ARDS	PA-22 Caribbean 150	D. V. Asher
	G-ARDT	PA-22 Tri-Pacer 160	A. A. Whiter
	G-ARDV	PA-22 Tri-Pacer 160	M. R. Angood
	G-ARDY	T.66 Nipper 2	P. F. Bennison
	G-ARDZ	Jodel D.140A	W. R. Dryden
	G-AREA	D.H.104 Dove 8	British Aerospace
	G-AREB	Cessna 175B Skylark	R. J. Postlethwaite & ptnrs.
	G-AREE	PA-23 Aztec 250	Montague Travel Ltd.
	G-AREF	PA-23 Aztec 250	Tricolore Trading Co. Ltd.
	G-AREH	D.H.82A Tiger Moth	T. Pate
	G-AREI	Auster 3	R. Alliker & ptnrs.
	G-AREJ	Beech 95 Travel Air	D. Huggett
	G-AREL	PA-22 Caribbean 150	H. H. Cousins
	G-AREO	PA-18 Super Cub 150	Lasham Gliding Soc. Ltd.
	G-ARET	PA-22 Tri-Pacer 160	P. & V. Slatterey
	G-AREV	PA-22 Tri-Pacer 160	Echo Victor Group
	G-AREX	Aeronca 15AC Sedan	A. E. Poulson
	G-AREZ	D.31 Turbulent	Gloster Aero Group
	G-ARFB	PA-22 Caribbean 150	Borrowash Estates Ltd.
	G-ARFD	PA-22 Tri-Pacer 160	C. Fergusson & ptnrs.
	G-ARFG	Cessna 175A Skylark	C. S. & Mrs. B. A. Frost
	G-ARFH	PA-24 Comanche 250	L. M. Walton
	G-ARFL	Cessna 175B Skylark	P. M. & R. G. Rees
	G-ARFM	Cessna 175B Skylark	Foyle Aviation Ltd.
	G-ARFO	Cessna 150A	R. J. Penwarden & T. Thornton
	G-ARFS	PA-22 Caribbean 150	M. H. Armstrong & C. McFadden
	G-ARFT	Jodel D.R. 1050	D. Howard
	G-ARFV	T.66 Nipper 2	D. J. Hockings
	G-ARGB	Auster 6A	A. M. Witt
	G-ARGG	D.H.C.1 Chipmunk 22	Air Navigation & Trading Co. Ltd.
	G-ARGK	Cessna 210	G. H. K. Rogers
	G-ARGO	PA-22 Colt 108	B. E. Goodman
	G-ARGR	V.708 Viscount	Inter City Airlines
	G-ARGV	PA-18 Super Cub 150	Deeside Gliding Club (Aberdeenshire) Ltd.
	G-ARGY	PA-22 Tri-Pacer 160	D. H. Tanner & I. J. Enoch
	G-ARGZ	D.31 Turbulent	C. Sewell
	G-ARHB	Forney F-1A Aircoupe	J. T. Mountain
	G-ARHC	Forney F-1A Aircoupe	W. F. Barnes
	G-ARHF	Forney F-1A Aircoupe	R. A. Nesbitt-Dufort
	G-ARHI	PA-24 Comanche 180	B.C.F.G. Ltd.
	G-ARHL	PA-23 Aztec 250	J. J. Freeman & Co. Ltd.
	G-ARHM	Auster 6A	D. Hollowell & ptnrs.
	G-ARHN	PA-22 Caribbean 150	P. H. Pickford
	G-ARHP	PA-22 Tri-Pacer 160	W. Wardle
	G-ARHR	PA-22 Caribbean 150	J. A. Hargraves
	G-ARHT	PA-22 Caribbean 150	J. S. Lewery
	G-ARHU	PA-22 Tri-Pacer 160	G. K. Hare & G. W. Worley
	G-ARHW	D.H.104 Dove 8	British Aerospace
	G-ARHZ	D.62 Condor	D. H. Wilson-Spratt
	G-ARIA	Bell 47G	Decca Navigator Co. Ltd.
	G-ARID	Cessna 172B	N. Law
	G-ARIE	PA-24 Comanche 250	W. Radwanski
	G-ARIF	O-H7 Minor Coupe	A. W. J. G. Ord-Hume
	G-ARIH	Auster 6A	B. D. Husband
	G-ARIK	PA-22 Caribbean 150	C. J. Berry
	G-ARIL	PA-22 Caribbean 150	G. N. Richardson Motors
	G-ARIN	PA-24 Comanche 250	Fisher Douglas Aviation Ltd.
	G-ARIR	V.708 Viscount	Inter City Airlines
	G-ARIU	Cessna 172B Skylark	P. A. Howell

Reg.	Type	Owner or Operator	Notes
G–ARIV	Cessna 172B	J. A. Hood	
G–ARIW	CP.301B Emeraude	C. Franklin	
G–ARJB	D.H.104 Dove 8	J. C. Bamford (Excavators) Ltd.	
G–ARJC	PA-22 Colt 108	C. E. Holland	
G–ARJE	PA-22 Colt 108	G. W. Doran	
G–ARJF	PA-22 Colt 108	E. G. Gilbert	
G–ARJG	PA-22 Colt 108	Sqd. Ldr. G. R. Sharp	
G–ARJH	PA-22 Colt 108	A. Walmsley	
G–ARJR	PA-23 Apache 160	R. F. Wanbon	
G–ARJS	PA-23 Apache 160	Bencray Ltd.	
G–ARJT	PA-23 Apache 160	R. D. Dickson	
G–ARJU	PA-23 Apache 160	Aviation Enterprises	
G–ARJV	PA-23 Apache 160	Gordon King (Aviation) Ltd.	
G–ARJW	PA-23 Apache 160	Gordon King (Aviation) Ltd.	
G–ARJZ	D.31 Turbulent	N. H. Jones	
G–ARKG	J/5G Autocar	C. Thompson	
G–ARKJ	Beech N35 Bonanza	R. J. Guise	
G–ARKK	PA-22 Colt 108	A. W. Baxter	
G–ARKM	PA-22 Colt 108	L. E. Usher	
G–ARKN	PA-22 Colt 108	J. H. Underwood & A. J. F. Tabenor	
G–ARKP	PA-22 Colt 108	C. J. & J. Freeman	
G–ARKR	PA-22 Colt 108	H. L. Crawley	
G–ARKS	PA-22 Colt 108	D. W. Mickleburgh	
G–ARLB	PA-24 Comanche 250	Stakehill Engineering Ltd.	
G–ARLD	H-395 Super Courier	P. F. Hall	
G–ARLG	Auster D.4/108	R. D. Hilliar-Symons	
G–ARLI	PA-23 Apache 150	K. A. Learmonth	
G–ARLK	PA-24 Comanche 250	M. Walker & C. Robinson	
G–ARLL	PA-24 Comanche 250	E. J. Spiers	
G–ARLP	A.61 Terrier	J. M. Jones	
G–ARLR	A.61 Terrier	G. Griffith	
G–ARLT	Cessna 172B Skyhawk	A. R. German & Sons	
G–ARLU	Cessna 172B Skyhawk	Hanoflite Ltd.	
G–ARLV	Cessna 172B Skyhawk	Shoreham Flight Simulation	
G–ARLW	Cessna 172B Skyhawk	T. J. Evason & ptnrs.	
G–ARLX	Jodel D.140B	Gladaircraft Ltd.	
G–ARLY	J/5P Autocar	J. Alva & ptnrs.	
G–ARLZ	D.31A Turbulent	R. W. Rushton	
G–ARMA	PA-23 Apache 160	Oxford Air Training School Ltd.	
G–ARMB	D.H.C.1 Chipmunk 22A	College of Air Training	
G–ARMC	D.H.C.1 Chipmunk 22A	W. London Aero Services Ltd.	
G–ARMD	D.H.C.1 Chipmunk 22A	College of Air Training	
G–ARMF	D.H.C.1 Chipmunk 22A	College of Air Training	
G–ARMG	D.H.C.1 Chipmunk 22A	College of Air Training	
G–ARMH	PA-23 Aztec 250	Royle Aviation Co. Ltd.	
G–ARMI	PA-23 Apache 160	Stapleford Flying Club Ltd.	
G–ARMJ	Cessna 185 Skywagon	British Skysports	
G–ARML	Cessna 175B Skylark	Woolmer Aircraft Ltd.	
G–ARMN	Cessna 175B Skylark	H. A. Claireaux & T. D. Gray	
G–ARMO	Cessna 172B Skyhawk	Sangria Designs Ltd. & BRM Plastics Ltd.	
G–ARMP	Cessna 172B	Southport & Merseyside Aero Club (1979) Ltd.	
G–ARMR	Cessna 172B Skyhawk	J. Braithwaite	
G–ARMW	H.S.748 Srs. 1	Dan-Air Services Ltd.	
G–ARMX	H.S.748 Srs. 1	Dan-Air Services Ltd.	
G–ARMZ	D.31 Turbulent	Frederick A. Shepherd	
G–ARNA	Mooney M.20B	R. Travers	
G–ARNB	J/5G Autocar	M. T. Jeffrey	
G–ARND	PA-22 Colt 108	Richard Rimington Ltd.	
G–ARNE	PA-22 Colt 108	T. D. L. Bowden	
G–ARNG	PA-22 Colt 108	W. London Aero Services Ltd.	
G–ARNI	PA-22 Colt 108	T. Rundle	
G–ARNJ	PA-22 Colt 108	MKM Flying Group	
G–ARNK	PA-22 Colt 108	D. P. Golding	
G–ARNL	PA-22 Colt 108	Mr. J. A. & Miss J. A. Dodsworth	
G–ARNN	GC-1B Swift	K. E. Sword	
G–ARNO	A.61 Terrier	M. B. Hill.	
G–ARNP	A.109 Airedale	T. A. J. Luther & ptnrs.	
G–ARNY	Jodel D.117	A. D. Henderson	
G–ARNZ	D.31 Turbulent	P. L. Cox & ptnrs.	
G–AROA	Cessna 172B Skyhawk	D. E. Partridge	
G–AROC	Cessna 175B Skylark	Yorkshire Flying Services Ltd.	

Notes	Reg.	Type	Owner or Operator
	G-AROD	Cessna 175B	Medical Co. Hospital Supplies Ltd.
	G-AROE	Aero 145	G. S. & Mrs. P. Galt
	G-AROF	L.40 Meta-Sokol	B. G. Barber
	G-AROJ	A.109 Airedale	D. J. Shaw
	G-AROK	Cessna 310F	S. E. Berry
	G-ARON	PA-22 Colt 108	M. A. Coward
	G-AROO	Forney F-1A Aircoupe	W. I. McMeekan
	G-AROR	Forney F-1A Aircoupe	Treswithick Air & Shipping Services Ltd.
	G-AROW	Jodel D.140B	Kent Gliding Club Ltd.
	G-AROY	Stearman A.75N.1	W. A. Jordan
	G-ARPD	*H.S.121 Trident 1C	CAA Fire School, Tees-side
	G-ARPH	H.S.121 Trident 1C	British Airways
	G-ARPK	H.S.121 Trident 1C	British Airways
	G-ARPL	H.S.121 Trident 1C	British Airways
	G-ARPN	H.S.121 Trident 1C	British Airways
	G-ARPO	H.S.121 Trident 1C	British Airways
	G-ARPP	H.S.121 Trident 1C	British Airways
	G-ARPR	*H.S.121 Trident 1C	CAA Fire School, Tees-side
	G-ARPW	H.S.121 Trident 1C	British Airways
	G-ARPX	H.S.121 Trident 1C	British Airways
	G-ARPZ	H.S.121 Trident 1C	British Airways
	G-ARRD	Jodel DR. 1050	N. L. E. Dupee
	G-ARRE	Jodel DR. 1050	E. H. Ellis
	G-ARRF	Cessna 150A	Cornwall Flying Club
	G-ARRH	Cessna 175B Skylark	R. A. Spiller
	G-ARRI	Cessna 175B Skylark	C. L. Thomas
	G-ARRL	J/1N Alpha	A. J. Brown
	G-ARRM	*Beagle B.206-X	Brighton Transport Museum
	G-ARRP	PA-28 Cherokee 160	Dyce Flying Group
	G-ARRS	CP-301A Emeraude	J. Y. Paxton
	G-ARRT	Wallis WA-116-1	K. H. Wallis
	G-ARRU	D.31 Turbulent	E. R. Gourd & ptnrs.
	G-ARRW	H.S.748 Srs. 1	Dan-Air Services Ltd.
	G-ARRY	Jodel D.140B	R. G. Andrews
	G-ARRZ	D.31 Turbulent	J. S. Barker
	G-ARSB	Cessna 150A	S. M. Upton
	G-ARSG	*Avro Triplane	Shuttleworth Trust
	G-ARSJ	CP.301-C2 Emeraude	G. G. Milton
	G-ARSL	A.61 Terrier	C. W. Thomas
	G-ARSN	D.H.104 Dove 8	Staravia Ltd.
	G-ARSP	L.40 Meta-Sokol	K. K. Johnson
	G-ARSU	PA-22 Colt 108	P. E. Palmer
	G-ARSW	PA-22 Colt 108	J. P. Smith
	G-ARSX	PA-22 Tri-Pacer 160	AF Aviation Ltd.
	G-ARTB	Mooney M.20B	R. E. Dagless
	G-ARTD	PA-23 Apache 160	Dr. D. A. Jones
	G-ARTF	D.31 Turbulent	L. J. Pennell
	G-ARTG	Hiller UH-12C	E. C. Francis
	G-ARTH	PA-12 Super Cruiser	D. Mallinson
	G-ARTL	D.H.82A Tiger Moth (T7281)	S. E. Marples
	G-ARTT	M.S.880B Rallye Club	J. Berry
	G-ARUE	D.H.104 Dove 7	Staravia Ltd.
	G-ARUG	J/5G Autocar	N. P. Biggs
	G-ARUH	Jodel DR. 1050	PFA Group
	G-ARUI	A.61 Terrier	D. C. Cullen
	G-ARUL	Cosmic Wind	J. Cull
	G-ARUM	D.H.104 Dove 8	National Coal Board
	G-ARUO	PA-24 Comanche 180	Uniform Oscar Group
	G-ARUR	PA-28 Cherokee 160	Falconash Ltd.
	G-ARUV	CP.301A Emeraude	J. Tanswell
	G-ARUY	J/1N Alpha	A. J. Brown
	G-ARUZ	Cessna 175C Skylark	J. E. Sansome & M. D. Faiers
	G-ARVF	V.1101 VC10	United Arab Emirates
	G-ARVJ	V.1101 VC10	British Airways
	G-ARVM	*V.1101 VC10	Aerospace Museum, Cosford
	G-ARVS	PA-28 Cherokee 160	Stapleford Flying Club Ltd.
	G-ARVT	PA-28 Cherokee 160	C. R. Knapton
	G-ARVU	PA-28 Cherokee 160	Palestar Aviation Ltd.
	G-ARVV	PA-28 Cherokee 160	R. J. Jackson
	G-ARVW	PA-28 Cherokee 160	Bolton Air Training School Ltd.
	G-ARVZ	D.62B Condor	O. G. Stuart-Lee
	G-ARWB	D.H.C.1 Chipmunk 200	Aero Bonner Co. Ltd.

Reg.	Type	Owner or Operator	Notes
G-ARWC	Cessna 150B	Worldwide Wheels Ltd.	
G-ARWH	Cessna 172C Skyhawk	Mrs. J. M. Kirk	
G-ARWM	Cessna 175C	Special Air Services	
G-ARWO	Cessna 172C Skyhawk	T. A. Cox & R. C. Jackman	
G-ARWR	Cessna 172C Skyhawk	W. of Scotland Flying Club Ltd.	
G-ARWS	Cessna 175C Skylark	J. Mudd	
G-ARWW	Bensen B.8M	B. McIntyre	
G-ARWX	Luton L.A.5A Major	A. G. Cameron	
G-ARWY	Mooney M.20A	P. J. Hatcher	
G-ARXC	A.109 Airedale	E. A. Wright	
G-ARXD	A.109 Airedale	D. Howden	
G-ARXF	PA-23 Aztec 250B	Weendy Aviation (UK)	
G-ARXG	PA-24 Comanche 250	F. Stewart	
G-ARXH	Bell 47G	A.C.C. Builders & Capricorn Studios [Ltd.	
G-ARXN	Tipsy Nipper 2	Griffon Flying Group	
G-ARXP	Luton L.A.4A Minor	W. C. Hymas	
G-ARXT	Jodel DR. 1050	A. W. Webster	
G-ARXU	Auster 6A	Bath & Wilts Gliding Club Ltd.	
G-ARXW	M.S.885 Super Rallye	E. W. Noakes	
G-ARXX	M.S.880B Rallye Club	M. S. Bird	
G-ARXY	M.S.880B Rallye Club	Horizon Flying Group	
G-ARYC	*H.S.125 Srs. 1	The Mosquito Aircraft Museum	
G-ARYF	PA-23 Aztec 250	I. J. T. Branson	
G-ARYH	PA-22 Tri-Pacer 160	Filtration (Water Treatment Engineers) Ltd.	
G-ARYI	Cessna 172C	J. T. Mirley & A. E. Wall	
G-ARYK	Cessna 172C	Mrs. K. M. & T. Hemsley	
G-ARYM	D.H.104 Dove 8	Canbaria (Aircraft Spares) Ltd.	
G-ARYR	PA-28 Cherokee 180	Glenochill Engineering	
G-ARYS	Cessna 172C Skyhawk	K. J. Squires	
G-ARYV	PA-24 Comanche 250	P. Meeson	
G-ARYZ	A.109 Airedale	J. D. Reid	
G-ARZA	Wallis WA.116 Srs. 1	N. D. Z. de Ferranti	
G-ARZB	Wallis WA.116 Srs. 1	K. H. Wallis	
G-ARZD	Cessna 172C	T. Poole & P. C. Thompson	
G-ARZF	Cesna 150B	J. E. F. Aviation Ltd.	
G-ARZM	D.31 Turbulent	N. H. Jones	
G-ARZN	Beech N35 Bonanza	Beech Aircraft Ltd.	
G-ARZP	A.109 Airedale	G. R. Evans	
G-ARZW	Currie Wot	D. F. Faulkener-Byrant	
G-ARZX	Cessna 150B	E. T. Wicks	
G-ASAA	Luton L.A.4A. Minor	Four Counties Flying Syndicate	
G-ASAI	A.109 Airedale	A. C. Watt	
G-ASAJ	A.61 Terrier 2	R Skingley & P. G. Bolton	
G-ASAK	A.61 Terrier 2	Rochford Hundred Flying Group	
G-ASAL	SAL Bulldog 120	British Aerospace	
G-ASAM	D.31 Turbulent	Tiger Club Ltd.	
G-ASAN	A.61 Terrier 2	M. R. E. & M. V. Stillingfleet	
G-ASAT	M.S.880B Rallye Club	J. N. Burke	
G-ASAU	M.S.880B Rallye Club	W. J. Armstrong	
G-ASAV	M.S.880B Rallye Club	McAully Flying Group	
G-ASAX	A.61 Terrier 2	G. Strathdee	
G-ASAZ	Hiller UH-12 E4	Morland Beazley Helicopters Ltd.	
G-ASBA	Currie Wot	M. A. Kaye	
G-ASBB	Beech 23 Musketeer	D. Silver	
G-ASBD	Hughes 269A	C. R. McNeil	
G-ASBG	HPR-7 Herald 203	Air UK	
G-ASBH	A.109 Airedale	Pyrochem (UK) Ltd.	
G-ASBS	C.P.301A Emeraude	D. M. Upfield	
G-ASBU	A.61 Terrier 2	G. Strathdee	
G-ASBY	A.109 Airedale	A. Farrell	
G-ASCC	Beagle E.3 AOP Mk. II	M. D. N. & A. C. Fisher	
G-ASCH	A.61 Terrier 2	Enstone Eagles Flying Group	
G-ASCJ	PA-24 Comanche 250	Telspec Ltd.	
G-ASCM	Isaacs Fury II (K2050)	Russwick Ltd.	
G-ASCU	PA-18A-150 Super Cub	Farm Aviation Services Ltd.	
G-ASCZ	CP.310A Emeraude	Hylton Flying Group	
G-ASDA	Beech 65–80 Queen Air	Parker & Heard Ltd.	
G-ASDD	D.H.104 Dove 5	Haywards Aviation Ltd.	
G-ASDL	A.61 Terrier 2	C. P. Lockyer & C. E. Mason	
G-ASDN	PA-24 Comanche 250	J. W. Hodgson	
G-ASDO	Beech 95A.55 Baron	Executive Aviation Ltd.	

Notes	Reg.	Type	Owner or Operator
	G-ASDY	Wallis WA-116/F	K. H. Wallis
	G-ASEA	Luton L.A.4A Minor	Toon Ghose Aviation Ltd.
	G-ASEB	Luton L.A.4A Minor	R. K. Lynn
	G-ASEE	J/IN Alpha	H. C. J. & Sara L. G. Williams
	G-ASEG	A.61 Terrier	J. T. Hogben
	G-ASEO	PA-24 Comanche 250	A. van Daalen
	G-ASEP	PA-23 Apache 235	G. R. Selbert
	G-ASEU	D.62A Condor	W. Grant & D. McNicholl
	G-ASEV	PA-23 Aztec 250	Selexpress Ltd.
	G-ASFA	Cessna 172D	R. A. Marven
	G-ASFD	L-200A Morava	D. Simon
	G-ASFG	PA-23 Aztec 250	Air Conditioning (Transport) Ltd.
	G-ASFJ	Beech P.35 Bonanza	Lockwood Brooks & Co. Ltd.
	G-ASFK	J/5G Autocar	Orman (Carrolls Farm) Ltd.
	G-ASFL	PA-28 Cherokee 180	K. Winfield & ptnrs.
	G-ASFR	Bo 208 Junior	Sunderland Flying Club Ltd.
	G-ASFX	D.31 Turbulent	E. F. Clapham & W. B. S. Dobie
	G-ASGC	*V.1151 Super VC10	Imperial War Museum, Duxford
	G-ASHA	Cessna F.172D	R. L. Fogg & Co. Ltd. & R. Soar
	G-ASHB	Cessna 182F	RN & R Marines Sport Parachute Association
	G-ASHH	PA-23 Aztec 250	G. Everington
	G-ASHJ	Brantly B.2B	A. G. Dean
	G-ASHK	Brantly B.2A	S. N. Cole
	G-ASHR	Beech B35–33 Debonair	C. M. Fraser & E. A. Perry
	G-ASHS	Stampe SV.4B	Tiger Club Ltd.
	G-ASHT	D.31 Turbulent	D. L. Riley
	G-ASHU	PA-15 Vagabond	G. J. Romanes
	G-ASHV	PA-E23 Aztec 250	Haywards Aviation Ltd.
	G-ASHW	D. H.104 Dove 8	L. de la Hay (Fishing & Marine) Salvage Ltd.
	G-ASHX	PA-28 Cherokee 180	R. Lee & D. Morris
	G-ASIB	Cessna F.172D	K. D. Horton
	G-ASII	PA-28 Cherokee 180	Worldwide Wheels Ltd. & ptnrs.
	G-ASIJ	PA-28 Cherokee 180	B. H. Pearce
	G-ASIL	PA-28 Cherokee 180	F. W. Shaw & Sons (Worthing) Ltd.
	G-ASIP	Auster 6A	Bristol Gliding Club (Pty) Ltd.
	G-ASIS	Jodel D.112 Club	E. F. Hazell
	G-ASIT	Cessna 180	A. J. Spiller (Gt. Staughton) Ltd.
	G-ASIY	PA-25 Pawnee 235	A.D.S. (Aerial) Ltd.
	G-ASJL	Beech H.35 Bonanza	P. M. Coulten
	G-ASJM	PA-30 Twin Comanche 160	The Derek Pointon Group
	G-ASJO	Beech B.23 Musketeer	Laxtonbridge Ltd.
	G-ASJU	Aero Commander 520	Interflight Ltd.
	G-ASJV	V.S.361 Spitfire IX (MH434)	Airborne Taxi Service Ltd.
	G-ASJY	GY-80 Horizon 160	A. D. Hemley
	G-ASJZ	Jodel D.117A	Wolverhampton Ultra-light Flying Group
	G-ASKC	*D.H.98 Mosquito 35 (TA719)	Skyfame Collection
	G-ASKH	D.H.98 Mosquito T.3 (RR299)	Hawker Siddeley Aviation
	G-ASKJ	A.61 Terrier I	Bristol & Gloucestershire Gliding Club
	G-ASKK	HPR-7 Herald 211	Air UK
	G-ASKL	Jodel D.150A	J. M. Graty
	G-ASKM	Beech B.65–80 Queen Air	H. Williams & ptnrs.
	G-ASKP	D.H.82A Tiger Moth	Tiger Club Ltd.
	G-ASKS	Cessna 336 Skymaster	M. J. Godwin
	G-ASKT	PA-28 Cherokee 180	Capel & Co. (Printers) Ltd.
	G-ASKV	PA-25 Pawnee 235	Westwick Distributors Ltd.
	G-ASLA	PA-25 Pawnee 235	Westwick Distributors Ltd.
	G-ASLE	PA-30 Twin Comanche 160	Baldocks of Wivelsfield Ltd.
	G-ASLF	Bensen B.7	S. R. Hughes
	G-ASLH	Cessna 182F	Celahurst Ltd.
	G-ASLK	PA-25 Pawnee 235	Westwick Distributors Ltd.
	G-ASLN	Forney F.1A Aircoupe	Cornwall Flying Club
	G-ASLR	Agusta-Bell 47J-2	David Barriball Ltd.
	G-ASLV	PA-28 Cherokee 235	C.S.E. (Aircraft Services) Ltd.
	G-ASLX	CP.301A Emeraude	K. C. Green
	G-ASMA	PA-30 Twin Comanche 160	M. G. Edmunds
	G-ASMC	P.56 Provost T.1.	W. Walker
	G-ASME	Bensen B.7M	C. R. Pepper & A. J. Tabenor
	G-ASMF	Beech D.95A Travel Air	R. M. & C. T. Harlock (Thorney) Ltd.
	G-ASMG	D.H.104 Dove 8	British Aerospace

Reg.	Type	Owner or Operator	Notes
G–ASMH	PA-30 Twin Comanche 160	MLP Aviation Ltd.	
G–ASMJ	Cessna F.172E	J. B. Stocks & J. E. Tribe	
G–ASML	Luton L.A.4A Minor	R. L. E. Horrell	
G–ASMM	D.31 Turbulent	Kenneth Browne	
G–ASMN	PA-23 Apache 160	W. London Aero Services Ltd.	
G–ASMO	PA-23 Apache 160	Aviation Enterprises	
G–ASMS	Cessna 150A	K. R. & T. W. Davies	
G–ASMT	Fairtravel Linnet 2	J. H. Hartley	
G–ASMU	Cessna 150D	Stapleford Flying Club Ltd.	
G–ASMV	CP1310-C3 Super Emeraude	P. F. D. Waltham	
G–ASMW	Cessna 150D	Yorkshire Light Aircraft Ltd.	
G–ASMY	PA-23 Apache 160	Thurston Aviation Ltd.	
G–ASMZ	A.61 Terrier 2	R. B. Humphries	
G–ASNA	PA-23 Aztec 250	Margate Motors Plant & Aircraft Hire [Ltd.	
G–ASNB	Auster 6A	M. Pocock & ptnrs.	
G–ASNC	Beagle D.5/180 Husky	S. J. Riddington	
G–ASND	PA-23 Aztec 250	Survair Ltd.	
G–ASNE	PA-28 Cherokee 180	J. L. Dexter	
G–ASNF	Ercoupe 415-CD	Charles Robertson (Developments) Ltd.	
G–ASNH	PA-23 Aztec 250	Derek Crouch (Contractors) Ltd.	
G–ASNI	CP1310-C3 Super Emeraude	D. Chapman	
G–ASNK	Cessna 205	Transgap Ltd.	
G–ASNL	Sikorsky S-61N	British Airways Helicopters Ltd.	
G–ASNN	Cessna 182F Skylane	Banbury Plant Hire Ltd.	
G–ASNP	Mooney M.20C Mark 21	W. Woodrum	
G–ASNU	H.S.125 Srs. 1	Flintgrange Ltd.	
G–ASNV	Agusta-Bell 47J-2	S. W. Electricity Board	
G–ASNW	Cessna F.172E	S. W. Biroth	
G–ASNY	Bensen B.8M	D. L. Wallis	
G–ASNZ	Bensen B.7M	W. H. Turner	
G–ASOB	PA-30 Twin Comanche 160	M. A. Grayburn	
G–ASOC	Auster 6A	Aquila Gliding Club	
G–ASOH	Beech B.55A Baron	Steer Aviation Ltd.	
G–ASOI	A.61 Terrier 2	R. H. Jowett	
G–ASOK	Cessna F.172E	Okay Flying Group	
G–ASON	PA-30 Twin Comanche 160	Roundham Garage Ltd.	
G–ASOO	PA-30 Twin Comanche 160	John Bisco (Cheltenham) Ltd. & ptnrs	
G–ASOV	PA-25 Pawnee 235	A.D.S. (Aerial) Ltd.	
G–ASOX	Cessna 205A	Halfpenny Green Skydiving Club	
G–ASPA	D.H.104 Dove 8	Staravia Ltd.	
G–ASPF	Jodel D.120	W. S. Howell	
G–ASPI	Cessna F.172E	A. M. Castleton & ptnrs.	
G–ASPK	PA-28 Cherokee 140	R. T. Love	
G–ASPP	*Bristol Boxkite	Shuttleworth Trust	
G–ASPS	Piper J-3C-65 Cub	A. J. Chalkley	
G–ASPT	Cessna 172D	C. H. Parker	
G–ASPU	D.31 Turbulent	I. Maclennen	
G–ASPV	D.H.82A Tiger Moth	B. S. Charters	
G–ASPX	Bensen B-8S	L. D. Goldsmith	
G–ASRB	D.62B Condor	Tiger Club Ltd.	
G–ASRC	D.62B Condor	Tiger Club Ltd.	
G–ASRE	PA-23 Aztec 250	S.P.T. Aircraft Ltd.	
G–ASRF	Jenny Wren	G. W. Gowland	
G–ASRH	PA-30 Twin Comanche 160	IOM & General Life Assurance Co. Ltd. & Bigland Holdings Ltd.	
G–ASRI	PA-23 Aztec 250	Meridian Airmaps Ltd.	
G–ASRK	A.109 Airedale	Atlantic & Caribbean Aviation Ltd.	
G–ASRO	PA-30 Twin Comanche 160	A. G. Perkins	
G–ASRP	Jodel DR 1050	D. J. M. Edmondston	
G–ASRR	Cessna 182G	John V. White Ltd.	
G–ASRT	Jodel D.150	H. M. Kendall	
G–ASRW	PA-28 Cherokee 180	K. R. Deering	
G–ASRX	Beech 65 A80 Queen Air	Seismograph Service (England) Ltd. & Parker & Heard Ltd.	
G–ASSB	PA-30 Twin Comanche 160	E. Berks Boat Co. Ltd.	
G–ASSE	PA-22 Colt 108	J. B. King	
G–ASSF	Cessna 182G Skylane	A. Newsham	
G–ASSI	H.S.125 Srs. 1	TBF (Transport) Ltd.	
G–ASSO	Cessna 150D	W. H. & J. Rogers Group Ltd.	
G–ASSP	PA-30 Twin Comanche 160	The Mastermix Engineering Co. Ltd.	
G–ASSR	PA-30 Twin Comanche 160	Direct Air Ltd.	
G–ASSS	Cessna 172E	D. H. N. Squires	

Notes	Reg.	Type	Owner or Operator
	G-ASST	Cessna 150D	F. R. H. Parker
	G-ASSU	CP 301 A Emeraude	R. W. Millward
	G-ASSW	PA-28 Cherokee 140	C. J. Plummer
	G-ASSY	D.31 Turbulent	G-ASSY Group
	G-ASTA	D.31 Turbulent	D. J. R. Williams
	G-ASTD	PA-23 Aztec 250	Peregrine Air Services Ltd.
	G-ASTG	Nord 1002	L. M. Walton
	G-ASTI	Auster 6A	M. Pocock
	G-ASTL	*Fairey Firefly I (Z2033)	Skyfame Collection
	G-ASTM	Hiller UH-12B	Bristow Helicopters Ltd.
	G-ASTP	Hiller UH-12C	L. Goddard
	G-ASTR	Hiller UH-12B	Bristow Helicopters Ltd.
	G-ASTZ	Hughes 269B	Twyford Moors (Helicopters) Ltd.
	G-ASUB	Mooney M.20E Super 21	T. J. Pigott
	G-ASUD	PA-28 Cherokee 180	H. J. W. Ellison
	G-ASUE	Cessna 150D	D. Huckle
	G-ASUG	*Beech E18S	Royal Scottish Museum
	G-ASUH	Cessna F.172E	G. H. Willson & E. Shipley
	G-ASUI	A.61 Terrier 2	Britrange Ltd.
	G-ASUL	Cessna 182G Skylane	G. H. Reeve
	G-ASUP	Cessna F.172E	W. T. Jenkins & ptnrs.
	G-ASUR	Dornier Do 28A-1	Sheffair Ltd.
	G-ASUS	Jurca MJ.2B Tempete	D. G. Jones
	G-ASVG	CP.301B Emeraude	K. R. Jackson
	G-ASVH	Hiller UH-12B	P. W. Hicks
	G-ASVI	Hiller UH-12B	Bristow Helicopters Ltd.
	G-ASVK	Hiller UH-12B	Bristow Helicopters Ltd.
	G-ASVL	Hiller UH-12B	Bristow Helicopters Ltd.
	G-ASVM	Cessna F.172E	J. White
	G-ASVN	Cessna U.206 Super Skywagon	British Skysports
	G-ASVO	HPR-7 Herald 214	British Air Ferries
	G-ASVP	PA-25 Pawnee 235	A.D.S. (Aerial) Ltd.
	G-ASVV	Cessna 3101	A. R. J. Propafloor
	G-ASVZ	PA-28 Cherokee 140	A. D. Henley
	G-ASWA	PA-28 Cherokee 140	Southend Light Aviation Centre Ltd.
	G-ASWB	A.109 Airedale	Gainsborough Cars
	G-ASWF	A.109 Airedale	D. W. Wastell
	G-ASWG	PA-25 Pawnee 235	A.D.S. (Aerial) Ltd.
	G-ASWH	Luton L.A.5A Major	N. P. Gething
	G-ASWJ	*Beagle 206 Srs. I (8449M)	RAF Halton
	G-ASWL	Cessna F.172F	C. Wilson
	G-ASWN	Bensen B.8M	D. R. Shepherd
	G-ASWO	Cessna 210D	DK Investments Ltd.
	G-ASWP	Beech A.23 Musketeer	Tenair Ltd.
	G-ASWT	Aero 145 Series 20	A. C. Frost
	G-ASWW	PA-30 Twin Comanche 160	Bristol & Wessex Flying Club Ltd.
	G-ASWX	PA-28 Cherokee 180	H. I. Jones (Whitby Garage) Ltd.
	G-ASXB	D.H.82A Tiger Moth	G. W. Bisshopp
	G-ASXC	SIPA 901	Waterside Flying Group
	G-ASXD	Brantly B.2B	Brantly Enterprises
	G-ASXF	Brantly 305	Express Aviation Services Ltd.
	G-ASXI	T.66 Nipper 2	D. Shrimpton
	G-ASXJ	Luton L.A.4A Minor	P. D. Lea & E. A. Lingard
	G-ASXR	Cessna 210	Dolphin Wetsuits Ltd.
	G-ASXS	Jodel DR.1050	J. P. I. Lloyd-Bostock
	G-ASXT	G.159 Gulfstream I	The Ford Motor Co. Ltd.
	G-ASXU	Jodel D.120A	R. W. & Mrs. J. Thompsett
	G-ASXX	*Avro 683 Lancaster 7 (NX611)	RAF Scampton Gate Guard
	G-ASXY	Jodel D.117A	P. A. Davies & ptnrs.
	G-ASXZ	Cessna 182G Skylane	P. M. Robertson
	G-ASYD	BAC One-Eleven 670	British Aerospace
	G-ASYG	A.61 Terrier 2	B. J. Guest
	G-ASYJ	Beech D.95A Travel Air	Crosby Aviation Ltd.
	G-ASYK	PA-30 Twin Comanche 160	Heath Street Car Hirings Ltd.
	G-ASYL	Cessna 150E	British Skysports
	G-ASYP	Cessna 150E	T. S. Quirk
	G-ASYV	Cessna 310G	Chromecourt Ltd.
	G-ASYW	Bell 47G-2	Bristow Helicopters Ltd.
	G-ASYZ	Victa Airtourer 100	J. B. Cave
	G-ASZB	Cessna 150E	H. J. Cox
	G-ASZD	Bo 208A2 Junior	F. P. Hall
	G-ASZE	A.61 Terrier 2	P. J. Moore
	G-ASZF	Boeing 707-336C	British Airways

Reg.	Type	Owner or Operator	Notes
G–ASZG	Boeing 707-336C	British Airways	
G–ASZJ	S.C.7 Skyvan 3A-100	Short Bros Ltd.	
G–ASZR	Fairtravel Linnet	H. C. D. & F. J. Garner	
G–ASZS	GY.80 Horizon 160	I. B. Willis	
G–ASZU	Cessna 150E	C. C. Dobson	
G–ASZV	T.66 Nipper 2	R. L. Mitcham	
G–ASZX	A.61 Terrier	W. D. Hill	
G–ASZY	FRED Srs. 2	E. Clutton & A. J. F. Tabenor	
G–ASZZ	Cessna 310J	European Steel Sheets Ltd.	
G–ATAA	PA-28 Cherokee 180	Brendair	
G–ATAD	Mooney M.20C	H. W. Walker	
G–ATAF	Cessna F.172F	G. Bush	
G–ATAG	Jodel DR. 1050	P. T. K. Roberts & D. Jones	
G–ATAH	Cessna 336 Skymaster	Alderney Air Charter	
G–ATAI	D.H.104 Dove 8	Centrax Ltd.	
G–ATAO	PA-24 Comanche 260	C. V. Jones	
G–ATAS	PA-28 Cherokee 180	D. R. Wood	
G–ATAT	Cessna 150E	The Derek Pointon Group	
G–ATAU	D.62B Condor	A. J. Abbey	
G–ATAV	D.62C Condor	Lasham Gliding Soc. Ltd.	
G–ATAW	A.109 Airedale	Jean Dalton	
G–ATBF	*F-86E Sabre 4 (XB733)	Historic Aircraft Preservation Soc.	
G–ATBG	Nord 1002	L. M. Walton	
G–ATBH	Aero 145	Colpak Aviation Ltd.	
G–ATBI	Beech A.23 Musketeer	R. F. G. Dent	
G–ATBJ	Sikorsky S-61N	British Airways Helicopters Ltd.	
G–ATBK	Cessna F.172F	F. W. Sherrell	
G–ATBL	D.H.60G Moth	M. E. Vaisey	
G–ATBN	PA-28 Cherokee 140	M. R. McGregor	
G–ATBP	Fournier RF-3	C. Jacques & ptnrs.	
G–ATBS	D.31 Turbulent	J. A. Thomas	
G–ATBU	A.61 Terrier 2	P. R. Anderson	
G–ATBV	PA-23 Aztec 250	Cabair Ltd.	
G–ATBW	T.66 Nipper 2	D. Marshall	
G–ATBX	PA-20 Pacer 135	M. R. Smith	
G–ATBZ	W.S-58 Wessex 60	Bristow Helicopters Ltd.	
G–ATCC	A.109 Airedale	A. Robinson	
G–ATCD	D.5/180 Husky	Oxford Flying & Gliding Group	
G–ATCE	Cessna U.206	J. J. Aviation Ltd.	
G–ATCI	Victa Airtourer 100	B. &. C. Building Materials (Canvey Island) Ltd.	
G–ATCJ	Luton L.A.4A Minor	R. M. Sharphouse	
G–ATCL	Victa Airtourer 100	D. Alexander	
G–ATCN	Luton L.A.4A Minor	J. C. Gates & C. Neilson	
G–ATCR	Cessna 310	Staverton Apache Ltd.	
G–ATCU	Cessna 337	University of Cambridge	
G–ATCX	Cessna 182H Skylane	I. H. Sugden	
G–ATCY	PA-23 Aztec 250	A.I.C. Security Services Ltd.	
G–ATDA	PA-28 Cherokee 160	D. E. Siviter (Motors) Ltd.	
G–ATDB	Nord 1101 Noralpha	J. B. Jackson	
G–ATDC	PA-23 Aztec 250	Edinburgh Flying Services Ltd.	
G–ATDN	A.61 Terrier 2	J. F. Moore	
G–ATDO	Bo 208C Junior	H. Swift	
G–ATDS	HPR.7 Herald 209	Express Air Services (CI) Ltd.	
G–ATDV	PA-24 Comanche 400	Ugrain Aviation (Jersey) Ltd.	
G–ATDZ	Z-326 Trener Master	M. D. Popoff	
G–ATED	Hiller UH-12E	North Scottish Helicopters Ltd.	
G–ATEF	Cessna 150E	M. R. Nichols & ptnrs.	
G–ATEG	Cessna 150E	Northair Aviation Ltd.	
G–ATEM	PA-28 Cherokee 180	G. Wyles & W. Adams	
G–ATEP	EAA Biplane	E. L. Martin	
G–ATES	PA-32 Cherokee Six 260	Aeroscot Ltd.	
G–ATEV	Jodel DR. 1050	B. A. Mills & G. W. Payne	
G–ATEW	PA-30 Twin Comanche 160	Air Northumbria Group	
G–ATEX	Victa Airtourer 100	D. C. Giles & ptnrs.	
G–ATEZ	PA-28 Cherokee 140	J. A. Burton	
G–ATFA	Bensen B-8	J. Butler	
G–ATFD	Jodel DR.1050	H. Fawcett & ptnrs.	
G–ATFF	PA-23 Aztec 250	Topflight Aviation Co. Ltd.	
G–ATFG	Brantly B.2B	R. J. Chapman Ltd.	
G–ATFK	PA-30 Twin Comanche 160	L. J. Martin	
G–ATFL	Cessna F.172F	R. L. Beverley	

Notes	Reg.	Type	Owner or Operator
	G–ATFM	Sikorsky S-61N	British Airways Helicopters Ltd.
	G–ATFR	PA-25 Pawnee 150	J. F. Pelham-Born & R. V. Miller
	G–ATFU	D.H.85 Leopard Moth	A. H. Carrington & C. D. Duthy-James
	G–ATFV	Agusta-Bell 47J-2A	GSM Helicopters Ltd.
	G–ATFW	Luton L.A.4A Minor	A. J. Hammond
	G–ATFX	Cessna F.172G	Birmingham Windscreens Ltd.
	G–ATFY	Cessna F.172G	E. Cure Ltd.
	G–ATGC	Victa Airtourer 100	Barnside Flying Group
	G–ATGE	Jodel DR.1050	J. R. Roberts
	G–ATGF	M.S.892A Rallye Commodore 150	E. G. Bostock & R. A. Punter
	G–ATGG	M.S.885 Super Rallye	Southend Flying Club Ltd.
	G–ATGH	Brantly B.2B	R. Crook
	G–ATGO	Cessna F.172G	Bristol & Wessex Aeroplane Club Ltd.
	G–ATGP	Jodel DR.1050	W. M. Haley
	G–ATGY	GY.80 Horizon	P. W. Gibberson
	G–ATGZ	GH-4 Gyroplane	G. Griffiths
	G–ATHA	PA-23 Apache 235	A. Tucker
	G–ATHD	D.H.C.1 Chipmunk 22	Spartan Flying Group Ltd.
	G–ATHF	Cessna 150F	Cambridge Aero Club Ltd.
	G–ATHG	Cessna 150F	G. T. Williams
	G–ATHH	Nord 1101 Noralpha	J. G. Kay
	G–ATHJ	PA-23 Aztec 250	B. J. & A. K. Whitemore
	G–ATHK	Aeronca 7AC Champion	A. Corran
	G–ATHL	Wallis WA-116/F	G. V. Wallis
	G–ATHM	Wallis WA-116 Srs. 1	Wallis Autogyros Ltd.
	G–ATHN	Nord 1101 Noralpha	E. L. Martin
	G–ATHR	PA-28 Cherokee 180	Britannia Airways Ltd.
	G–ATHT	Victa Airtourer 115	Southend District Flying Club
	G–ATHU	A.61 Terrier 1	Marquess of Salisbury & Hon. C. E. V. Cecil
	G–ATHV	Cessna 150F	A. W. Pyle
	G–ATHW	Mooney Mk. 20E	F. J. L. Aran
	G–ATHX	Jodel DR. 100A	J. W. West
	G–ATHZ	Cessna 150F	Solitair Flight Management Ltd.
	G–ATIA	PA-24 Comanche 260	A. P. Holmes & ptnrs.
	G–ATIB	Bensen B.8M	K. J. Atkins
	G–ATIC	Jodel DR.1050	R. J. Hurstone & G. D. Kinnie
	G–ATID	Cessna 337	C. S. Bishop Ltd.
	G–ATIE	Cessna 150F	Staverton Flying School Ltd.
	G–ATIG	HPR-7 Herald 214	Brymon Aviation Ltd.
	G–ATIN	Jodel D.117	Mrs. M. J. Underhill
	G–ATIR	Stampe SV.4C	Mitchell Aviation
	G–ATIS	PA-28 Cherokee 160	P. H. Gurney
	G–ATIX	Nord 1101 Noralpha	A. E. Hutton
	G–ATIZ	Jodel D.117	N. Chandler
	G–ATJA	Jodel DR.1050	Lesley Dee Fashions (Groby) Ltd.
	G–ATJC	Victa Airtourer 100	G. F. Richardson & G. J. Cresswell
	G–ATJD	PA-28 Cherokee 140	Novair Ltd.
	G–ATJF	PA-28 Cherokee 140	Alpha Flying Club
	G–ATJG	PA-28 Cherokee 140	A. Nayyar & C. Lindesay
	G–ATJL	PA-24 Comanche 260	M. W. Webb
	G–ATJN	Jodel D.119	A. L. Wickens
	G–ATJP	PA-23 Apache 160	DMR Computer Ltd.
	G–ATJR	PA-E23 Aztec 250	Bonaire
	G–ATJT	GY.80 Horizon 160	M. Chamberlain
	G–ATJU	Cessna 150F	Sunderland Flying Club Ltd.
	G–ATJV	PA-32 Cherokee Six 260	Skytrade
	G–ATJW	Nord 1101 Noralpha	H. W. Elkin
	G–ATJX	Bu 131 Jungmann (AT + JX)	J. E. Fricker & G. H. A. Bird
	G–ATJZ	PA-E23 Aztec 250	New Guarantee Trust Ltd.
	G–ATKC	Stampe S.V.4B	Tiger Club Ltd.
	G–ATKD	Cessna 150F	Cambridge Aero Club Ltd.
	G–ATKE	Cessna 150F	Skegness Air Taxi Services Ltd.
	G–ATKF	Cessna 150F	Cambridge Aero Club Ltd.
	G–ATKG	Hiller UH-12B	Bristow Helicopters Ltd.
	G–ATKH	Luton L.A.4A Minor	L. Hepper
	G–ATKI	Piper J-3C-65 Cub	A. C. Netting
	G–ATKS	Cessna F.172G	Derek Crouch (Contractors) Ltd.
	G–ATKT	Cessna F.172G	N. Y. Souster
	G–ATKU	Cessna F.172G	S. E. Ward & Sons (Engineers) Ltd.
	G–ATKX	Jodel D.140C	Tiger Club Ltd.

G–AOHV (Top) V.802 Viscount of British Air Ferries/*A. S. Wright*

G–AOSU (Centre) D.H.C.1 Chipmunk 22 (Lycoming)

G-ATJX (Bottom) Bucker Bu 131 Jungmann

Reg.	Type	Owner or Operator	Notes
G–ATKY	Cessna 150F	R. L. Beverley	
G–ATKZ	T.66-2 Nipper	M. W. Knights	
G–ATLA	Cessna 182J Skylane	Shefford Transport Engineers Ltd.	
G–ATLB	Jodel DR.1050-M1	Tiger Club Ltd.	
G–ATLC	PA-23 Aztec 250	Alderney Air Charter Ltd.	
G–ATLD	Cessna E-310K	Centreline Air Services Ltd.	
G–ATLG	Hiller UH-12B	Bristow Helicopters Ltd.	
G–ATLH	Fewsdale Gyro-Glider	F. Fewsdale	
G–ATLM	Cessna F.172G	Yorkshire Flying Services Ltd.	
G–ATLN	Cessna F.172G	Ceejay Aviation	
G–ATLO	Brantly 305	Freemans of Bewdley (Aviation) Ltd.	
G–ATLP	Bensen B.8M	C. D. Julian	
G–ATLR	Cessna F.172G	A. Wood & R. F. Patmore	
G–ATLT	Cessna U-206A	North Denes Aerodrome Ltd.	
G–ATLW	PA-28 Cherokee 180	Links Systems Ltd.	
G–ATMB	Cessna F.150F	J. R. Gray	
G–ATMC	Cessna F.150F	Cambridge Aero Club Ltd.	
G–ATMG	M.S.893 Rallye Commodore 180	F. W. Fay & ptnrs.	
G–ATMH	Beagle D.5/180 Husky	Devon & Somerset Gliding Club Ltd.	
G–ATMI	H.S.748 Srs. 2A	Dan-Air Services Ltd.	
G–ATMJ	H.S.748 Srs. 2A	Dan-Air Services Ltd.	
G–ATMM	Cessna F.150F	Cambridge Technical Developments (Leasing) Ltd.	
G–ATMN	Cessna F.150F	Routair Flying Services Ltd.	
G–ATMP	Cessna 210F	J. L. Way	
G–ATMT	PA-30 Twin Comanche 160	D. H. T. Bain	
G–ATMU	PA-23 Apache 160	Randy International Flying Associates Ltd.	
G–ATMW	PA-28 Cherokee 140	Bencray Ltd.	
G–ATMX	Cessna F.150F	H. M. Synge	
G–ATMY	Cessna 150F	S. E. Fellows	
G–ATNB	PA-28 Cherokee 180	Chaplin Auto Preparation Ltd.	
G–ATNE	Cessna F.150F	Bristol & Wessex Aeroplane Club Ltd.	
G–ATNI	Cessna F.150F	Rolim Ltd.	
G–ATNK	Cessna F.150F	Lothian Air Hire Ltd.	
G–ATNL	Cessna F.150F	R. & Mrs. P. R. Budd	
G–ATNU	Cessna 182A	A. Bennett & W. W. Willis	
G–ATNV	PA-24 Comanche 260	Self-Fly Europe	
G–ATNX	Cessna F.150F	Mooney Aviation Ltd.	
G–ATOA	PA-23 Apache 160	Aviation Enterprises	
G–ATOD	Cessna F.150F	W. T. Tummon	
G–ATOE	Cessna F.150F	Argo Air Services Ltd.	
G–ATOH	D.62B Condor	G. H. Daniels	
G–ATOI	PA-28 Cherokee 140	O. & E. Flying Ltd.	
G–ATOJ	PA-28 Cherokee 140	O. T. Kernahan	
G–ATOK	PA-28 Cherokee 140	Link-Hampson Ltd.	
G–ATOL	PA-28 Cherokee 140	Tamar Flying Group	
G–ATOM	PA-28 Cherokee 140	I. T. Dalingwater & ptnrs.	
G–ATON	PA-28 Cherokee 140	Newcastle-upon-Tyne Aero Club Ltd.	
G–ATOO	PA-28 Cherokee 140	W. Davies	
G–ATOP	PA-28 Cherokee 140	T. R. & Mrs. E. A. Wiltshire	
G–ATOR	PA-28 Cherokee 140	J. P. Taylor	
G–ATOS	PA-28 Cherokee 140	AFT Craft Ltd.	
G–ATOT	PA-28 Cherokee 180	Roy Peplow & Co.	
G–ATOU	Mooney M.20E Super 21	Charleslock Motors Ltd.	
G–ATOY	*PA-24 Comanche 260	Museum of Flight, E. Fortune	
G–ATPD	H.S. 125 Srs 1B	Juliett Aviation Ltd.	
G–ATPE	H.S. 125-1B	Moseley Group (PSV) Ltd.	
G–ATPJ	BAC One-Eleven 301	Dan-Air Services Ltd.	
G–ATPL	BAC One-Eleven 301	Dan-Air Services Ltd.	
G–ATPM	Cessna F.150F	Dan-Air Flying Club	
G–ATPN	PA-28 Cherokee 140	D. A. Thompson & L. Martin	
G–ATPT	Cessna 182J Skylane	Western Models Ltd.	
G–ATPV	JB.01 Minicab	S. Russell	
G–ATRC	Beech B.95A Travel Air	Rogers Aviation Sales Ltd.	
G–ATRG	PA-18 Super Cub 150	Lasham Gliding Soc. Ltd.	
G–ATRI	Bo 208C Junior	W. H. Jones	
G–ATRK	Cessna F.150F	A. Ditheridge	
G–ATRL	Cessna F.150F	R. J. Jackson	
G–ATRM	Cessna F.150F	J. W. C. A. Coulcutt	
G–ATRN	Cessna F.150F	J. Gregson & L. Chiappi	
G–ATRO	PA-28 Cherokee 140	390th Flying Group	

Notes	Reg.	Type	Owner or Operator
	G–ATRP	PA-28 Cherokee 140	T. R. E. Cook
	G–ATRR	PA-28 Cherokee 140	W. B. J. & A. M. Davis
	G–ATRU	PA-28 Cherokee 180	Britannia Airways Ltd.
	G–ATRW	PA-32 Cherokee Six 260	Kent Messenger Ltd.
	G–ATRX	PA-32 Cherokee Six 260	R. F. Gibbs
	G–ATRY	Alon A-2 Aircoupe	B. W. George
	G–ATSB	PA-E23 Aztec 250	A. M. Stol Air Ltd. (G–ATZJ)
	G–ATSI	Bo 208C Junior	T. M. H. Paterson
	G–ATSL	Cessna F.172G	H. G. Le Cheminant
	G–ATSM	Cessna 337A	Tremletts (Skycraft) Ltd.
	G–ATSR	Beech M.35 Bonanza	Alstan Aviation Ltd.
	G–ATST	M.S.893A Rallye Commodore	Blackwood Tugs Ltd.
	G–ATSU	Jodel D.140B	J. S. Burnett Ltd.
	G–ATSX	Bo 208C Junior	Cotswold Aero Club Ltd.
	G–ATSY	Wassmer WA41 Super Baladou IV	Gilbey Warren Co. Ltd.
	G–ATSZ	PA-30 Twin Comanche 160	Air Peterborough
	G–ATTB	Wallis WA.116-1	D. A. Wallis
	G–ATTD	Cessna 182J Skylane	Hanro Aviation Ltd.
	G–ATTF	PA-28 Cherokee 140	Mooney Aviation Ltd.
	G–ATTG	PA-28 Cherokee 140	J. F. Thurlow & J. H. Pickering
	G–ATTI	PA-28 Cherokee 140	A. S. Dredge & A. Gaydon
	G–ATTK	PA-28 Cherokee 140	Andrewsfield Flying Club Ltd.
	G–ATTM	Jodel DR. 250-160	R. W. Tomkinson
	G–ATTP	BAC One-Eleven 207	Dan-Air Services Ltd.
	G–ATTR	Bo 208C Junior 3	S. Luck
	G–ATTU	PA-28 Cherokee 140	Leith Air Ltd.
	G–ATTV	PA-28 Cherokee 140	Somerville Heighes
	G–ATTX	PA-28 Cherokee 180	Violet M. Lambeth
	G–ATTY	PA-32 Cherokee Six 260	L. A. Dingemans & D. J. Everett
	G–ATUA	PA-25 Pawnee 235	A.D.S. (Aerial) Ltd.
	G–ATUB	PA-28 Cherokee 140	British Airways
	G–ATUC	PA-28 Cherokee 140	Airways Aero Associations Ltd.
	G–ATUD	PA-28 Cherokee 140	British Airways
	G–ATUF	Cessna F.150F	Devair Aviation Services Ltd.
	G–ATUG	D.62B Condor	A. M. Hazell & C. B. Marsh
	G–ATUH	T.66 Nipper	V. H. Hallam
	G–ATUI	Bo 208C Junior	Cotswold Aero Club
	G–ATUL	PA-28 Cherokee 180	R. J. Roberts & Sons
	G–ATVF	D.H.C.1 Chipmunk 22	RAFGSA
	G–ATVG	Hiller UH-12E	Management Aviation Ltd.
	G–ATVH	BAC One-Eleven 207	Dan-Air Services Ltd. *City of Newcastle-upon-Tyne*
	G–ATVI	SIPA 903	J. Martin
	G–ATVK	PA-28 Cherokee 140	E. A. Clack
	G–ATVL	PA-28 Cherokee 140	West London Aero Services
	G–ATVO	PA-28 Cherokee 140	E. C. Andrews
	G–ATVP	*F.B.5 Gunbus (2345)	RAF Museum
	G–ATVS	PA-28 Cherokee 180	Marshalls Woodflakes Ltd.
	G–ATVW	D.62B Condor	J. R. Stainer & D. W. Evernden
	G–ATVX	Bo 208C Junior	G. W. Stanmore
	G–ATWA	Jodel DR.1050	Llamedoes Flying Group
	G–ATWE	M.S.892A Rallye Commodore	D. I. Murray
	G–ATWG	PA-30 Twin Comanche 160	L. R. Davies
	G–ATWJ	Cessna F.172F	C. J. & J. Freeman
	G–ATWL	Jodel D.120	T. A. S. Rosie
	G–ATWO	PA-28 Cherokee 180	Vectaphone Manufacturing Ltd.
	G–ATWP	Alon A-2 Aircoupe	F. Bolton
	G–ATWR	PA-30 Twin Comanche 160	Lubenham Fidelities & Investments Co. Ltd.
	G–ATWS	Luton L.A.4A Minor	K. J. Hazelwood
	G–ATWV	Boeing 707-336C	British Airways
	G–ATWZ	M.S.892 Rallye Commodore	R. G. Allen & K. F. Olivier
	G–ATXA	PA-22 Tri-Pacer 150	J. W. Holmes
	G–ATXD	PA-30 Twin Comanche 160	Southwark Estates Ltd.
	G–ATXF	GY-80 Horizon 150	A. I. Milne
	G–ATXM	PA-28 Cherokee 180	J. Khan
	G–ATXN	Mitchell-Proctor Kittiwake	D. W. Kent
	G–ATXO	SIPA 903	M. Hillam
	G–ATXR	AFB 1 gas balloon	Mrs. C. M. Bulmer *Omega One*
	G–ATXZ	Bo 208C Junior	J. I. Visser
	G–ATYA	PA-25 Pawnee 235	Skegness Air Taxi Services Ltd.
	G–ATYM	Cessna F.150G	J. T. Tyer

Reg.	Type	Owner or Operator	Notes
G-ATYN	Cessna F.150G	Skegness Air Taxi Services Ltd.	
G-ATYS	PA-28 Cherokee 180	E. K. Chalke	
G-ATYV	Bell 47G	Two Mile Oak Garage	
G-ATYZ	M.S.880B Rallye Club	Nylo Flying Group	
G-ATZA	Bo 208C Junior	W. C. Roberts	
G-ATZB	Hiller UH-12B	Bristow Helicopters Ltd.	
G-ATZC	Boeing 707-365C	British Caledonian Airways *Loch Katrine*	
G-ATZD	Boeing 707-365C	British Airways	
G-ATZG	AFB2 gas balloon	F./Lt. S. Cameron *Aeolis*	
G-ATZK	PA-28 Cherokee 180	MGF Racing Sidecars	
G-ATZM	Piper J-3C-65 Cub	R. W. Davison	
G-ATZO	Beagle B.206 Srs. 1	Arnos Exporting Co. Ltd.	
G-ATZS	Wassmer WA41 Super IV Baladou	J. R. MacAlpine-Dounie & P. A. May	
G-ATZU	PA-30 Twin Comanche 160	Routair Flying Services Ltd.	
G-ATZV	PA-30 Twin Comanche 160	Aviation International (CI) Ltd.	
G-ATZY	Cessna F.150G	Air Service Training Ltd.	
G-ATZZ	Cessna F.150G	Argo Air Services Ltd.	
G-AUTO	Cessna 441 Conquest	Automobile Association Ltd.	
G-AVAA	Cessna F.150G	Argo Air Services Ltd.	
G-AVAI	H.S.125 Srs. 3B	Aravco Ltd.	
G-AVAJ	Hiller UH-12B	Bristow Helicopters Ltd.	
G-AVAO	PA-30 Twin Comanche 160	Heath Street Car Hirings Ltd.	
G-AVAP	Cessna F.150G	Seawing Flying Club Ltd.	
G-AVAR	Cessna F.150G	West Wales Flying Club	
G-AVAS	Cessna F.172H	Birmingham Aviation Ltd.	
G-AVAU	PA-30 Twin Comanche 160	L. Batin	
G-AVAW	D.62B Condor	Avato Flying Group	
G-AVAX	PA-28 Cherokee 180	College of Air Training	
G-AVAY	PA-28 Cherokee 180	College of Air Training	
G-AVAZ	PA-28 Cherokee 180	College of Air Training	
G-AVBA	PA-28 Cherokee 180	College of Air Training	
G-AVBB	PA-28 Cherokee 180	College of Air Training	
G-AVBC	PA-28 Cherokee 180	College of Air Training	
G-AVBE	PA-28 Cherokee 180	College of Air Training	
G-AVBG	PA-28 Cherokee 180	College of Air Training	
G-AVBH	PA-28 Cherokee 180	College of Air Training	
G-AVBL	PA-30 Twin Comanche 160	Oakden Investments Ltd.	
G-AVBP	PA-28 Cherokee 140	Bristol & Wessex Aeroplane Club [Ltd.	
G-AVBS	PA-28 Cherokee 180	I. J. James	
G-AVBT	PA-28 Cherokee 180	Cardinal Engineering Ltd.	
G-AVBU	PA-32 Cherokee Six 260	Tempus Aviation (Holdings) Ltd.	
G-AVBW	BAC One-Eleven 320	Laker Airways	
G-AVBX	BAC One-Eleven 320	Laker Airways	
G-AVBY	BAC One-Eleven 320	Laker Airways	
G-AVBZ	Cessna F.172H	J. Seville	
G-AVCA	Brantly B.2B	M. J. & Mrs. G. M. Page	
G-AVCC	Cessna F.172H	Mercia Estates Ltd.	
G-AVCD	Cessna F.172H	Toon Ghose Aviation Ltd.	
G-AVCE	Cessna F.172H	Cleco Electrical Industries Ltd.	
G-AVCM	PA-24 Comanche 260	F. Smith & Sons Ltd.	
G-AVCS	A.61 Terrier 1	L. M. Farrell & A. R. C. Hunter	
G-AVCT	Cessna F.150G	Sierra Aviation Services	
G-AVCU	Cessna F.150G	P. R. Moss	
G-AVCV	Cessna 182J Skylane	J. A. Moores	
G-AVCW	PA-30 Twin Comanche 160	Manro Transport Ltd.	
G-AVCX	PA-30 Twin Comanche 160	F. J. Stevens	
G-AVCY	PA-30 Twin Comanche 160	T. S. Grimshaw Ltd.	
G-AVDA	Cessna 182K Skylane	J. W. Grant	
G-AVDE	Turner Gyroglider Mk. 1	J. S. Smith	
G-AVDF	*Beagle Pup 100	Brighton Transport Museum	
G-AVDG	Wallis WA-116 Srs. 1	K. H. Wallis	
G-AVDR	Beech B80 Queen Air	Air Camelot	
G-AVDS	Beech B80 Queen Air	Air Camelot	
G-AVDT	Aeronca 7AC Champion	W. R. Prescott	
G-AVDW	D.62B Condor	Essex Aviation	
G-AVDX	H.S.125 Srs. 3B	Civil Aviation Authority	
G-AVDY	Luton L.A.4A Minor	D. E. Evans & ptnrs.	
G-AVDZ	PA-25 Pawnee 235	Skegness Air Taxi Services Ltd.	
G-AVEB	Morane MS 230 Et 2	Hon. P. Lindsay	

Notes	Reg.	Type	Owner or Operator
	G–AVEC	Cessna F.172H	W. H. Ekin (Engineering) Co. Ltd.
	G–AVEF	Jodel D.150	Tiger Club Ltd.
	G–AVEG	SIAI-Marchetti S.205	Hanway Car Sales Ltd.
	G–AVEH	SIAI-Marchetti S.205	K. D. Gomm
	G–AVEM	Cessna F.150G	Phoenix Aviation Ltd.
	G–AVEN	Cessna F.150G	Glenrothes Flying School Ltd.
	G–AVEO	Cessna F.150G	Phoenix Aviation Ltd.
	G–AVER	Cessna F.150G	B. I. Chapman
	G–AVET	Beech C55 Baron	Spline Gauges Ltd.
	G–AVEU	Wassmer WA.41 Baladou	Baladou Flying Group
	G–AVEX	D.62B Condor	Cotswold Roller Hire Ltd.
	G–AVEY	Currie Super Wot	A. Eastelow
	G–AVEZ	HPR-7 Herald 210	Air UK
	G–AVFA	H.S.121 Trident 2E	British Airways
	G–AVFB	H.S.121 Trident 2E	British Airways
	G–AVFC	H.S.121 Trident 2E	British Airways
	G–AVFD	H.S.121 Trident 2E	British Airways
	G–AVFE	H.S.121 Trident 2E	British Airways
	G–AVFF	H.S.121 Trident 2E	British Airways
	G–AVFG	H.S.121 Trident 2E	British Airways
	G–AVFH	H.S.121 Trident 2E	British Airways
	G–AVFI	H.S.121 Trident 2E	British Airways
	G–AVFJ	H.S.121 Trident 2E	British Airways
	G–AVFK	H.S.121 Trident 2E	British Airways
	G–AVFL	H.S.121 Trident 2E	British Airways
	G–AVFM	H.S.121 Trident 2E	British Airways
	G–AVFN	H.S.121 Trident 2E	British Airways
	G–AVFO	H.S.121 Trident 2E	British Airways
	G–AVFP	PA-28 Cherokee 140	H. D. Vince Ltd.
	G–AVFR	PA-28 Cherokee 140	P. J. Sellar & B. M. O'Brien
	G–AVFS	PA-32 Cherokee Six 300	Headcorn Parachute Club Ltd.
	G–AVFU	PA-32 Cherokee Six 300	S. L. H. Construction Ltd.
	G–AVFW	PA-30 Twin Comanche 160	Woodlands Investments Ltd.
	G–AVFX	PA-28 Cherokee 140	J. E. Lawson
	G–AVFY	PA-28 Cherokee 140	F. Spencer-Jones
	G–AVFZ	PA-28 Cherokee 140	Keenair Services Ltd.
	G–AVGA	PA-24 Comanche 260	A. Rennie
	G–AVGB	PA-28 Cherokee 140	G. Abbot
	G–AVGC	PA-28 Cherokee 140	B. A. Bennett
	G–AVGD	PA-28 Cherokee 140	A. D. Wren
	G–AVGE	PA-28 Cherokee 140	A. G. Branch Contractors Ltd.
	G–AVGH	PA-28 Cherokee 140	Mooney Aviation Ltd.
	G–AVGI	PA-28 Cherokee 140	Bencray Ltd.
	G–AVGJ	Jodel DR.1050	H. A. N. Crocker
	G–AVGK	PA-28 Cherokee 180	Liverpool Aero Club Ltd.
	G–AVGL	Cessna F.150G	F .A. E. Pyle
	G–AVGN	PA-24 Comanche 260	Viscount Chelsea
	G–AVGP	BAC One-Eleven 408	British Airways
	G–AVGU	Cessna F.150G	J. A. & Mrs. J. M. C. Pothecary
	G–AVGV	Cessna F.150G	British Skysports
	G–AVGY	Cessna 182K Skylane	H. P. Nicholls
	G–AVGZ	Jodel DR.1050	W. H. Gilchrist & A. T. Howarth
	G–AVHF	Beech A.23 Musketeer	R. W. Neale
	G–AVHH	Cessna F.172H	R. T. Pritchard
	G–AVHJ	Wassmer WA.41 Baladou	D. G. Pickering & ptnrs.
	G–AVHL	Jodel DR.105A	G. L. Winterbourne
	G–AVHM	Cessna F.150G	M. Tosh
	G–AVHN	Cessna F.150G	Bristol and Wessex Aero Club Ltd.
	G–AVHT	Auster AOP.9 (WZ711)	M. Somerton-Rayner
	G–AVHY	Fournier RF.4D	R. Swinn & J. Conolly
	G–AVHZ	PA-30 Twin Comanche 160	P. S. Bubbear & J. M. Glanville
	G–AVIA	Cessna F.150G	Laarbruch Flying Club
	G–AVIB	Cessna F.150G	Warwickshire Aero Club
	G–AVIC	Cessna F.172H	Pembrokeshire Air
	G–AVID	Cessna 182J	Inch Farming Co. Ltd.
	G–AVIE	Cessna F.172H	N. Denes Aerodrome Ltd.
	G–AVIG	A-B 206B JetRanger	Bristow Helicopters Ltd.
	G–AVII	A-B 206A JetRanger	Bristow Helicopters Ltd.
	G–AVIL	Alon A.2 Aircoupe	D. W. Vernon
	G–AVIN	M.S.880B Rallye Club	D. R. F. Sapte
	G–AVIO	M.S.880B Rallye Club	A. F. Bullock
	G–AVIP	Brantly B.2B	Cosworth Engineering Ltd.
	G–AVIR	Cessna F.172H	W. Lancashire Aero Club Ltd.

Reg.	*Type*	*Owner or Operator*	*Notes*
G–AVIS	Cessna F.172H	Jon Paul Photography Ltd.	
G–AVIT	Cessna F.150G	Shropshire Aero Club Ltd.	
G–AVIZ	Scheibe SF.25A Motorfalke	D. C. Pattison & D. A. Wilson	
G–AVJB	V.815 Viscount	British Air Ferries	
G–AVJE	Cessna F.150G	L. R. Churchill & P. R. Green	
G–AVJF	Cessna F.172H	W. R. & Mrs. B. M. Young	
G–AVJG	Cessna 337B	P. R. Moss	
G–AVJH	D.62 Condor	Lleyn Flying Group	
G–AVJI	Cessna F.172H	Royal Artillery Aero Club Ltd.	
G–AVJJ	PA-30 Twin Comanche 160	Peregrine Air Services Ltd.	
G–AVJK	Jodel DR.1050 M.1	J. H. B. Urmston	
G–AVJN	Brantly B.2B	John Berry Ltd.	
G–AVJO	Fokker E.111 Replica (422)	Personal Plane Services Ltd.	
G–AVJU	PA-24 Comanche 260	Anchor Finance Ltd.	
G–AVJV	Wallis WA-117 Srs. 1	K. H. Wallis (G–ATCV)	
G–AVJW	Wallis WA-118 Srs. 2	K. H. Wallis (G–ATPW)	
G–AVKB	MB.50 Pipistrelle	R. K. Haldenby & T. S. Warren	
G–AVKD	Fournier RF.4D	Lasham RF4 Group	
G–AVKE	*Gadfly HDW.1	British Rotorcraft Museum	
G–AVKG	Cessna F. 172H	W. Lancs Aero Club Ltd.	
G–AVKI	Nipper T.66 Srs. 3	R. J. Corbett & ptnrs.	
G–AVKJ	Nipper T.66 Srs. 3	P. W. Hunter	
G–AVKM	D.62B Condor	H. R. Rowley & ptnrs.	
G–AVKN	Cessna 401	Barline Aviation Ltd.	
G–AVKP	A.109 Airedale	British Gliding Association Ltd.	
G–AVKR	Bo 208C Junior	D. F. Barley & D. A. Bishop	
G–AVKS	Bell 47G-2	Bristow Helicopters Ltd.	
G–AVKX	Hiller UH-12E	Management Aviation Ltd.	
G–AVKY	Hiller UH-12E	Agricopters Ltd.	
G–AVKZ	PA-23 Aztec 250	Hunnable Holdings Ltd.	
G–AVLA	PA-28 Cherokee 140	Mrs. J. I. & P. Sprent	
G–AVLB	PA-28 Cherokee 140	D. C. Tyne & Co.	
G–AVLC	PA-28 Cherokee 140	F. C. V. Hopkins	
G–AVLD	PA-28 Cherokee 140	P. V. Evans	
G–AVLE	PA-28 Cherokee 140	P. A. Johnstone & E. J. Morgan	
G–AVLF	PA-28 Cherokee 140	W. London Aero Services Ltd.	
G–AVLG	PA-28 Cherokee 140	D. Golding & P. J. Pearce	
G–AVLH	PA-28 Cherokee 140	T. L. Wilkinson	
G–AVLI	PA-28 Cherokee 140	R. H. Neale	
G–AVLJ	PA-28 Cherokee 140	Barbers Animal Products Ltd. & E. D. Hughes	
G–AVLN	B.121 Pup 2	C. B. G. Masefield	
G–AVLO	Bo 208C Junior	P. J. Thompson & R. G. Williams	
G–AVLP	PA-23 Aztec 250	B.K.S. Surveys Ltd.	
G–AVLR	PA-28 Cherokee 140	G. H. & Mrs. K. M. Wylde	
G–AVLS	PA-28 Cherokee 140	K. Dando	
G–AVLT	PA-28 Cherokee 140	E. A. Clack & M. T. Pritchard	
G–AVLU	PA-28 Cherokee 140	London Transport (Central Road Services) Sports Association Flying Club	
G–AVLW	Fournier RF 4D	T. M. W. Webster	
G–AVLY	Jodel D.120A	R. Arbon & ptnrs.	
G–AVMA	GY.80 Horizon 180	B. R. & S. Hildick	
G–AVMB	D.62B Condor	Tiger Club Ltd.	
G–AVMD	Cessna 150G	K. J. Jarvis	
G–AVMF	Cessna F. 150G	J. D. Palfreman	
G–AVMH	BAC One-Eleven 510	British Airways	
G–AVMI	BAC One-Eleven 510	British Airways	
G–AVMJ	BAC One-Eleven 510	British Airways	
G–AVMK	BAC One-Eleven 510	British Airways	
G–AVML	BAC One-Eleven 510	British Airways	
G–AVMM	BAC One-Eleven 510	British Airways	
G–AVMN	BAC One-Eleven 510	British Airways	
G–AVMO	BAC One-Eleven 510	British Airways	
G–AVMP	BAC One-Eleven 510	British Airways	
G–AVMR	BAC One-Eleven 510	British Airways	
G–AVMS	BAC One-Eleven 510	British Airways	
G–AVMT	BAC One-Eleven 510	British Airways	
G–AVMU	BAC One-Eleven 510	British Airways	
G–AVMV	BAC One-Eleven 510	British Airways	
G–AVMW	BAC One-Eleven 510	British Airways	
G–AVMX	BAC One-Eleven 510	British Airways	
G–AVMY	BAC One-Eleven 510	British Airways	

Notes	Reg.	Type	Owner or Operator
	G–AVMZ	BAC One-Eleven 510	British Airways
	G–AVNB	Cessna F.150G	G. A. J. Bowles
	G–AVNC	Cessna F.150G	Northampton Aviation Services Ltd.
	G–AVNG	Beech A80 Queen Air	Parker & Heard Ltd.
	G–AVNI	PA-30 Twin Comanche	D.P. Aviation
	G–AVNL	PA-23 Aztec 250	Cabair Ltd.
	G–AVNM	PA-28 Cherokee 180	College of Air Training
	G–AVNN	PA-28 Cherokee 180	College of Air Training
	G–AVNO	PA-28 Cherokee 180	College of Air Training
	G–AVNP	PA-28 Cherokee 180	College of Air Training
	G–AVNR	PA-28 Cherokee 180	College of Air Training
	G–AVNS	PA-28 Cherokee 180	College of Air Training
	G–AVNT	PA-28 Cherokee 180	College of Air Training
	G–AVNU	PA-28 Cherokee 180	College of Air Training
	G–AVNV	PA-28 Cherokee 180	College of Air Training
	G–AVNW	PA-28 Cherokee 180	College of Air Training
	G–AVNX	Fournier RF-4D	O. C. Harris & C. G. Masterman
	G–AVNY	Fournier RF-4D	A. N. Mavrogordato
	G–AVNZ	Fournier RF-4D	Cobra Group
	G–AVOA	Jodel DR.1050	I. MacPherson
	G–AVOD	Beagle D5/180 Husky	J. Hockton
	G–AVOH	D.62B Condor	N. H. Jones
	G–AVOI	H.S.125 Srs. 3B	Markheath Securities Ltd.
	G–AVOM	Jodel DR.221	M. A. Mountford
	G–AVON	Luton LA.5A Major	G. R. Mee
	G–AVOO	PA-18-150 Super Cub	T. A. McMullin
	G–AVOR	Lockspeiser Land Development Aircraft	D. Lockspeiser
	G–AVOZ	PA-28 Cherokee 180	Downley Garages Ltd.
	G–AVPA	Sopwith Pup Replica	C. J. Warrilow
	G–AVPB	Boeing 707-336C	British Airtours Ltd.
	G–AVPC	D.31 Turbulent	J. Sharp
	G–AVPD	D.9 Bebe	S. W. McKay
	G–AVPE	H.S.125 Srs. 3B	British Aerospace
	G–AVPH	Cessna F.150G	W. Lancashire Aero Club
	G–AVPI	Cessna F.172H	R. Jones & J. Chancellor
	G–AVPJ	D.H.82A Tiger Moth	The Barnstormers Ltd.
	G–AVPK	M.S.892A Rallye Commodore	Papa Kilo Flying Group
	G–AVPL	M.S.892 Rallye Commodore	J. H. Atkinson
	G–AVPM	Jodel D.117	J. Houghton
	G–AVPN	HPR.7 Herald 213	Air UK
	G–AVPR	PA-30 Twin Comanche 160	Cold Storage (Jersey) Ltd.
	G–AVPS	PA-30 Twin Comanche 160	Russell Foster Holdings Ltd.
	G–AVPT	PA-18 Super Cub 150	Tiger Club Ltd.
	G–AVPU	PA-18 Super Cub 150	Scottish Gliding Union Ltd.
	G–AVPV	PA-28 Cherokee 180	E. A. Clack
	G–AVPX	Taylor JT.1 Monoplane	S. M. Smith
	G–AVPY	PA-25 Pawnee 235	Farm Aviation Services Ltd.
	G–AVRF	H.S.125 Srs. 3B	BAC (Holdings) Ltd.
	G–AVRG	H.S.125 Srs. 3B	Shell Aircraft Ltd.
	G–AVRK	PA-28 Cherokee 180	Dollar Air Services Ltd.
	G–AVRL	Boeing 737-204	Britannia Airways Ltd. *Sir Ernest Shackleton*
	G–AVRM	Boeing 737-204	Britannia Airways Ltd. *James Watt*
	G–AVRN	Boeing 737-204	Britannia Airways Ltd. *Capt. James Cook*
	G–AVRO	Boeing 737-204	Britannia Airways Ltd. *Sir Francis Drake*
	G–AVRP	PA-28 Cherokee 140	J. T. Turner & R. A. B. Perkins
	G–AVRS	GY.80 Horizon 180	Rogers Autos Ltd.
	G–AVRT	PA-28 Cherokee 140	F. Clarke
	G–AVRU	PA-28 Cherokee 180	Fenland Tractors Ltd.
	G–AVRW	GY-20 Minicab	R. B. Pybus
	G–AVRX	PA-23 Aztec 250	The Cronite Group Ltd.
	G–AVRY	PA-28 Cherokee 180	Roses Flying Group
	G–AVRZ	PA-28 Cherokee 180	Briglea Engineering Ltd.
	G–AVSA	PA-28 Cherokee 180	Alliance Aviation Ltd.
	G–AVSB	PA-28 Cherokee 180	White House Garage Ltd.
	G–AVSC	PA-28 Cherokee 180	W. London Aero Services Ltd.
	G–AVSD	PA-28 Cherokee 180	Tracair Ltd.
	G–AVSE	PA-28 Cherokee 180	S E Aviation Ltd.
	G–AVSF	PA-28 Cherokee 180	Goodwood Terrena Ltd.
	G–AVSH	PA-28 Cherokee 180	Elken Ltd.

Reg.	Type	Owner or Operator	Notes
G-AVSI	PA-28 Cherokee 140	E. P. van Mechelen	
G-AVSP	PA-28 Cherokee 180	Trig Engineering Ltd.	
G-AVSR	Beagle D 5/180 Husky	A. L. Young	
G-AVTB	Nipper T.66 Srs. 3	S. Stride & J. Hobday	
G-AVTE	Bell 206A JetRanger	W. R. Finance Ltd.	
G-AVTJ	PA-32 Cherokee Six 260	G. Jacobs	
G-AVTK	PA-32 Cherokee Six 260	Mannix Aviation Ltd.	
G-AVTM	Cessna F.150H	J. K. Bamrah & J. B. Warwicker	
G-AVTO	Cessna F.150H	M. W. Jacobs	
G-AVTP	Cessna F.172H	Western Air Training Ltd.	
G-AVTT	Ercoupe 415D	L. C. Bourne	
G-AVTV	M.S.893A Rallye Commodore	A. Lister	
G-AVTX	Taylor JT.1 Monoplane	P. Lockwood	
G-AVUA	Cessna F.172H	Recreational Flying Centre (Popham) Ltd.	
G-AVUD	PA-30 Twin Comanche 160B	F. M. Aviation	
G-AVUG	Cessna F.150H	Dukeries Aviation Ltd.	
G-AVUH	Cessna F.150H	Sunderland Flying Club Ltd.	
G-AVUI	Cessna F.150H	Dukeries Aviation Ltd.	
G-AVUJ	F.8L Falco 4	M. Shield	
G-AVUK	Enstrom F-28A	Spooner Aviation Ltd.	
G-AVUL	Cessna F.172H	D. H. Stephens & J. A. Blythe	
G-AVUM	Hughes 269B	P. Berriman	
G-AVUN	PA-30 Twin Comanche 160B	Southshine Securities Ltd.	
G-AVUS	PA-28 Cherokee 140	R. Matthews	
G-AVUT	PA-28 Cherokee 140	W. B. Bateson	
G-AVUU	PA-28 Cherokee 140	Goodwood Estate Co. Ltd.	
G-AVUV	Cessna 310N	Airwork Services Ltd.	
G-AVUZ	PA-32 Cherokee Six 300	P. & Mrs. M. E. Biggins	
G-AVVB	H.S.125 Srs. 3B	Brown & Root (UK) Ltd.	
G-AVVC	Cessna F.172H	Kestrel Air Ltd.	
G-AVVE	Cessna F.150H	A. J. McDonald & H. P. Vox	
G-AVVF	D.H.104 Dove 8	Martin Baker (Engineering) Ltd.	
G-AVVG	PA-28 Cherokee 180	604 Squadron Flying Club	
G-AVVI	PA-30 Twin Comanche 160B	Steepletone Products Ltd.	
G-AVVJ	M.S.893A Rallye Commodore	F. A. O. Gaze	
G-AVVL	Cessna F.150H	R. J. & Mrs. B. Ford-Sagers	
G-AVVM	Jodel D.117	R. R. Corker	
G-AVVN	D.62C Condor	Avato Flying Group	
G-AVVO	*Avro 652A Anson 19 (VL348)	Newark Air Museum	
G-AVVS	Hughes 269B	W. Holmes	
G-AVVT	PA-23 Aztec 250	Kayglynn Investment Holdings Ltd.	
G-AVVU	Beech A.23A Musketeer	Tony McCann Motors Ltd.	
G-AVVV	PA-28 Cherokee 180	C. B. Healey	
G-AVVW	Cessna F.150H	Bristol & Wessex Aeroplane Club [Ltd.	
G-AVVX	Cessna F.150H	J. C. & Mrs. M. Abberley	
G-AVVY	Cessna F.150H	W. S. Davies	
G-AVWA	PA-28 Cherokee 140	W. London Aero Services Ltd.	
G-AVWD	PA-28 Cherokee 140	Manchester School of Flying Ltd.	
G-AVWE	PA-28 Cherokee 140	W. C. C. Meyer	
G-AVWF	PA-28 Cherokee 140	Liverpool Aero Club Ltd.	
G-AVWG	PA-28 Cherokee 140	Bencray Ltd.	
G-AVWH	PA-28 Cherokee 140	B. P. W. Faithfull	
G-AVWI	PA-28 Cherokee 140	L. M. Veitch	
G-AVWJ	PA-28 Cherokee 140	E.F.G. Flying Services Ltd.	
G-AVWL	PA-28 Cherokee 140	Bristol & Wessex Aeroplane Club Ltd.	
G-AVWM	PA-28 Cherokee 140	E. A. Clack & M. T. Pritchard	
G-AVWN	PA-28R Cherokee Arrow 180	L. Cooksey	
G-AVWO	PA-28R Cherokee Arrow 180	C & S Controls Ltd.	
G-AVWR	PA-28R Cherokee Arrow 180	Mono-Construction Ltd.	
G-AVWT	PA-28R Cherokee Arrow 180	G. W. Barker & ptnrs.	
G-AVWU	PA-28R Cherokee Arrow 180	Horizon Flyers Ltd.	
G-AVWV	PA-28R Cherokee Arrow 180	Mapair Ltd.	
G-AVWW	Mooney M.20F	A. & J. & G. Cullen	
G-AVWY	Fournier RF-4D	T. G. Hoult	
G-AVXA	PA-25 Pawnee 235	Howard Avis (Aviation) Ltd.	
G-AVXB	PL-1 Gyroplane	C. Mowat	
G-AVXC	Nipper T.66 Srs. 3	W. G. Wells & ptnrs.	
G-AVXD	Nipper T.66 Srs. 3	J. M. Murrie	
G-AVXF	PA-28R Cherokee Arrow 180	Rogers Aviation Sales Ltd.	
G-AVXI	H.S.748 Srs. 2	Civil Aviation Authority	
G-AVXJ	H.S.748 Srs. 2	Civil Aviation Authority	

Notes	Reg.	Type	Owner or Operator
	G–AVXK	H.S.125 Srs. 3B-RA	Overseas Business Service Ltd.
	G–AVXV	Bleriot XI (BAPC 104)	L. D. Goldsmith
	G–AVXW	D.62B Condor	Medway Flying Group Ltd.
	G–AVXX	Cessna FR.172E	Hadrian Flying Group
	G–AVXY	Auster AOP.9 (XK417)	R. Windley
	G–AVYE	*H.S.121 Trident 1E-140	Science Museum Store, Wroughton
	G–AVYF	Beech A.23-24 Musketeer	Wearside Aviation Group
	G–AVYK	A.61 Terrier 3	Airways Aero Associations Ltd.
	G–AVYL	PA-28 Cherokee 180	Miller Aerial Spraying Ltd.
	G–AVYM	PA-28 Cherokee 180	Border Piper Aviation Ltd.
	G–AVYO	PA-28 Cherokee 140	Goodwood Estate Co. Ltd.
	G–AVYP	PA-28 Cherokee 140	T. D. Reid (Braids) Ltd.
	G–AVYR	PA-28 Cherokee 140	E. Drew (Construction & Plant Hire) Ltd.
	G–AVYS	PA-28R Cherokee Arrow 180	Leisure Sport Ltd.
	G–AVYT	PA-28R Cherokee Arrow 180	H. Stephenson
	G–AVYV	Jodel D.120	J. B. J. Berrow & M. A. Kaye
	G–AVYX	AB-206A JetRanger	S.W. Electricity Board
	G–AVYZ	BAC One-Eleven 320L	Laker Airways
	G–AVZA	IMCO Callair A-9	Arable & Bulb Chemicals Ltd.
	G–AVZB	Aero Z-37 Cmelak	ADS (Aerial) Ltd.
	G–AVZC	Hughes 269B	Ocean Rangers Charters Ltd.
	G–AVZE	D.62B Condor	A. J. M. Trowbridge & J. Harris
	G–AVZI	Bo 208C Junior	C. F. Rogers
	G–AVZM	Beagle B.121 Pup 1	ARAZ Group
	G–AVZN	Beagle B.121 Pup 1	W. E. Cro & Sons Ltd.
	G–AVZO	Beagle B.121 Pup 1	Dungenair Ltd.
	G–AVZP	Beagle B.121 Pup 1	T. A. White
	G–AVZR	PA-28 Cherokee 180	G. Knowles & ptnrs.
	G–AVZS	Cessna 310B	J & B's Aviation Ltd.
	G–AVZT	PA-31 Navajo	Cabair Ltd.
	G–AVZU	Cessna F.150H	Miss I. Sammons
	G–AVZV	Cessna F.172H	Hull Aero Club
	G–AVZW	EAA Model P Biplane	R. G. Maidment & G. R. Edmundson
	G–AVZX	M.S.880B Rallye Club	L. Clutton & ptnrs.
	G–AVZY	M.S.880B Rallye Club	R. McLindsay
	G–AWAA	M.S.880B Rallye Club	P. A. Cairns
	G–AWAC	GY-80 Horizon 180	A. E. Chapman
	G–AWAD	Beech D 55 Baron	College of Air Training
	G–AWAE	Beech D 55 Baron	College of Air Training
	G–AWAF	Beech D 55 Baron	College of Air Training
	G–AWAG	Beech D 55 Baron	College of Air Training
	G–AWAH	Beech D 55 Baron	College of Air Training
	G–AWAI	Beech D 55 Baron	College of Air Training
	G–AWAJ	Beech D 55 Baron	College of Air Training
	G–AWAK	Beech D 55 Baron	College of Air Training
	G–AWAL	Beech D 55 Baron	College of Air Training
	G–AWAM	Beech D 55 Baron	College of Air Training
	G–AWAN	Beech D 55 Baron	College of Air Training
	G–AWAO	Beech D 55 Baron	College of Air Training
	G–AWAP	SA 318B Alouette Astazou	Helicopter Hire Ltd.
	G–AWAT	D.62B Condor	Tiger Club Ltd.
	G–AWAU	*Vickers F.B.27A Vimy (replica) (F8614)	RAF Museum
	G–AWAV	Cessna F.150F	J. F. Thurlow & J. H. Pickering
	G–AWAW	Cessna F.150F	S.B. International Aviation Ltd.
	G–AWAX	Cessna 150D	Cambridge Technical Developments (Leasing) Ltd.
	G–AWAZ	PA-28R Cherokee Arrow 180	M. I. Edwards (Engineering) Ltd.
	G–AWBA	PA-28R Cherokee Arrow 180	W. London Aero Services Ltd.
	G–AWBB	PA-28R Cherokee Arrow 180	Brian Neale Ltd.
	G–AWBC	PA-28R Cherokee Arrow 180	Astrapoint Ltd.
	G–AWBG	PA-28 Cherokee 140	EFG Flying Services Ltd.
	G–AWBH	PA-28 Cherokee 140	R. C. A. Mackworth
	G–AWBJ	Fournier RF.4D	The BJ Group
	G–AWBL	BAC One-Eleven 416	British Airways
	G–AWBM	D.31 Turbulent	J. T. S. Lewis
	G–AWBN	PA-30 Twin Comanche 160	Stourfield Investments Ltd.
	G–AWBP	Cessna 182L Skylane	A. H. & Mrs. P. M. Butcher
	G–AWBS	PA-28 Cherokee 140	W. London Aero Services Ltd.
	G–AWBT	PA-30 Twin Comanche 160	R. M. English & Son Ltd.
	G–AWBU	Morane-Saulnier N (replica) (M.S.50)	Personal Plane Services Ltd.

Reg.	Type	Owner or Operator	Notes
G–AWBV	Cessna 182L Skylane	Hunting Surveys & Consultants Ltd.	
G–AWBW	*Cessna F.172H	Brunel Technical College, Lulsgate	
G–AWBX	Cessna F.150H	R. Featherstone	
G–AWCD	CEA DR.253	D. H. Smith	
G–AWCH	Cessna F.172H	M. Bua	
G–AWCJ	Cessna F.150H	Transknight Ltd.	
G–AWCK	Cessna F.150H	Wycombe Air Centre Ltd.	
G–AWCL	Cessna F.150H	Chalkair Ltd.	
G–AWCM	Cessna F.150H	Peterborough Aero Club Ltd.	
G–AWCN	Cessna FR.172E	LEC Refrigeration Ltd.	
G–AWCO	Cessna F.150H	K. A. Learmonth	
G–AWCP	Cessna F.150H (tailwheel)	Herefordshire Aero Club Ltd.	
G–AWCW	Beech E.95 Travel Air	H. W. Astor	
G–AWCY	PA-32 Cherokee Six 260	Robinson & Carr Ltd.	
G–AWDA	Nipper T.66 Srs. 3	Felthorpe Tipsy Group	
G–AWDD	Nipper T.66 Srs. 3	T. D. G. Roberts	
G–AWDH	D.31 Turbulent	J. H. Tetley	
G–AWDI	PA-23 Aztec 250	Air Foyle Ltd.	
G–AWDL	PA-25 Pawnee 235	Peter Charles (Airfarmers) Ltd.	
G–AWDO	D.31 Turbulent	R. Watling-Greenwood	
G–AWDP	PA-28 Cherokee 180	Brian Ilston Ltd.	
G–AWDR	Cessna FR.172E	Largoair Ltd.	
G–AWDU	Brantly B.2B	S. N. Cole	
G–AWDW	Bensen B.8	M. R. Langton	
G–AWDX	Beagle B.121 Pup I	J. Pearse & O. D. West	
G–AWED	PA-31-310 Navajo	Wheal Reath Properties Ltd.	
G–AWEF	Stampe SV.4B	Tiger Club Ltd.	
G–AWEG	Cessna 172G	H. Lawson	
G–AWEI	D.62B Condor	Tiger Club Ltd.	
G–AWEL	Fournier RF.4D	A. B. Clymo	
G–AWEM	Fournier RF.4D	B. J. Griffin	
G–AWEN	Jodel DR.1050	L. G. Earnshaw & ptnrs.	
G–AWEO	Cessna F.150H	Banbury Plant Hire Ltd.	
G–AWEP	JB-01 Minicab	P. H. Dyson & N. B. Gibbons	
G–AWER	PA-23 Aztec 250	Woodgate Aviation	
G–AWET	PA-28 Cherokee 180	Broadland Flying Group Ltd.	
G–AWEU	PA-28 Cherokee 140	Liverpool Aero Club Ltd.	
G–AWEV	PA-28 Cherokee 140	J. T. Garrod	
G–AWEX	PA-28 Cherokee 140	Northampton Heating & Ventilating Ltd.	
G–AWEZ	PA-28R Cherokee Arrow 180	P. H. de Havilland	
G–AWFB	PA-28R Cherokee Arrow 180	Luke Aviation Ltd.	
G–AWFC	PA-28R Cherokee Arrow 180	J. A. Clarke	
G–AWFD	PA-28R Cherokee Arrow 180	D. E. Roberts & J. G. Fisher	
G–AWFE	Jodel D.140E	Airscooters Ltd.	
G–AWFF	Cessna F.150H	Fowler Aviation Ltd.	
G–AWFH	Cessna F.150H	Norfolk & Norwich Aero Club Ltd.	
G–AWFJ	PA-28R Cherokee Arrow 180	R. Watt	
G–AWFK	PA-28R Cherokee Arrow 180	J. A. Rundle (Holdings) Ltd.	
G–AWFN	D.62B Condor	J. Guy	
G–AWFO	D.62B Condor	Cornwall Flying Group	
G–AWFP	D.62B Condor	Wingsouth Ltd.	
G–AWFR	D.31 Turbulent	S. W. Usherwood	
G–AWFT	Jodel D.9 Bebe	W. H. Cole	
G–AWFW	Jodel D.117	F. H. Greenwell	
G–AWFX	Sikorsky S-61N	British Airways Helicopters Ltd.	
G–AWFY	SA.318C Alouette Astazou	N. Hutchings Ltd.	
G–AWFZ	Beech A23 Musketeer	R. Sweet & B. D. Corbett	
G–AWGA	A.109 Airedale	A. C. W. Day	
G–AWGC	Cessna F.172H	R. J. Ford-Sagers	
G–AWGD	Cessna F.172H	T. J. W. Hood	
G–AWGE	Cessna F.172H	R. A. Gray	
G–AWGJ	Cessna F.172H	J. & C. J. Freeman	
G–AWGK	Cessna F.150H	G. R. Brown	
G–AWGM	Mitchell-Procter Kittiwake 2	RAF Halton Flying Club Ltd.	
G–AWGN	Fournier RF.4D	P. J. Foreman	
G–AWGP	Cessna T210H	Gledhill Water Storage Ltd.	
G–AWGR	Cessna F.172H	P. Bushell	
G–AWGU	AB-206B JetRanger II	British Airways Helicopters Ltd.	
G–AWGX	Cessna F.172H	Aberdeen Aero Club	
G–AWGY	Cessna F.150H	Exeter Flying Club Ltd.	
G–AWGZ	Taylor JT.1 Monoplane	G. Jones	
G–AWHB	*CASA 2.111 (6J+PR)	Historic Aircraft Museum, Southend	

Notes	Reg.	Type	Owner or Operator
	G–AWHU	Boeing 707-379C	British Airways
	G–AWHV	Rollason Beta B.2A	Tiger Club Ltd.
	G–AWHW	Rollason Beta B.2A	C. E. Bellhouse
	G–AWHX	Rollason Beta B.2	J. J. Cooke
	G–AWIF	Brookland Mosquito	R. Watson
	G–AWIG	Jodel D.112	K. R. Nunn
	G–AWII	V.S.349 Spitfire VC (AR501)	Shuttleworth Trust
	G–AWIJ	V.S.329 Spitfire IIA (P7350)	RAF Battle of Britain Historic Aircraft Flight
	G–AWIK	Beech 23 Musketeer	Resource Investors Management Ltd.
	G–AWIN	Campbell-Bensen B.8MC	M. J. Cuttel & J. Deane
	G–AWIO	Brantly B.2B	G. J. Ward & E. J. Roche
	G–AWIP	Luton L.A.4A Minor	J. Houghton
	G–AWIR	Midget Mustang	K. E. Sword
	G–AWIT	PA-28 Cherokee 180	C. D. Linton & ptnrs.
	G–AWIV	Storey TSR.3	C. J. Jesson
	G–AWIW	Stampe SV.4B	Historic Aircraft Museum, Southend
	G–AWIY	PA-23 Aztec 250	Queen's University of Belfast
	G–AWJA	Cessna 182L Skylane	Mercia Aviation
	G–AWJC	Brighton gas balloon	P. D. Furlong *Slippery William*
	G–AWJE	Nipper T.66 Srs. 3	Jubilee Group
	G–AWJF	Nipper T.66 Srs. 3	R. Wilcock
	G–AWJI	M.S.880B Rallye Club	Thames Estaury Flying Services
	G–AWJO	Tigercraft Tiger Mk. II	K. Aziz
	G–AWJP	Tigercraft Tiger Mk. III	Frederick Fewsdale
	G–AWJR	Tigercraft Tiger Mk. I	Frederick Fewsdale
	G–AWJS	Tigercraft Mosquito Mk. I	Frederick Fewsdale
	G–AWJT	Tigercraft Tiger Mk. I	J. H. Turner
	G–AWJV	*D.H.98 Mosquito TT Mk. 35 (TA634)	Mosquito Aircraft Museum
	G–AWJW	AB-206B JetRanger II	Flairair
	G–AWJX	Z.526 Akrobat	Aerobatics International Ltd.
	G–AWJY	Z.526 Akrobat	Ello Manufacturing Co.
	G–AWKB	M.J.5 Siroceo F2/39	G. D. Claxton
	G–AWKD	PA-17 Vagabond	A. T. & Mrs. M. R. Dowie
	G–AWKM	B.121 Pup I	D. M. G. Jenkins
	G–AWKO	B.121 Pup I	S. G. Bailey & G. H. Willson
	G–AWKP	Jodel DR.253	R. C. Chandless
	G–AWKS	M.S.880B Rallye Club	A. Thomas
	G–AWKT	M.S.880B Rallye Club	D. C. Strain
	G–AWKW	PA-24 Comanche 180	F. J. Bellamy
	G–AWKX	Beech A65 Queen Air	Andrew Edie Aviation
	G–AWKZ	PA-23 Apache 160	E. A. Clack & T. Pritchard
	G–AWLA	Cessna F.150H	S. M. Bent & P. Thompson
	G–AWLB	D.31 Turbulent	A. E. Shouler
	G–AWLE	Cessna F.172H	Sunderland Flying Club Ltd.
	G–AWLF	Cessna F.172H	Clement Spring Co. Ltd.
	G–AWLG	SIPA 903	S. W. Markham
	G–AWLI	PA-22 Tri-Pacer 150	D. T. Cheetham
	G–AWLJ	Cessna F.150H	D. S. Watts
	G–AWLL	AB-206B JetRanger 2	Gleneagles Helicopters Ltd.
	G–AWLM	Bensen B.8MS	C. J. E. Ashby
	G–AWLO	Boeing N2 S-5 Kaydet	Warbirds of GB, Blackbushe
	G–AWLP	Mooney M.20F	Siminco Ltd.
	G–AWLR	Nipper T.66 Srs. 3	J. D. Lowther
	G–AWLS	Nipper T.66 Srs. 3	D. W. Griffiths
	G–AWLW	Hawker Hurricane IIB (P3308)	Davis Trust, Strathallan
	G–AWLY	Cessna F.150H	Birmingham Aviation Ltd.
	G–AWLZ	Fournier RF.4D	E. V. Goodwin & C. R. Williamson
	G–AWMC	Campbell-Bensen B.8MS	M. E. Sykes-Hankinson
	G–AWMD	Jodel D.11	D. A. Lord
	G–AWMF	PA-18-150 Super Cub	Airways Aero Associations Ltd.
	G–AWMI	Glos-Airtourer 115	V. C. Birch
	G–AWMK	AB-206A JetRanger	Bristow Helicopters Ltd.
	G–AWMM	M.S.893A Rallye Commodore 180	Callow Aviation
	G–AWMN	Luton L.A.4A Minor	R. E. R. Wilks
	G–AWMP	Cessna F.172H	W. Rennie-Roberts
	G–AWMR	D.31 Turbulent	P. R. M. Bowlan
	G–AWMT	Cessna F.150H	J. A. Wright

Reg.	Type	Owner or Operator	Notes
G–AWMU	Cessna F.172H	Tees-side Transport Commercial Services Ltd.	
G–AWNA	Boeing 747-136	British Airways *Sir Richard Grenville*	
G–AWNB	Boeing 747-136	British Airways *City of Newcastle*	
G–AWNC	Boeing 747-136	British Airways *City of Belfast*	
G–AWND	Boeing 747-136	British Airways *Christopher Marlowe*	
G–AWNE	Boeing 747-136	British Airways *Sir Francis Drake*	
G–AWNF	Boeing 747-136	British Airways	
G–AWNG	Boeing 747-136	British Airways *City of London*	
G–AWNH	Boeing 747-136	British Airways *Sir Walter Raleigh*	
G–AWNJ	Boeing 747-136	British Airways *John Donne*	
G–AWNL	Boeing 747-136	British Airways *William Shakespeare*	
G–AWNM	Boeing 747-136	British Airways	
G–AWNN	Boeing 747-136	British Airways *Sebastian Cabot*	
G–AWNO	Boeing 747-136	British Airways *Sir Francis Bacon*	
G–AWNP	Boeing 747-136	British Airways *Sir John Hawkins*	
G–AWNT	BN-2A Islander	B.K.S. Surveys Ltd.	
G–AWOA	M.S.880B Rallye Club	G. C. Taylor	
G–AWOE	Aero Commander 680E	J. M. Houlder	
G–AWOF	PA-15 Vagabond	D. S. Morgan	
G–AWOH	PA-17 Vagabond	K. M. Bowen	
G–AWOL	Bell 206B JetRanger 2	Gleneagles Helicopters Ltd.	
G–AWOT	Cessna F.150H	Coventry Air Training School Ltd.	
G–AWOU	Cessna 170B	Red Fir Aviation Ltd.	
G–AWPA	D.31A Turbulent	J. T. Heaton	
G–AWPH	P.56 Provost T Mk. I	J. A. D. Bradshaw	
G–AWPJ	Cessna F.150H	M. L. Sarjeant	
G–AWPK	PA-23 Aztec 250	Air Atlantique Ltd.	
G–AWPL	Bensen B.8	N. F. Higgins	
G–AWPN	Shield Xyla	T. Worrall	
G–AWPP	Cessna F.150H	D. Williams	
G–AWPS	PA-28 Cherokee 140	J. D. Widdicombe & ptnrs.	
G–AWPU	Cessna F.150J	Light Planes (Lancashire) Ltd.	
G–AWPW	PA-12 Super Cruiser	P. Ligertwood	
G–AWPX	Cessna 150E	W. R. Emberton	
G–AWPY	Bensen B.8M	J. M. Deane	
G–AWPZ	Andreasson BA-4B	S. A. W. Becker	
G–AWRB	B.121 Pup I	P. O. P. Pulvermacher	
G–AWRK	Cessna F.150J	T. D. Close	
G–AWRL	Cessna F.172H	M. A. Cooper	
G–AWRP	Grasshopper CR.LTH-I	Cierva Rotorcraft Ltd.	
G–AWRS	*Avro 19 Srs. 2	N. E. Aircraft Museum	
G–AWRT	Glos-Airtourer 115	Hollybush Investments Ltd.	
G–AWRY	*P.56 Provost T.I (XF836)	Shuttleworth Trust	
G–AWRZ	Bell 47G-5	Heliwork Finance Ltd.	
G–AWSA	*Avro 652A Anson 19 (N5054)	Norfolk & Suffolk Aviation Museum	
G–AWSD	Cessna F.150J	Felthorpe Flying Group Ltd.	
G–AWSH	Z.526 Akrobat	Aerobatics International Ltd.	
G–AWSK	Agusta-Bell 47G-2	Bristow Helicopters Ltd.	
G–AWSL	PA-28 Cherokee 180D	Fascia Ltd.	
G–AWSM	PA-28 Cherokee 235	Colton Aviation Services Ltd.	
G–AWSN	D.62B Condor	J. Leader	
G–AWSO	D.62B Condor	N. H. Jones	
G–AWSP	D.62B Condor	R. Q. & A. S. Bond	
G–AWSR	D.62B Condor	N. H. Jones	
G–AWSS	D.62C Condor	J. L. Kinch & R. J. Leitch	
G–AWST	D.62B Condor	Humberside Aviation	
G–AWSU	F.8L Falco Srs. 4	M. Slazenger	
G–AWSV	Skeeter 12 (XM553)	Maj. M. Somerton-Rayner	
G–AWSY	Boeing 737-204	Britannia Airways Ltd. *General James Wolfe*	
G–AWSZ	M.S.894A Rallye Minerva 220	D. Quinn & J. McCloskey	
G–AWTA	Cessna 310N	A. H. Wiltshire	
G–AWTJ	Cessna F.150J	Metropolitan Police Flying Club	
G–AWTL	PA-28 Cherokee 180D	Leston Aviation	
G–AWTM	PA-28 Cherokee 140	Keenair Services Ltd.	
G–AWTR	Beech A.23 Musketeer	J. & P. Donoher	
G–AWTU	Beech A.23 Musketeer	Air Haven	
G–AWTV	Beech A.23 Musketeer	D. J. Johnson	
G–AWTW	Beech B.55 Baron	Worldwide Wheels Ltd.	
G–AWTX	Cessna F.150J	Tango X-Ray Flying Group	
G–AWUA	Cessna P.206D	G. B. Grant & Sons (Farmers) Ltd.	
G–AWUB	GY.201-Minicab	H. P. Burrill	

Notes	Reg.	Type	Owner or Operator
	G–AWUE	Jodel DR.1050	S. Bichan
	G–AWUF	H.S.125 Srs. 1B	MAM Aviation Ltd.
	G–AWUG	Cessna F.150H	Norfolk & Norwich Aero Club Ltd.
	G–AWUH	Cessna F.150H	M. A. Barrass & G. Scott
	G–AWUJ	Cessna F.150H	R. J. Jones & J. M. Allen
	G–AWUL	Cessna F.150H	W. Linskill
	G–AWUM	Cessna F.150H	Western Air Training Ltd.
	G–AWUN	Cessna F.150H	Northamptonshire School of Flying Ltd.
	G–AWUO	Cessna F.150H	W. Todd
	G–AWUP	Cessna F.150H	R. H. Timmis
	G–AWUR	Cessna F.150J	K. A. Learmonth
	G–AWUS	Cessna F.150J	Practavia Ltd.
	G–AWUT	Cessna F.150J	T. I. Murtough
	G–AWUU	Cessna F.150J	Enniskillen Flying Club
	G–AWUW	Cessna F.172H	E. Shipley & H. Wilson
	G–AWUX	Cessna F.172H	J. D. A. Shields & ptnrs.
	G–AWUY	Cessna F.172H	J. & B. Powell (Printers) Ltd.
	G–AWUZ	Cessna F. 172H	K. Wickenden
	G–AWVA	Cessna F. 172H	R. G. F. Allwright
	G–AWVB	Jodel D.117	C. M. & T. R. C. Griffin
	G–AWVC	B.121 Pup 1	Middleton Music Stores Ltd.
	G–AWVE	Jodel DR.1050M.1	E. A. Taylor
	G–AWVF	P.56 Provost T.1 (XF877)	Rural Flying Corps
	G–AWVH	Glos-Airtourer 115	Mardenair Ltd.
	G–AWVK	H.P.137 Jetstream	Decca Navigator Co. Ltd.
	G–AWVN	Aeronca 7AC Champion	Bowker Air Services Ltd.
	G–AWVP	Brookland Hornet	Brookland Rotorcraft Ltd.
	G–AWVS	Cessna 337D	W. H. Crispe & Sons Ltd.
	G–AWVW	PA-23 Aztec 250D	Heath Street Car Hirings Ltd.
	G–AWVZ	Jodel D.112	D. C. Stokes
	G–AWWE	B.121 Pup 2	G. J. Bunting
	G–AWWF	B.121 Pup 1	J. Pothecary
	G–AWWI	Jodel D.117	J. H. Kirkham
	G–AWWL	H.S.125 Srs. 3B-RA	Euro Airfinance Ltd.
	G–AWWM	GY-201 Minicab	J. S. Brayshaw
	G–AWWN	Jodel DR.1051	Jodel Flying Group
	G–AWWO	Jodel DR.1050	D. R. Gray & ptnrs.
	G–AWWP	Woody Pusher III	M. S. Bird & Mrs. R. D. Bird
	G–AWWS	SC.7 Skyvan Srs. 3	Vernair Transport Services
	G–AWWT	D.31 Turbulent	J. G. Alderton & J. Martendale
	G–AWWU	Cessna FR.172F	Dowdeswell Engineering Co. Ltd.
	G–AWWV	Cessna FR.172F	I. R. Hamilton & J. P. M. Stewart
	G–AWWW	Cessna 401	Westair Flying Services Ltd.
	G–AWWX	BAC One-Eleven 509	Dan-Air Services Ltd.
	G–AWWZ	BAC One-Eleven 509	Monarch Airlines Ltd.
	G–AWXA	Cessna 182M Skylane	R. E. & Mrs. U. C. Markelow
	G–AWXH	Cessna F.150H	Bristol & Wessex Aeroplane Club Ltd.
	G–AWXO	H.S.125 Srs. 400B	McAlpine Aviation Ltd.
	G–AWXR	PA-28 Cherokee 180D	J. D. Williams
	G–AWXS	PA-28 Cherokee 180D	Rayhenro Flying Group
	G–AWXU	Cessna F.150J	Hornet Aviation Ltd.
	G–AWXV	Cessna F.172H	D. N. Forrest
	G–AWXW	PA-23 Aztec 250D	Thurston Aviation Ltd.
	G–AWXX	Wessex Mk, 60 Srs. 1	Bristow Helicoptors Ltd.
	G–AWXY	M.S.885 Super Rallye	J. & B. Fowler
	G–AWXZ	SNCAN SV-4C	Personal Plane Services Ltd.
	G–AWYB	Cessna FR.172F	C. W. Larkin
	G–AWYE	H.S.125 Srs. 1B	Rolls-Royce Ltd.
	G–AWYF	G.159 Gulfstream 1	Ford Motor Co. Ltd.
	G–AWYH	Aero Commander 200D	W. J. D. Roberts
	G–AWYJ	B.121 Pup 2	D. Calabritto
	G–AWYL	Jodel DR.253B	Clarville Ltd.
	G–AWYO	B.121 Pup 1	B. R. C. Wild
	G–AWYR	BAC One-Eleven 501	British Caledonian Airways *Isle of Tiree*
	G–AWYS	BAC One-Eleven 501	British Caledonian Airways *Isle of Bute*
	G–AWYT	BAC One-Eleven 501	British Caledonian Airways *Isle of Barra*
	G–AWYU	BAC One-Eleven 501	British Caledonian Airways *Isle of Colonsay*

Reg.	Type	Owner or Operator	Notes
G–AWYV	BAC One-Eleven 501	British Caledonian Airways *Isle of Harris*	
G–AWYX	M.S.880B Rallye Club	Joy M. L. Edwards	
G–AWYY	T.57 Camel replica (C1701)	Leisure Sport Ltd.	
G–AWYZ	H.S.121 Trident 3B	British Airways	
G–AWZA	H.S.121 Trident 3B	British Airways	
G–AWZB	H.S.121 Trident 3B	British Airways	
G–AWZC	H.S.121 Trident 3B	British Airways	
G–AWZD	H.S.121 Trident 3B	British Airways	
G–AWZE	H.S.121 Trident 3B	British Airways	
G–AWZF	H.S.121 Trident 3B	British Airways	
G–AWZG	H.S.121 Trident 3B	British Airways	
G–AWZH	H.S.121 Trident 3B	British Airways	
G–AWZI	H.S.121 Trident 3B	British Airways	
G–AWZJ	H.S.121 Trident 3B	British Airways	
G–AWZK	H.S.121 Trident 3B	British Airways	
G–AWZL	H.S.121 Trident 3B	British Airways	
G–AWZM	H.S.121 Trident 3B	British Airways	
G–AWZN	H.S.121 Trident 3B	British Airways	
G–AWZO	H.S.121 Trident 3B	British Airways	
G–AWZP	H.S.121 Trident 3B	British Airways	
G–AWZR	H.S.121 Trident 3B	British Airways	
G–AWZS	H.S.121 Trident 3B	British Airways	
G–AWZU	H.S.121 Trident 3B	British Airways	
G–AWZV	H.S.121 Trident 3B	British Airways	
G–AWZW	H.S.121 Trident 3B	British Airways	
G–AWZX	H.S.121 Trident 3B	British Airways	
G–AWZZ	H.S.121 Trident 3B	British Airways	
G–AXAB	PA-28 Cherokee 140	Bencray Ltd.	
G–AXAK	M.S.880B Rallye Club	R. L. & Mrs. C. Stewart	
G–AXAN	D.H.82A Tiger Moth	A. J. Cheshire	
G–AXAO	Omega 56 balloon	P. D. Furlong *Renatus Cartesius*	
G–AXAS	Wallis WA-116T	K. H. Wallis (G–AVDH)	
G–AXAT	Jodel D.117A	J. F. Barber	
G–AXAU	PA-30 Twin Comanche 160C	Bartcourt Ltd.	
G–AXAV	PA-30 Twin Comanche 160C	Nottingham Aviation Ltd.	
G–AXAW	Cessna 421A	Bembridge Air Hire Ltd.	
G–AXAX	PA-23 Aztec 250D	Euroair Transport Ltd.	
G–AXAZ	PA-31 Navajo	Meridian Airmaps Ltd.	
G–AXBB	BAC One-Eleven 409	Air UK *Island Entente*	
G–AXBD	PA-25 Pawnee 235C	Farm Aviation Services Ltd.	
G–AXBG	Bensen B.8M	R. Curtis	
G–AXBH	Cessna F.172H	Farrowcrest Ltd.	
G–AXBJ	Cessna F.172H	D. Burbidge & R. Evans	
G–AXBW	D.H.82A Tiger Moth (T5879)	R. Venning	
G–AXBY	Cessna 401A	Hawtal Whiting (Design & Engineering) Co. Ltd.	
G–AXBZ	D.H.82A Tiger Moth	D. H. McWhir	
G–AXCA	PA-28R Cherokee Arrow 200	J. A. Tooth	
G–AXCC	Bell 47G-2	P. Lancaster	
G–AXCD	Agusta-Bell 47G-2	Bristow Helicopters Ltd.	
G–AXCF	Agusta-Bell 47G-2	Bristow Helicopters Ltd.	
G–AXCG	Jodel D.117	J. W. Hollingsworth & M. J. Doherty	
G–AXCI	*Bensen B.8M	Loughborough & Leicester Aircraft Museum	
G–AXCK	BAC One-Eleven 401	Dan-Air Services Ltd.	
G–AXCL	M.S.880B Rallye Club	G. W. Tietjer & C. Welch	
G–AXCM	M.S.880B Rallye Club	P. A. & W. A. Benjamin	
G–AXCN	M.S.880B Rallye Club	O. G. Baum	
G–AXCP	BAC One-Eleven 401	Dan Air Services Ltd.	
G–AXCX	B.121 Pup 2	Deltair Ltd.	
G–AXCY	Jodel D.117	R. M. Rennoldson	
G–AXCZ	Stampe SV-4C	Keenair Services Ltd.	
G–AXDB	Piper L-4 Cub	N. D. Norman	
G–AXDC	PA-23 Aztec 250D	Trago Mills (South Devon) Ltd.	
G–AXDE	Bensen B.8	T. J. Hartwell	
G–AXDH	BN-2A Islander	Parachute Regiment Freefall Club	
G–AXDI	Cessna F.172H	Jim Russell International Racing Drivers Ltd.	
G–AXDK	Jodel DR.315	W. B. Wright & Sons Ltd.	
G–AXDL	PA-30 Twin Comanche 160C	Northern Executive Aviation Ltd.	
G–AXDM	H.S.125 Srs. 400B	Ferranti Ltd.	

Notes	Reg.	Type	Owner or Operator
	G–AXDN	*BAC-Sud Concorde 01	Duxford Aviation Soc.
	G–AXDU	B.121 Pup 2	Deltair Ltd.
	G–AXDV	B.121 Pup 1	R. A. Chappell
	G–AXDW	B.121 Pup 1	Cranfield Institute of Technology
	G–AXDY	Falconar F-11	G. K. Ellis
	G–AXDZ	Cassutt Racer Srs. 111M	A. Chadwick
	G–AXEB	Cassutt Racer Srs. 111M	G. E. Horder
	G–AXEC	Cessna 182M	Mendring Ltd.
	G–AXED	PA-25 Pawnee 235	Sprayfields (Scothern) Ltd.
	G–AXEI	*Ward Gnome	Lincolnshire Aviation Museum
	G–AXEO	Scheibe SF.25B Falke	D. Collinson
	G–AXER	PA-30 Twin Comanche 160C	Astrojet Ltd. & Danro Dental Products Ltd.
	G–AXES	B.121 Pup 2	D. A. Lowe & A. Molesworth
	G–AXET	B.121 Pup 2	J. C. Furneaux
	G–AXEU	B.121 Pup 2	Wings Flying Group
	G–AXEV	B.121 Pup 2	B. Richardson
	G–AXEW	B.121 Pup 1	C. J. Spicer & A. A. Gray
	G–AXEX	B.121 Pup 1	Lubair (Transport Services) Ltd.
	G–AXFA	PA-23 Aztec 250D	Stapleford Flying Club Ltd.
	G–AXFB	BN-3 Nymph	D. McIntyre
	G–AXFD	PA-25 Pawnee 235	J.E.F. Aviation Ltd.
	G–AXFE	Beech B.90 King Air	GKN Contractors Ltd.
	G–AXFG	Cessna 337D	Alfred Smith & Son (Penzance) Ltd.
	G–AXFH	D.H.114 Heron 1B/C	Hurst Rent-a-Car Ltd.
	G–AXFM	Servotec Grasshopper 3	Cierva Rotorcraft Ltd.
	G–AXFN	Jodel D.119	P. D. Wheatland
	G–AXGA	PA-19 Super Cub 95	Felthorpe Flying Group Ltd.
	G–AXGC	M.S.880B Rallye Club	K. M. & H. Bowen
	G–AXGD	M.S.880B Rallye Club	J. G. R. Read & ptnrs.
	G–AXGE	M.S.880B Rallye Club	A. R. T. Banks
	G–AXGG	Cessna F.150J	G. W. G. C. Sudlow
	G–AXGP	Piper L-4B Cub	W. K. Butler
	G–AXGR	Luton L.A.4A Minor	T. M. W. Webster
	G–AXGS	D.62B Condor	Tiger Club Ltd.
	G–AXGT	D.62B Condor	P. Simpson & ptnrs.
	G–AXGU	D.62B Condor	Tiger Club Ltd.
	G–AXGV	D.62B Condor	C. N. Bonneywell
	G–AXGW	Boeing 707-336C	British Airways
	G–AXGX	Boeing 707-336C	British Airways
	G–AXGZ	D.62B Condor	P. W. Johnson & J. T. Hayes
	G–AXHA	Cessna 337A	E. A. Pitcher
	G–AXHC	Stampe SV-4C	K. L. Hawse & ptnrs.
	G–AXHE	BN-2A Islander	Peterborough Parachute Centre Ltd.
	G–AXHG	M.S.880B Rallye Club	D. M. Leonard
	G–AXHI	M.S.880B Rallye Club	Mona Services
	G–AXHO	B.121 Pup 2	Harvair Ltd.
	G–AXHP	Piper L-4H Cub	W. F. Barnes & R. W. Griffin
	G–AXHR	Piper L-4H Cub (329601)	D. E. Elphick
	G–AXHS	M.S.880B Rallye Club	R. Allan
	G–AXHT	M.S.880B Rallye Club	J. L. Osbourne & A. M. Sutton
	G–AXHV	Jodel D.117A	D. M. Cashmore & K. R. Payne
	G–AXHX	M.S.892A Rallye Commodore	G. V. Blakeway
	G–AXIA	B.121 Pup 1	Cranfield Institute of Technology
	G–AXIE	B.121 Pup 2	I. J. Ross & ptnrs.
	G–AXIF	B.121 Pup 2	D. W. Goodenough & Nick Alder Farms Ltd.
	G–AXIG	B.125 Bulldog 104	George House (Holdings) Ltd.
	G–AXIH	Bu 133 Jungmeister	M. W. Stow
	G–AXIO	PA-28 Cherokee 140B	W. London Aero Services Ltd.
	G–AXIR	PA-28 Cherokee 140B	C. T. Brinson
	G–AXIT	M.S.893A Rallye Commodore 180	South Wales Gliding Club Ltd.
	G–AXIW	Scheibe SF.25B Falke	Herefordshire Gliding Club Ltd.
	G–AXIX	Glos-Airtourer 150	R. Gilkes
	G–AXIY	Bird Gyrocopter	Gerald Bird
	G–AXJB	Omega 84 balloon	Hot-air Group *Jester*
	G–AXJH	B.121 Pup 2	J. Pearce
	G–AXJI	B.121 Pup 2	E. A. Clack
	G–AXJJ	B.121 Pup 2	P. Hirst & ptnrs.
	G–AXJK	BAC One-Eleven 501	British Caledonian Airways *Isle of Staffa*

Reg.	*Type*	*Owner or Operator*	*Notes*
G–AXJM	BAC One-Eleven 501	British Caledonian Airways *Isle of Islay*	
G–AXJN	B.121 Pup 2	Toon Ghose Aviation Ltd.	
G–AXJO	B.121 Pup 2	J. A. D. Bradshaw	
G–AXJR	Scheibe SF.25B Falke	M. J. Munday & N. E. M. Coombe	
G–AXJV	PA-28 Cherokee 140B	Mona Aviation Ltd.	
G–AXJW	PA-28 Cherokee 140B	I. Goodchild	
G–AXJX	PA-28 Cherokee 140B	Manchester School of Flying Ltd.	
G–AXJY	Cessna U-206D Super Skywagon	Hereford Parachute Club Ltd.	
G–AXKD	PA-23 Aztec 250D	Jones & Bailey Contractors Ltd.	
G–AXKH	Luton L.A.4A Minor	M. E. Vaisey	
G–AXKI	Jodel D.9 Bebe	T. A. Hodges	
G–AXKJ	Jodel D.9 Bebe	C. H. Morris	
G–AXKK	Westland Bell 47G-4A	Bristow Helicopters Ltd.	
G–AXKL	Westland Bell 47G-4A	Bristow Helicopters Ltd.	
G–AXKM	Westland Bell 47G-4A	Bristow Helicopters Ltd.	
G–AXKN	Westland Bell 47G-4A	Bristow Helicopters Ltd.	
G–AXKO	Westland Bell 47G-4A	Bristow Helicopters Ltd.	
G–AXKR	Westland Bell 47G-4A	Bristow Helicopters Ltd.	
G–AXKS	Westland Bell 47G-4A	Bristow Helicopters Ltd.	
G–AXKU	Westland Bell 47G-4A	Bristow Helicopters Ltd.	
G–AXKV	Westland Bell 47G-4A	Bristow Helicopters Ltd.	
G–AXKW	Westland Bell 47G-4A	Bristow Helicopters Ltd.	
G–AXKX	Westland Bell 47G-4A	Bristow Helicopters Ltd.	
G–AXKY	Westland Bell 47G-4A	Bristow Helicopters Ltd.	
G–AXKZ	Westland Bell 47G-4A	Bristow Helicopters Ltd.	
G–AXLA	Westland Bell 47G-4A	Bristow Helicopters Ltd.	
G–AXLG	Cessna 310K	Smiths (Outdrives) Ltd.	
G–AXLI	Nipper T.66 Srs. 3	P. W. Thomas & K. Richter	
G–AXLS	Jodel DR.105A	T. L. Giles	
G–AXLZ	PA-19 Super Cub 95	J. C. Quantrell	
G–AXMA	PA-24 Comanche 180	Tegrel Products Ltd.	
G–AXMB	Slingsby Motor Cadet	I. G. Smith	
G–AXMD	Omega 20 balloon	Nimble Bread Ltd. *Nimble*	
G–AXMG	BAC One-Eleven 518	Monarch Airlines Ltd.	
G–AXMM	Bell 206A JetRanger	R. G. Woodward	
G–AXMN	J/5B Autocar	I. R. F. Hammond	
G–AXMP	PA-28 Cherokee 180	Concorde Garage (Elmsawell) Ltd.	
G–AXMR	PA-31-300 Navajo	Airmore Aviation Ltd.	
G–AXMS	PA-30 Twin Comanche 160C	Ernest Green International Ltd.	
G–AXMT	Bu 133 Jungmeister	S. R. Flack	
G–AXMU	BAC One-Eleven 432	Air UK *Island Esprit*	
G–AXMW	B.121 Pup 1	DJP Engineering (Knebworth) Ltd.	
G–AXMX	B.121 Pup 2	Susan A. Jones	
G–AXMY	PA-30 Twin Comanche 160C	Leisair Avionics Ltd.	
G–AXNA	Boeing 737-204C	Britannia Airways Ltd. *Robert Clive of India*	
G–AXNB	Boeing 737-204C	Britannia Airways Ltd. *Charles Darwin*	
G–AXNC	Boeing 737-204	Britannia Airways Ltd. *Sir Frederick Handley Page*	
G–AXNJ	Wassmer Jodel D.120	Clive Flying Group	
G–AXNK	Cessna F.150J	Mona Aviation Ltd.	
G–AXNL	B.121 Pup 1	Mike Bennett Ltd.	
G–AXNM	B.121 Pup 1	Majabay Ltd.	
G–AXNN	B.121 Pup 2	Knights Aviation Ltd.	
G–AXNP	B.121 Pup 2	Deltair Ltd.	
G–AXNR	B.121 Pup 2	Specialised Mouldings Ltd. & G. Broadley	
G–AXNS	B.121 Pup 2	S. J. Figures & N. Fields	
G–AXNW	SNCAN SV-4C	E. N. Grace	
G–AXNX	Cessna 182M	Air Tows Ltd.	
G–AXNY	Fixter Pixie	J. van Geest	
G–AXNZ	Pitts S.1C Special	W. A. Jordan	
G–AXOG	PA-23 Aztec 250D	R. W. Diggens	
G–AXOH	M.S.894 Rallye Minerva	Bristol Cars Ltd.	
G–AXOI	Jodel D.9 Bebe	P. W. Thomas	
G–AXOJ	B.121 Pup 2	TM Air Ltd.	
G–AXOL	Currie Wot	D. R. Campbell & A. Kay	
G–AXOR	PA-28 Cherokee 180D	P. D. Allum & K. F. Atkins	
G–AXOS	M.S.894A Rallye Minerva	Seven Flying Group	

Notes	Reg.	Type	Owner or Operator
	G–AXOT	M.S.893 Rallye Commodore 180	Col. J. F. Williams-Wynne
	G–AXOV	Beech B55A Baron	S. Brod
	G–AXOX	BAC One-Eleven 432	Air UK *Island Endeavour*
	G–AXOZ	B.121 Pup 1	Tees-side Aviation Ltd.
	G–AXPB	B.121 Pup 1	D. Smith
	G–AXPD	B.121 Pup 1	Surrey & Kent Flying Club Ltd.
	G–AXPF	Cessna F.150K	Westair Flying Services Ltd.
	G–AXPG	Mignet HM-293	W. H. Cole (Historic Aircraft Museum)
	G–AXPM	B.121 Pup 1	D. Taylor
	G–AXPN	B.121 Pup 2	Starline Elms Coaches
	G–AXPZ	Campbell Cricket	W. R. Partridge
	G–AXRA	Campbell Cricket	L. E. Schnurr
	G–AXRB	Campbell Cricket	J. C. P. Thomas
	G–AXRC	Campbell Cricket	K. W. Hayr
	G–AXRD	Campbell Cricket	Glyndwr Rees
	G–AXRK	Practavia Sprite 115	E. G. Thale
	G–AXRL	PA 28 Cherokee 160	T. W. Clark
	G–AXRO	PA-30 Twin Comanche 160C	Havelet Aviation
	G–AXRP	SNCAN SV-4C	M. D. Tweedie & ptnrs.
	G–AXRR	Auster AOP.9 (XR241)	Shuttleworth Trust
	G–AXRS	Boeing 707-355C	Monarch Airlines Ltd.
	G–AXRT	Cessna FA.150K (tailwheel)	W. H. Milner
	G–AXRU	Cessna FA.150K	H. A. Knight
	G–AXSC	B.121 Pup 1	J. Hawkins
	G–AXSD	B.121 Pup 1	Surrey & Kent Flying Club Ltd.
	G–AXSF	Nash Petrel	Nash Aircraft Ltd.
	G–AXSG	PA-28 Cherokee 180	Shropshire Aero Club Ltd.
	G–AXSH	PA-28 Cherokee 140B	Brailey & Co. (Aviation) Ltd.
	G–AXSJ	Cessna FA.150K	Practavia Ltd.
	G–AXSM	Jodel DR.1051	C. Cousten
	G–AXSV	Jodel DR.340	Beamville Ltd.
	G–AXSW	Cessna FA.150K	Furness Aviation Ltd.
	G–AXSX	Beech C.23 Musketeer	D. M. Balfour
	G–AXSZ	PA-28 Cherokee 140B	N. Cureton & R. B. Cheek
	G–AXTA	PA-28 Cherokee 140B	Carlisle Aviation Co. Ltd.
	G–AXTC	PA-28 Cherokee 140B	Airways Aero Associations Ltd.
	G–AXTD	PA-28 Cherokee 140B	Vincent-Walker Engineering Ltd.
	G–AXTE	PA-28 Cherokee 140B	N. Clayton
	G–AXTG	PA-28 Cherokee 140B	M. J. Cowham
	G–AXTH	PA-28 Cherokee 140B	W. London Aero Services Ltd.
	G–AXTI	PA-28 Cherokee 140B	LT (Central Road Services) Sports Association
	G–AXTJ	PA-28 Cherokee 140B	TK Aero Enterprises Ltd.
	G–AXTK	PA-28 Cherokee 140B	Andrewsfield Flying Club Ltd.
	G–AXTL	PA-28 Cherokee 140B	L. Williams
	G–AXTM	PA-28 Cherokee 140B	Cormack (Aircraft Services) Ltd.
	G–AXTO	PA-24 Comanche 260	Wolverhouse Ltd.
	G–AXTP	PA-28 Cherokee 180	D. I. L. Butler
	G–AXTX	Jodel D.112	J. J. Penney
	G–AXUA	B.121 Pup 1	F. R. Blennerhassett & C. Wedlake
	G–AXUB	BN-2A Islander	Headcorn Parachute Club
	G–AXUC	PA-12 Super Cruiser	V. N. Mukaloff
	G–AXUE	Jodel DR.105A	Carlton Flying Group
	G–AXUF	Cessna FA.150K	Airwork Services Ltd.
	G–AXUI	H.P.137 Jetstream 1	Cranfield Institute of Technology
	G–AXUJ	J/1 Autocrat	R. G. Earp & J. W. H. Lee
	G–AXUK	Jodel DR.1050	R. Pidcock & ptnrs.
	G–AXUM	H.P.137 Jetstream 1	Cranfield Institute of Technology
	G–AXUV	Cessna F.172H	F. A. & Mrs. E. M. Smith
	G–AXUW	Cessna FA.150K	Coventry Air Training School
	G–AXUX	Beech B95 Travel Air	J. H. Southern
	G–AXUZ	Practavia Sprite 125	C. B. Healey
	G–AXVB	Cessna F.172H	C. Gabbitas
	G–AXVC	Cessna FA.150K	Rob Hughes Garages Ltd.
	G–AXVG	H.S.748 Srs. 2	Dan-Air Services Ltd.
	G–AXVK	Campbell Cricket	Campbell Gyroplanes Ltd.
	G–AXVM	Campbell Cricket	D. M. Organ
	G–AXVN	McCandless M.4	R. McCandless
	G–AXVS	Jodel DR.1050	F. W. Tilley
	G–AXVU	Omega 84 balloon	Brede Balloons Ltd. *Henry VII*
	G–AXVV	Piper L-4H Cub	J. MacCarthy

Reg.	Type	Owner or Operator	Notes
G-AXVW	Cessna F.150K	R. Stubbings & M. Driscoll	
G-AXVX	Cessna F.172H	D'Austere Ltd.	
G-AXWA	Auster AOP 9	T. Platt	
G-AXWB	Omega 65 balloon	A. Robinson & M. J. Moore *Ezekiel*	
G-AXWD	Jurca MJ 10	F.P.A. Group	
G-AXWE	Cessna F.150K	Light Planes (Lancashire) Ltd.	
G-AXWF	Cessna F.172H	Redfir Aviation Ltd.	
G-AXWG	BN-2A Islander	Bristow Helicopters Ltd.	
G-AXWP	BN-2A Islander	Aurigny Air Services	
G-AXWR	BN-2A Islander	Aurigny Air Services	
G-AXWT	Jodel D.11	R. Owen	
G-AXWV	Jodel DR.253	Murray Motors	
G-AXWZ	PA-28R Cherokee Arrow 200	Melbourns Brewery Ltd.	
G-AXXA	PA-28 Cherokee 180E	Sicon Hydraulics Ltd.	
G-AXXC	CP.301B Emeraude	J. R. R. Gale & J. Tetley	
G-AXXG	BN-2A Islander	Air Orkney	
G-AXXH	BN-2A Islander	Westward Airways	
G-AXXJ	BN-2A Islander	Haywards Aviation Ltd.	
G-AXXM	CP.301A Emeraude	Mrs. P. H. Wren	
G-AXXN	WHE Airbuggy	W. H. Beevers	
G-AXXR	Beech 95-B55A Baron	Firth Plant Ltd.	
G-AXXV	D.H.82A Tiger Moth (DE992)	L. B. Jefferies	
G-AXXW	Jodel D.117	G. Staples	
G-AXXY	Boeing 707-336B	British Airways	
G-AXXZ	Boeing 707-336B	British Airways	
G-AXYA	PA-31-300 Navajo	Air Foyle Ltd.	
G-AXYD	BAC One-Eleven 509	Dan-Air Services Ltd.	
G-AXYK	Taylor JT.1 Monoplane	C. Oakins	
G-AXYU	Jodel D.9 Bebe	D. P. Jones & S. R. Sissons	
G-AXYV	Luton Beta Srs. 2	D. G. Wiggins	
G-AXYX	WHE Airbuggy	R. T. Ginn	
G-AXYY	WHE Airbuggy	M. P. Chetwynd-Talbot	
G-AXYZ	WHE Airbuggy	W. H. Ekin	
G-AXZA	WHE Airbuggy	J. K. Davies	
G-AXZB	WHE Airbuggy	J. D. Jewitt	
G-AXZD	PA-28 Cherokee 180E	College of Air Training	
G-AXZE	PA-28 Cherokee 180E	College of Air Training	
G-AXZF	PA-28 Cherokee 180E	College of Air Training	
G-AXZJ	Cessna F.172H	Smiths' Aviation	
G-AXZM	Nipper Mk. III	S. J. Booth & D. A. Young	
G-AXZO	Cessna 180	R.S.A. Parachute Club Ltd.	
G-AXZP	PA-23 Aztec 250	White House Garage, Ashford Ltd.	
G-AXZR	Taylor JT.2 Titch	A. J. Fowler & D. E. Evans	
G-AXZT	Jodel D.117	H. W. Baines	
G-AXZU	Cessna 182N	R. Taylor & ptnrs.	
G-AYAA	PA-28 Cherokee 180E	College of Air Training	
G-AYAB	PA-28 Cherokee 180E	College of Air Training	
G-AYAC	PA-28R Cherokee Arrow 200	Steer Aviation Ltd.	
G-AYAD	PA-30 Twin Comanche 160C	Merlin Service Stations Ltd.	
G-AYAE	Bell 47G-4A	Helicopter Hire Ltd.	
G-AYAF	PA-30 Twin Comanche 160C	Arrow Air Services (Charter) Ltd.	
G-AYAI	Fournier RF-5	Exeter RF Group	
G-AYAJ	Cameron O-84 balloon	E. T. Hall *Flaming Pearl*	
G-AYAK	Yak-11 (14)	A. E. Hutton	
G-AYAL	Omega 56 balloon	Nimble Bread Ltd. *Nimble II*	
G-AYAN	Slingsby Motor Cadet Mk. III	I. Stevenson	
G-AYAO	Cessna F.172H	Transmatic Fyllan Ltd.	
G-AYAP	PA-28 Cherokee 180E	R. L. Taylor & A. Jahanfar	
G-AYAR	PA-28 Cherokee 180E	College of Air Training	
G-AYAS	PA-28 Cherokee 180E	College of Air Training	
G-AYAT	PA-28 Cherokee 180E	College of Air Training	
G-AYAU	PA-28 Cherokee 180E	College of Air Training	
G-AYAV	PA-28 Cherokee 180E	College of Air Training	
G-AYAW	PA-28 Cherokee 180E	College of Air Training	
G-AYBD	Cessna F.150K	R. P. Cole & ptnrs.	
G-AYBG	Scheibe SF.25B Falke	Doncaster Sailplane Services	
G-AYBK	PA-28 Cherokee 180E	College of Air Training	
G-AYBO	PA-23 Aztec 250D	Twinguard Leasing Ltd.	
G-AYBP	Jodel D.112	F. H. French	
G-AYBT	PA-28 Cherokee 180E	College of Air Training	
G-AYBU	Western 84 balloon	D. R. Gibbons	
G-AYBV	Chasle Tourbillon	B. A. Mills	

Notes	Reg.	Type	Owner or Operator
	G–AYCC	Campbell Cricket	K. W. E. Denson
	G–AYCE	CP.301C Emeraude	K. Webb
	G–AYCF	Cessna FA.150K	E. J. Atkins
	G–AYCG	SNCAN SV-4C	N. Bignall
	G–AYCJ	Cessna TP.206D	Brymon Aviation Ltd.
	G–AYCM	Bell 206A JetRanger	W.R. Finance Ltd.
	G–AYCN	Piper L-4H Cub	F. J. Cox
	G–AYCO	CEA DR.360	L. M. Gould
	G–AYCP	Jodel D.112	W. Hutchings
	G–AYCT	Cessna F.172H	Kontrox Ltd.
	G–AYDG	M.S.894A Rallye Minerva	R. Vaughan & F. T. Skipper (Electronics) Ltd.
	G–AYDI	D.H.82A Tiger Moth	R. B. Woods
	G–AYDJ	Campbell Cricket	A. van Preussen
	G–AYDR	SNCAN SV-4C	R. A. Phillips
	G–AYDU	AJEP Tailwind	AJEP Development Ltd.
	G–AYDV	Coates SA.11-1 Swalesong	J. R. Coates
	G–AYDW	A.61 Terrier 2	J. S. Harwood
	G–AYDX	A.61 Terrier 2	R. G. Ford
	G–AYDY	Luton L.A.4A Minor	C. R. Scott
	G–AYDZ	Jodel DR.200	Don Martin (Car Sales) Ltd.
	G–AYEB	Jodel D.112	F. W. & G. F. T. Taylor
	G–AYEC	CP.301A Emeraude	A. P. Docherty & J. S. Barker
	G–AYED	PA-24 Comanche 260	Patgrove Ltd.
	G–AYEE	PA-28 Cherokee 180E	College of Air Training
	G–AYEF	PA-28 Cherokee 180E	College of Air Training
	G–AYEG	Falconer F-9	A. G. Thelwall
	G–AYEH	Jodel DR.1050	R. O. F. Harper & P. R. Skeels
	G–AYEI	PA-31-300 Navajo	Keller, Bryant & Co. Ltd.
	G–AYEJ	Jodel DR.1050	G. Weaver
	G–AYEK	Jodel DR.1050	I. Shaw & B. Hanson
	G–AYEL	Bell 47G-5	Dollar Air Services Ltd.
	G–AYEN	Piper L-4H Cub	P. Warde & C. F. Morris
	G–AYER	H.S.125 Srs. 403B	MAM Aviation Ltd.
	G–AYES	M.S.892A Rallye Commodore 150	Waveney Flying Group
	G–AYET	M.S.892A Rallye Commodore 150	Cleanacres Ltd.
	G–AYEU	Brookland Hornet	J. B. Verney
	G–AYEV	Jodel DR.1050	M. R. Ireland
	G–AYEW	Jodel DR.1051	D. G. Hammersley & R. E. Kendal
	G–AYEX	Boeing 707-355C	British Caledonian Airways *Loch Leven*
	G–AYEY	Cessa F.150K	Kerry Type
	G–AYFA	SA Twin Pioneer 3	Flight One Ltd.
	G–AYFC	D.62B Condor	J. B. Randle
	G–AYFD	D.62B Condor	Tiger Club Ltd.
	G–AYFE	D.62C Condor	R. R. Harris
	G–AYFF	D.62B Condor	A. F. S. Caldecourt
	G–AYFG	D.62C Condor	Wolds Gliding Club
	G–AYFH	D.62B Condor	B. J. Collins
	G–AYFJ	M.S.880B Rallye Club	K. Hutson & ptnrs.
	G–AYFM	H.S.125 Srs. 403B	Ford Motor Co.
	G–AYFP	Jodel D.140	S. K. Minocha
	G–AYFS	Brookland Hornet	Brookland Rotorcraft Ltd.
	G–AYFT	PA-39 Twin Comanche C/R	F. Kirby
	G–AYFV	Crosby BA-4B	I. N. Jennison
	G–AYFX	AA-1 Yankee	D. W. F. Willard & B. Refson
	G–AYFY	EAA Biplane	H. Kuehling
	G–AYFZ	PA-31-300 Navajo	Anglo-Normandy Aviation Ltd.
	G–AYGA	Jodel D.117	E. J. Baxter & ptnrs.
	G–AYGB	Cessna 310Q	Airwork Services Ltd.
	G–AYGC	Cessna F.150K	D. W. Barron
	G–AYGD	Jodel DR.1051	I. S. Walsh
	G–AYGE	SNCAN SV-4C	The Hon. A. M. J. Rothschild
	G–AYGG	Jodel D.120	J. E. Hobbs
	G–AYGN	Cessna 210K	J. W. O'Sullivan
	G–AYGX	Cessna FR.172G	Ranelagh Garage Ltd.
	G–AYGZ	Beech 58 Baron	General Engineering Co. (Ilford) Ltd.
	G–AYHA	AA-1 Yankee	P. Chambers
	G–AYHH	Campbell Cricket	A. E. Sawyer
	G–AYHI	Campbell Cricket	E. Tests
	G–AYHX	Jodel D.117A	L. J. E. Goldfinch

Reg.	Type	Owner or Operator	Notes
G–AYHY	Fournier RF-4D	Tiger Club Ltd.	
G–AYIA	Hughes 369HS	G. D. E. Bilton	
G–AYIB	Cessna 182N Skylane	Emair Bridlington Ltd.	
G–AYIF	PA-28 Cherokee 140C	C.S.E. (Aircraft Services) Ltd.	
G–AYIG	PA-28 Cherokee 140C	Harris Aviation Services Ltd.	
G–AYIH	PA-28 Cherokee 140C	Emilio Ferrari Ltd.	
G–AYII	PA-28R Cherokee Arrow 200	Devon Growers Ltd. & A. L. Bacon	
G–AYIJ	SNCAN SV-4B	R. J. Maxey & ptnrs.	
G–AYIL	Scheibe SF.25B Falke	S. Evans & ptnrs.	
G–AYIO	PA-28 Cherokee 140C	S. Grant & ptnrs.	
G–AYIP	PA-39 Twin Comanche C/R	P. D. Lees & G. Pinkus	
G–AYIT	D.H.82A Tiger Moth	R. L. H. Alexander & ptnrs.	
G–AYIU	Cessna 182N Skylane	Moatair Service Ltd.	
G–AYJA	Jodel DR. 1050	J. R. Surbey	
G–AYJB	SNCAN SV-4C	N. J. Robertson & ptnrs.	
G–AYJD	Alpavia-Fournier RF-3	C. Wren	
G–AYJP	PA-28 Cherokee 140C	RAF Brize Norton Flying Club Ltd.	
G–AYJR	PA-28 Cherokee 140C	RAF Brize Norton Flying Club Ltd.	
G–AYJS	PA-28 Cherokee 140C	Thames Estuary Flying Services Ltd.	
G–AYJT	PA-28 Cherokee 140C	Jennifer M. Whitaker	
G–AYJU	Cessna TP-206A	Berrard Aviation	
G–AYJW	Cessna FR.172G	Alpine Press Ltd.	
G–AYJY	Isaacs Fury II	A. V. Francis	
G–AYKA	Beech 95-B55A Baron	Flockvale Ltd.	
G–AYKC	D.H.82A Tiger Moth	G. Freeman & ptnrs.	
G–AYKD	Jodel DR.1050	R. G. E. Armfield	
G–AYKF	M.S.880B Rallye Club	Martin Ltd.	
G–AYKJ	Jodel D.117A	G. R. W. Monksfield	
G–AYKK	Jodel D.117	P. Cawkwell & ptnrs.	
G–AYKL	Cessna F.150L	Aero Group 78	
G–AYKS	Leopoldoff L-7	D. J. Elliott	
G–AYKT	Jodel D.117	J. P. Owen-Jones	
G–AYKU	PA-E23 Aztec 250D	Simulated Flight Training Ltd.	
G–AYKV	PA-28 Cherokee 140C	J. M. Whitaker	
G–AYKW	PA-28 Cherokee 140C	R. E. & R. A. Yallop	
G–AYKX	PA-28 Cherokee 140C	Dicoll Electronics Ltd.	
G–AYKZ	SAI KZ-8	R. E. Mitchell	
G–AYLA	Glos-Airtourer 115	Vagabond Flying Group	
G–AYLB	PA-39 Twin Comanche C/R	E. A. Radnall & Co. Ltd.	
G–AYLC	Jodel DR.1051	E. W. B. Trollope	
G–AYLE	M.S.880B Rallye Club	J. E. Stephenson	
G–AYLF	Jodel DR.1051	W. A. G. Willbond	
G–AYLG	H.S.125 Srs. 400B	British Steel Corporation	
G–AYLJ	PA-31 Navajo	Northern Executive Aviation Ltd.	
G–AYLK	Stampe SV-4C	R. W. & P. R. Budge	
G–AYLL	Jodel DR.1050	V. E. Hanning-Lee	
G–AYLO	AA-1 Yankee	M. Brown	
G–AYLP	AA-1 Yankee	D. Nairn & E. Y. Hawkins	
G–AYLT	Boeing 707-336C	—	
G–AYLU	Pitts S-1D Special	I. M. G. Senior & J. G. Harper	
G–AYLV	Jodel D.120	R. E. Wray	
G–AYLX	Hughes 269C	Feastlight Ltd.	
G–AYLY	PA-23 Aztec 250	Air UK	
G–AYLZ	Super Aero 45 Srs. 2	J. R. B. Aviation Ltd.	
G–AYMA	Stolp Starduster Too	K. D. Ballinger & A. R. T. Jones	
G–AYME	Fournier RF.5	R. D. Goodger	
G–AYMG	HPR-7 Herald 213	Air UK	
G–AYMK	PA-28 Cherokee 140C	The Piper Flying Group	
G–AYML	PA-28 Cherokee 140C	J. M. Bendle	
G–AYMM	Cessna 421B	Rogers Aviation Sales Ltd.	
G–AYMN	PA-28 Cherokee 140C	F.R. Aviation Ltd.	
G–AYMO	PA-23 Aztec 250	Goldstar Publications Ltd.	
G–AYMP	Currie Wot Special	H. F. Moffatt	
G–AYMR	Lederlin 380L Ladybug	J. S. Brayshaw	
G–AYMT	Jodel DR.1050	Merlin Flying Club Ltd.	
G–AYMU	Jodel D.112	R. M. White	
G–AYMV	Western 20 balloon	G. F. Turnbull & ptnrs. *Tinkerbelle*	
G–AYMW	Bell 206A JetRanger	Wykeham Helicopters Ltd.	
G–AYMX	Bell 206A JetRanger	W. Holmes	
G–AYMY	Bell 47G-5	Nelhams Productions Ltd.	
G–AYMZ	PA-28 Cherokee 140C	Carlisle Aviation Co. Ltd.	
G–AYNA	Currie Wot	D. G. Crew	
G–AYNB	PA-31-300 Navajo	B. Mendes	

Notes	Reg.	Type	Owner or Operator
	G-AYNC	Wessex Mk. 60 Srs. 1	Bristow Helicopters Ltd.
	G-AYND	Cessna 310Q	IBC Transport Containers Ltd.
	G-AYNF	PA-28 Cherokee 140C	N. P. Bendle
	G-AYNG	PA-28 Cherokee 140C	R. M. Norbury
	G-AYNJ	PA-28 Cherokee 140C	J. O. Carlisle
	G-AYNN	Cessna 185B Skywagon	Bencray Ltd.
	G-AYNP	Westland S.55 Srs. 3	Bristow Helicopters Ltd.
	G-AYNR	H.S.125 Srs. 400B	McAlpine Aviation Ltd.
	G-AYNS	Airmaster H2-B1	D. J. Fry
	G-AYOD	Cessna 172	J. Vicary
	G-AYOJ	H.S.125 Srs. 400B	Bakerloo Investments Co. Ltd.
	G-AYOL	GY-80 Horizon 180	J.B.D.R. Flying Group Ltd.
	G-AYOM	Sikorsky S-61N Mk. 2	British Airways Helicopters Ltd.
	G-AYOP	BAC One-Eleven 530	British Caledonian Airways *Isle of Hoy*
	G-AYOU	Cessna 401B	Zaar International Cinema & TV Programmes Ltd.
	G-AYOW	Cessna 182N Skylane	D. P. H. Lennox
	G-AYOX	V.814 Viscount	British Midland Airways Ltd.
	G-AYOY	Sikorsky S-61N Mk. 2	British Airways Helicopters Ltd.
	G-AYOZ	Cessna FA.150L	Exeter Flying Club Ltd.
	G-AYPA	Beech A-24R Sierra	D. Williamson
	G-AYPB	Beech C-23 Musketeer	B. F. Bloomfield
	G-AYPC	Beech 70 Queen Air	Vernair Transport Services
	G-AYPD	Beech 95 B.55 Baron	Sir W. S. Dugdale
	G-AYPE	Bo 209 Monsun	Papa Echo Ltd.
	G-AYPF	Cessna F.177RG	Paine Electrics Marine Ltd.
	G-AYPG	Cessna F.177RG	Smith Aviation
	G-AYPH	Cessna F.177RG	G. A. Witherington
	G-AYPI	Cessna F.177RG	Cardinal Aviation Ltd.
	G-AYPJ	PA-28 Cherokee 180E	M. J. F. Aviation Ltd.
	G-AYPM	PA-19 Super Cub 95	C. H. A. Bott
	G-AYPO	PA-19 Super Cub 95	Mrs. J. E. Mavrogordato
	G-AYPP	PA-19 Super Cub 95	McAully Flying Group
	G-AYPR	PA-19 Super Cub 95	T. E. C. Cushing Ltd.
	G-AYPS	PA-19 Super Cub 95	K. F. J. Gardner
	G-AYPT	PA-19 Super Cub 95	Laarbruch Flying Club
	G-AYPU	PA-28R Cherokee Arrow 200	Alpine Ltd.
	G-AYPV	PA-28 Cherokee 140D	Meeting Point Ltd.
	G-AYPZ	Campbell Cricket	A. Melody
	G-AYRA	Campbell Cricket	R. C. Thomas
	G-AYRB	Campbell Cricket	G. L. Clarke
	G-AYRC	Campbell Cricket	G. A. Coventry
	G-AYRE	Campbell Cricket	Campbell Aircraft Ltd.
	G-AYRF	Cessna F.150L	Northern Auto Salvage
	G-AYRG	Cessna F.172K	D. J. C. Turnbull
	G-AYRH	M.S.892A Rallye Commodore 150	J. D. Watt
	G-AYRI	PA-28R Cherokee Arrow 200	K.R. Tools & Delta Motor Co. Ltd.
	G-AYRK	Cessna 150J	K. A. Learmonth
	G-AYRL	Fournier SFS.31 Milan	W. A. L. Mitchell
	G-AYRM	PA-28 Cherokee 140D	E. S. Dignam
	G-AYRN	Schleicher ASK-14	V. J. F. Falconer
	G-AYRO	Cessna FA.150L Aerobat	Buddale Ltd.
	G-AYRP	Cessna FA.150L Aerobat	Whitefriar Ltd.
	G-AYRS	Jodel D.120A	J. H. Tetley & G. C. Smith
	G-AYRT	Cessna F.172K	Fly-Gay Ltd.
	G-AYRU	BN-2A-6 Islander	Joint Services Parachute Centre
	G-AYSA	PA-23 Aztec 250C	C. Young & J. E. Burt
	G-AYSB	PA-30 Twin Comanche 160C	Sandcliffe Aviation
	G-AYSG	Cessna F.172K	Coventry (Civil) Aviation Ltd.
	G-AYSK	Luton L.A.4A Minor	P. F. Bennison & ptnrs.
	G-AYSX	Cessna F.177RG	Nasaire Ltd.
	G-AYSY	Cessna F.177RG	Wells (Barrow) Ltd.
	G-AYSZ	Cessna FA.150L Aerobat	O. P. Edwards & M. B. Fletcher
	G-AYTA	M.S.880B Rallye Club	Willoughby Farms Ltd.
	G-AYTB	M.S.880B Rallye Club	Champ Flying Group
	G-AYTC	PA-E23 Aztec 250C	New Guarantee Trust Finance Ltd.
	G-AYTD	PA-23 Aztec 250C	Interland Air Services Ltd.
	G-AYTF	Bell 206B JetRanger 2	Group Lotus Car Co. Ltd.
	G-AYTH	Cessna FR.172H	Zonex Ltd.
	G-AYTJ	Cessna 207 Super Skywagon	Foxair
	G-AYTN	Cameron O-65 balloon	P. G. Hall & R. F. Jessett *Prometheus*

Reg.	Type	Owner or Operator	Notes
G–AYTP	PA-23E Aztec 250E	J. Traynor	
G–AYTR	CP.301A Emeraude	D. A. Cuttriss & A. H. Jefferson	
G–AYTT	Phoenix LA-4A Duet	Gp. Capt. A. S. Knowles	
G–AYTV	MJ.2A Tempete	P. Russell	
G–AYTY	Bensen Autogyro	J. H. Wood	
G–AYUB	CEA DR.253B	D. M. E. Rawling	
G–AYUC	Cessna F.150L	Lincoln Aero Club Ltd.	
G–AYUD	PA-25 Pawnee 235	Farmair Ltd.	
G–AYUF	PA-31-300 Navajo	Cabair Ltd.	
G–AYUH	PA-28 Cherokee 180F	M. S. Bayliss	
G–AYUI	PA-28 Cherokee 180	Routair Aviation Services Ltd.	
G–AYUJ	Evans VP.1 Volksplane	R. F. Selby	
G–AYUL	PA-23 Aztec 250E	Kattan (GB) Ltd.	
G–AYUM	Slingsby T.61 Falke	Doncaster & District Gliding Club	
G–AYUN	Slingsby T.61 Falke	C. W. Vigar & R. J. Watts	
G–AYUP	Slingsby T.61A Falke	Cranwell Gliding Club	
G–AYUR	Slingsby T.61 Falke	W. A. Urwin	
G–AYUS	Taylor JT.1 Monoplane	D. G. J. Barker	
G–AYUT	Jodel DR.1050	R. Norris	
G–AYUV	Cessna F.172H	Arch Motors Manufacturing Ltd.	
G–AYUX	D.H.82A Tiger Moth (PG651)	P. R. Harris	
G–AYUY	Cessna FA.150L Aerobat	J. H. Blake & S. G. D. Ritchie	
G–AYVA	Cameron O-84 balloon	A. Kirk *April Fool*	
G–AYVB	Cessna F.172K	Botsford & Willard Ltd.	
G–AYVC	PA-23 Aztec 250E	McAlpine Aviation Ltd.	
G–AYVF	H.S.121 Trident 3B	British Airways	
G–AYVI	Cessna T.210H	Trident Marine Ltd.	
G–AYVJ	PA-23 Aztec 250D	Kilby Bros. (Property) Ltd.	
G–AYVM	PA-31-300 Navajo	Casair Aviation Services Ltd.	
G–AYVO	Wallis WA120 Srs. 1	K. H. Wallis	
G–AYVP	Woody Pusher	J. R. Wraight	
G–AYVT	Brochet MB.84	Dunelm Flying Group	
G–AYVU	Cameron O-56 balloon	Shell-Mex & B.P. Ltd. *Hot Potato*	
G–AYVY	D.H.82A Tiger Moth (PG617)	G. Smith	
G–AYWA	*Avro 19 Srs. 2	Strathallan Aircraft Collection	
G–AYWD	Cessna 182N	Trans Para Aviation Ltd.	
G–AYWE	PA-28 Cherokee 140C	Symtec Systems Ltd.	
G–AYWF	PA-23 Aztec 250C	Peregrine Air Services Ltd.	
G–AYWG	PA-E23 Aztec 250C	David Knott (Plant) Ltd.	
G–AYWH	Jodel D.117A	J. M. Knapp & ptnrs.	
G–AYWI	BN-2A Mk. III-1 Trislander	Aurigny Air Services	
G–AYWL	Taylor JT.1 Monoplane	D. G. Wiggins	
G–AYWM	Glos-Airtourer Super 150	F. B. Miles	
G–AYWS	Beech C23 Musketeer	Waygrand Ltd.	
G–AYWT	Stampe SV-4C	B. K. Lecomber	
G–AYWU	Cessna 150G	C. L. Duke	
G–AYWW	PA-28R Cherokee Arrow 200D	J. A. Butterfield & ptnrs.	
G–AYXO	Luton L.A.5 Major	A. C. T. Broomcroft	
G–AYXP	Jodel D.117A	G. N. Davies	
G–AYXS	SIAI-Marchetti S205-18R	W. F. South	
G–AYXT	Westland S.55 Srs. 2	Autair International Ltd.	
G–AYXU	Champion 7KCAB Citabria	H. Fould & ptnrs.	
G–AYXV	Cessna FA.150L	Leo Designs	
G–AYXW	Evans VP.1 Volksplane	J. S. Penny	
G–AYXX	Cessna F.172H	Robinaire Ltd.	
G–AYXY	PA-39 Twin Comanche 160 C/R	James D. Peace & Co.	
G–AYYC	Taylor JT.1 Monoplane	F. J. Hoysted	
G–AYYD	M.S.894A Rallye Minerva	J. E. Dyson & T. E. O'Connors	
G–AYYF	Cessna F.150L	Falcon Aero Club	
G–AYYG	H.S.748 Srs. 2A	Dan-Air Services Ltd.	
G–AYYK	Slingsby T.61A Falke	Polish Flying Club Ltd.	
G–AYYL	Slingsby T.61A Falke	Airways Aero Associations Ltd.	
G–AYYN	PA-28R Cherokee Arrow 200B	International Ski-Sales Ltd. & W. & R. Leggott Ltd.	
G–AYYO	Jodel DR.1050/M1	Bustard Flying Club Ltd.	
G–AYYT	Jodel DR.1050/M1	T. S. Warren & ptnrs.	
G–AYYU	Beech C23 Musketeer	A. F. Clements	
G–AYYX	M.S.880B Rallye Club	D. Hall & V. Thompson	
G–AYYY	M.S.880B Rallye Club	G. R. Porter	
G–AYYZ	M.S.880B Rallye Club	Thames Estuary Flying Services Ltd.	
G–AYZC	PA-E23 Aztec 250D	Pennine Commercial Holdings Ltd.	

Notes	Reg.	Type	Owner or Operator
	G–AYZE	PA-39 Twin Comanche 160 C/R	J. R. Fuller
	G–AYZH	Taylor JT.2 Titch	K. J. Munro
	G–AYZI	Stampe SV-4C	F. M. Barrett
	G–AYZJ	*Westland Sikorsky S-55 (XM685)	Newark Air Museum
	G–AYZK	Jodel DR.1050/MI	G. S. Claybourn
	G–AYZN	PA-E23 Aztec 250	Central Air Services (Air Envoy) Ltd.
	G–AYZS	D.62B Condor	D. P. Horridge & C. P. F. Degen
	G–AYZT	D.62B Condor	R. R. Harris
	G–AYZU	Slingsby T.61A Falke	The Falcon Gliding Group
	G–AYZW	Slingsby T.61A Falke	J. A. Dandie & R. J. M. Clement
	G–AYZX	Fournier RF-5	R. D. Goodger
	G–AYZY	PA-39 Twin Comanche 160C/R	Euromotive Ltd.
	G–AZAB	PA-30 Twin Comanche 160	T. W. P. Sheffield
	G–AZAD	Jodel DR.1051	I. C. Young & J. S. Paget
	G–AZAG	AB 206A JetRanger	Heliwork Ltd.
	G–AZAJ	PA-28R Cherokee Arrow 200B	McKenzie & Tapp Ltd.
	G–AZAU	Grasshopper Type 02	Cierva Rotorcraft Ltd.
	G–AZAV	Cessna 337F	W. T. Johnson & Sons (Huddersfield) Ltd.
	G–AZAW	GY-80 Horizon 160	Scottish Electric Ltd.
	G–AZAZ	Bensen B.8M	FAA Museum
	G–AZBA	T.66 Nipper 3	E. N. Simmons
	G–AZBB	MBB Bo 209 Monsun 160FV	Cheyne Motors Ltd.
	G–AZBC	PA-39 Twin Comanche 160 C/R	Dennis Silver & Co.
	G–AZBE	Glos-Airtourer Super 150	T. C. Edwards
	G–AZBI	Jodel D.150	T. A. Rawson & K. H. Siorpaes
	G–AZBK	PA-E23 Aztec 250E	Qualitair Engineering Ltd.
	G–AZBL	Jodel D.9 Bebe	West Midlands Flying Group
	G–AZBN	AT-16 Harvard 2B (FT391)	Colt Car Co. Ltd.
	G–AZBT	Western O-65 balloon	D. J. Harris *Hermes*
	G–AZBW	PA-39 Twin Comanche 160 C/R	Rijory Ltd.
	G–AZBX	Western O-65 balloon	Jasper Balloon Group *Thursday's Child*
	G–AZCB	Stampe SV-4C	M. J. Coburn
	G–AZCE	Pitts S-1 Special	R. J. Oulton
	G–AZCF	Sikorsky S-61N	British Airways Helicopters Ltd.
	G–AZCH	H.S.125 Srs. 3B/RA	Leopard Aviation Ltd.
	G–AZCI	Cessna 320A Skyknight	Landsurcon (Air Survey) Ltd.
	G–AZCK	B.121 Pup 2	Wickenby Flying Club Ltd.
	G–AZCL	B.121 Pup 2	Cameron Rainwear Ltd.
	G–AZCP	B.121 Pup 1	M. M. Pepper
	G–AZCT	B.121 Pup 1	D. R. Rolfe
	G–AZCU	B.121 Pup 1	Surrey & Kent Flying Club
	G–AZCZ	B.121 Pup 2	A. J. Mustarde
	G–AZDA	B.121 Pup 1	G. H. G. Bishop & K. E. Fehrenbach
	G–AZDC	Sikorsky S-61N	Bristow Helicopters Ltd. *Dunnotar*
	G–AZDD	MBB Bo 209 Monsun 150FF	E. M. Emerson & N. Hughes-Narborough
	G–AZDE	PA-28R Cherokee Arrow 200B	Electro-Motion UK (Export) Ltd.
	G–AZDF	Cameron O-84 balloon	M. G. Abram *Hannibal*
	G–AZDH	PA-31-300 Navajo	Casair Aviation Services Ltd.
	G–AZDK	Beech B55 Baron	Burton Metal Fabrications Ltd.
	G–AZDW	PA-28 Cherokee 180F	DFS Aviation Ltd.
	G–AZDX	PA-28 Cherokee 180F	Anglo-Dansk Marine Engineering Co. Ltd.
	G–AZDY	D.H.82A Tiger Moth	B. A. Mills
	G–AZDZ	Cessna 172K	Stansted Fluid Power Ltd.
	G–AZEA	Cessna 182N	Forth Flying Group Ltd.
	G–AZED	BAC One-Eleven 414	Dan-Air Services Ltd.
	G–AZEE	M.S.880B Rallye Club	P. L. Clements
	G–AZEF	Jodel D.120	P. Cawkwell & G. Firth
	G–AZEG	PA-28 Cherokee 140D	J. W. Simmons
	G–AZER	Cameron O-42 balloon	M. P. Dokk-Olsen & P. L. Jaye *Shy Tot*
	G–AZEU	B.121 Pup 2	J. N. Russell
	G–AZEV	B.121 Pup 2	G. P. Martin
	G–AZEW	B.121 Pup 2	Deltair Ltd.
	G–AZFA	B.121 Pup 2	K. F. Plummer
	G–AZFB	Boeing 720–051B	Monarch Airlines Ltd.

Reg.	Type	Owner or Operator	Notes
G-AZFC	PA-28 Cherokee 140D	J. R. Wardle	
G-AZFE	PA-23 Aztec 250D	Air Charter (Scotland) Ltd.	
G-AZFF	Jodel D.112	M. K. Field	
G-AZFI	PA-28R Cherokee Arrow 200B	Hawksworth Garage Ltd.	
G-AZFM	PA-28R Cherokee Arrow 200B	Lincs Poultry Machinery Ltd.	
G-AZFO	PA-39 Twin Comanche 160 C/R	G. Firbank	
G-AZFP	Cessna F.177RG	Afronix (UK) Ltd.	
G-AZFR	Cessna 401B	Johnson Group Management Services Ltd.	
G-AZFS	Beech B80 Queen Air	Globetrotter Survey Co. Ltd.	
G-AZFZ	Cessna 414	C. H. Taylor & Co. Ltd.	
G-AZGA	Jodel D.120	M. A. Webb	
G-AZGB	PA-E23 Aztec 250D	Qualitair Engineering Ltd.	
G-AZGC	Stampe SV-4C (No. 120)	The Hon. Patrick Lindsay	
G-AZGE	Stampe SV-4A	M. R. L. Astor	
G-AZGF	B.121 Pup 2	J. A. Macreadie	
G-AZGG	Beech C90 King Air	Hay & Co. (Lerwick) Ltd.	
G-AZGH	M.S.880B Rallye Club	R. G. Moore	
G-AZGI	M.S.880B Rallye Club	G. E. M. Hallett & ptnrs.	
G-AZGJ	M.S.880B Rallye Club	S. C. Howes & E. T. French	
G-AZGL	M.S.894A Rallye Minerva	The Cambridge Aero Club Ltd.	
G-AZGY	CP.301B Emeraude	Rodingair Flying Group	
G-AZGZ	D.H.82A Tiger Moth	F. R. Manning	
G-AZHA	PA-E23 Aztec 250E	Air Charter (Scotland) Ltd.	
G-AZHB	Robin HR 100-200	W. H. Everett & Son Ltd.	
G-AZHC	Jodel D.112	J. A. Summer & A. Burton	
G-AZHD	Slingsby T.61A Falke	West Wales Gliding Co. Ltd.	
G-AZHF	Cessna 150L	Coventry Air Training School Ltd.	
G-AZHH	SA 102.5 Cavalier	D. W. Buckle *Time*	
G-AZHI	Glos-Airtourer Super 150	A. W. Jenner & ptnrs	
G-AZHJ	S.A. Twin Pioneer Srs. 3	Flight One Ltd.	
G-AZHK	Robin HR 100-200	E. A. & W. M. C. Payton	
G-AZHL	PA-31-300 Navajo	BAC Windows Ltd.	
G-AZHM	Cassutt Racer	J. A. H. Chadwick	
G-AZHO	Jodel DR. 1050	S. Alexander	
G-AZHR	Piccard Ax6 balloon	J. W. Moss *Happiness*	
G-AZHT	Glos-Airtourer T.3	D. C. Giles	
G-AZHU	Luton L.A.4A Minor	F. Didsbury	
G-AZIA	PA-39 Twin Comanche 160 C/R	Worldair Sales Ltd.	
G-AZIB	ST-10 Diplomate	Wilmslow Audio Ltd.	
G-AZID	Cessna FA.150L	Oldment Ltd.	
G-AZIE	PA-25 Pawnee 235	Aerocare Agricultural Services Ltd.	
G-AZIG	Fournier RF-4D	A. H. R. Stansfield	
G-AZIH	J/IN Alpha	L. A. & P. Groves	
G-AZII	Jodel D.117A	J. S. Brayshaw	
G-AZIJ	Jodel DR.360	K. H. Tostevin	
G-AZIK	PA-34-200 Seneca	C.S.E. (Aircraft Services) Ltd.	
G-AZIL	Slingsby T.61B Falke	I. Jamieson	
G-AZIM	PA-31 Navajo	IDS Aircraft Ltd.	
G-AZIO	SNCAN SV-4C	Rollason Aircraft & Engines Ltd.	
G-AZIP	Cameron O-65 balloon	Dante Balloon Group *Dante*	
G-AZIR	Stampe SV-4C	Rollason Aircraft & Engines Ltd.	
G-AZJA	BN-2A-1 Mk. III Trislander	Aurigny Air Services	
G-AZJB	PA-34-200 Seneca	W. S. Churchill	
G-AZJC	Fournier RF-5	A. Edie	
G-AZJD	AT-6D Harvard III	Gladaircraft Ltd.	
G-AZJE	JB-01 Minicab	J. B. Evans	
G-AZJI	Western O-65 balloon	W. Davison *Peek-a-Boo*	
G-AZJN	Robin DR 300/140	Wright Farm Eggs Ltd.	
G-AZJV	Cessna F.172L	The JV Group	
G-AZJW	Cessna F.150L	Seair	
G-AZJX	Cessna F.150L	Gordon King (Aviation) Ltd.	
G-AZJY	Cessna FRA.150L	Shropshire Aero Club Ltd.	
G-AZJZ	PA-23 Aztec 250E	Air Commuter Ltd.	
G-AZKA	M.S.880B Rallye Club	W. K. Anderson	
G-AZKC	M.S.880B Rallye Club	L. J. Martin	
G-AZKD	M.S.880B Rallye Club	C. B. Dew	
G-AZKE	M.S.880B Rallye Club	M. J. Powell	
G-AZKG	Cessna F.172L	Wycombe Air Centre Ltd.	
G-AZKH	Cessna F.177RG	F. B. Spriggs	
G-AZKI	Noorduyn Harvard 2B (FT229)	A. E. Hutton	

Notes	Reg.	Type	Owner or Operator
	G–AZKK	Cameron O-56 balloon	Gemini Balloon Group *Gemini*
	G–AZKM	Boeing 720-051B	Monarch Airlines Ltd.
	G–AZKN	Robin HR.100/200	P. T. Bolton
	G–AZKO	Cessna F.337F	S. G. Spindlow
	G–AZKP	Jodel D.117	C. M. Fitton
	G–AZKR	PA-24 Comanche 180	B. J. Boughton
	G–AZKS	AA-1A Trainer	G. A. P. N. Barlow
	G–AZKV	Cessna FRA.150L	Penguin Flight
	G–AZKW	Cessna F.172L	Banbury Plant Hire Ltd.
	G–AZKZ	Cessna F.172L	Sprowston Engineering Ltd.
	G–AZLA	Taylor JT.2 Titch	Jeffrey Chappell
	G–AZLE	Boeing N2S–5 Kaydet	A. E. Poulson
	G–AZLF	Jodel D.120	J. Brooks
	G–AZLH	Cessna F.150L	E. Midlands School of Flying Ltd.
	G–AZLJ	BN-2A-1 Mk. III Trislander	Aurigny Air Services
	G–AZLK	Cessna F.150L	Mercury Flying Club Ltd.
	G–AZLL	Cessna FRA.150L	Cleveland Flying School Ltd.
	G–AZLM	Cessna F.172L	J. F. Davis & Q. J. Rigby
	G–AZLN	PA-28 Cherokee 180F	D. H. L. Wigan
	G–AZLO	Cessna F.337F	Leasetec Ltd.
	G–AZLP	V.813 Viscount	British Midland Airways Ltd.
	G–AZLR	V.813 Viscount	British Midland Airways Ltd.
	G–AZLS	V.813 Viscount	British Midland Airways Ltd.
	G–AZLV	Cessna 172K	J. Braithwaite (Aerial Photography) Ltd.
	G–AZLY	Cessna F.150L	W. Linskill
	G–AZLZ	Cessna F.150L	J. Fricker
	G–AZMA	Jodel D.140B	W. A. Braim Ltd.
	G–AZMB	Bell 47G–3B	Helicopter Farming Ltd.
	G–AZMC	Slingsby T.61A Falke	Essex Gliding Club Ltd.
	G–AZMD	Slingsby T.61C Falke	P. J. Moss & ptnrs.
	G–AZME	PA-31-300 Navajo	All Seasons Contracting Co. Ltd.
	G–AZMF	BAC One-Eleven 530	British Caledonian Airways
	G–AZMH	Morane-Saulnier M.S.500 (7A+WN)	Hon. P. Lindsay
	G–AZMJ	AA-5 Traveler	A. A. Cansick
	G–AZMK	PA-23 Aztec 250	Frank Graham Aviation Ltd.
	G–AZMN	Glos-Airtourer T.5	R. G. Lowerson & T. Ellefson
	G–AZMO	PA-32 Cherokee Six 260	Dateline International Dating Systems Ltd. & Grangewood Press Ltd.
	G–AZMV	D.62C Condor	Ouse Gliding Club Ltd.
	G–AZMX	PA-28 Cherokee 140	Mooney Aviation Ltd.
	G–AZMY	SIAI-Marchetti SF-260	Miss W. M. Miller
	G–AZMZ	M.S.893A Rallye Commodore 150	John Hatswell Ltd.
	G–AZNA	V.813 Viscount	British Midland Airways Ltd.
	G–AZNB	V.813 Viscount	British Midland Airways Ltd.
	G–AZNC	V.813 Viscount	British Midland Airways Ltd.
	G–AZNF	Stampe SV-4C	H. J. Smith
	G–AZNI	S.A.315B Lama	Dollar Air Services Ltd (G–AWLC)
	G–AZNJ	M.S.880B Rallye Club	Miss J. G. White
	G–AZNK	Stampe SV-4A	P. Bond
	G–AZNL	PA-28R Cherokee Arrow 200D	C. R. Balls
	G–AZNO	Cessna 182P	Strathmarine Flying Group
	G–AZNT	Cameron O-84 balloon	Cameron Balloons Ltd. *Oberon*
	G–AZNX	Boeing 720-051B	Monarch Airlines Ltd.
	G–AZNY	PA-E23 Aztec 250E	P & B Investment Holdings Ltd.
	G–AZNZ	Boeing 737-222	Britannia Airways Ltd. *Henry Hudson*
	G–AZOA	MBB Bo 209 Monsun 150FF	Dr. G. R. Outwin
	G–AZOB	MBB Bo 209 Monsun 150FF	G. N. Richardson
	G–AZOD	PA-23 Aztec 250D	Peregrine Air Services Ltd.
	G–AZOE	Glos-Airtourer 115	P. C. Logsdon
	G–AZOF	Glos-Airtourer Super 150	Armstrong Equipment Leasing Ltd.
	G–AZOG	PA-28R Cherokee Arrow 200D	Winchfield Enterprises Ltd.
	G–AZOH	Beech 65-B90 Queen Air	Clyde Surveys Ltd.
	G–AZOL	PA-34-200 Seneca	Granpack Ltd.
	G–AZOM	MBB Bo 105D	B.E.A.S. Ltd.
	G–AZON	PA-34-200-2 Seneca	Alma Investments Ltd.
	G–AZOO	Western O-65 balloon	Southern Balloon Group *Carousel*
	G–AZOS	Jurca Sirocco	R. Wells

Reg.	Type	Owner or Operator	Notes
G–AZOT	PA-34-200 Seneca	L. G. Payne	
G–AZOU	Jodel DR.1051	T. W. Jones & ptnrs.	
G–AZOZ	Cessna FRA.150L	Airwork Services Ltd.	
G–AZPA	PA-25 Pawnee 235	Farm Aviation Services Ltd.	
G–AZPC	Slingsby T.61C Falke	B. C. Dixon	
G–AZPF	Fournier RF-5	R. Pye	
G–AZPH	Craft-Pitts S-1S Special	Aerobatics International Ltd.	
G–AZPV	Luton L.A.4A Minor	J. Scott	
G–AZPX	Western O-31 balloon	E. R. McCosh *Nessie*	
G–AZPZ	BAC One-Eleven 515	Dan-Air Services Ltd.	
G–AZRA	MBB Bo 209 Monsun	The BBC Club	
G–AZRC	Cessna 340	Margate Motors (Plant & Aircraft) Ltd.	
G–AZRD	Cessna 401B	John Finlan Ltd.	
G–AZRF	Sikorsky S-61N	Bristow Helicopters *Pitcaple*	
G–AZRG	PA-23 Aztec 250D	Woodgate Aviation (IOM) Ltd.	
G–AZRH	PA-28 Cherokee 140D	K9 Transit & Co.	
G–AZRI	Payne balloon	G. F. Payne *Shoestring*	
G–AZRK	Fournier RF-5	Strathtay Flying Group	
G–AZRL	PA-19 Super Cub 95	B. A. Dunlop	
G–AZRM	Fournier RF-5	Miss R. S. A. Lloyd-Bostock	
G–AZRN	Cameron O-84 balloon	M. Yarrow *Gravida II*	
G–AZRP	Glos-Airtourer 115	Torfaen Self Drive Hire Ltd.	
G–AZRR	Cessna 310Q	Ames Company (Transport) Ltd.	
G–AZRS	PA-22 Tri-Pacer 150	E. A. Harrhy	
G–AZRU	AB-206B JetRanger 2	Dollar Air Services Ltd.	
G–AZRV	PA-28R Cherokee Arrow 200B	S. G. Daniel	
G–AZRW	Cessna T.337C	A.D.S. (Aerial) Ltd.	
G–AZRX	GY-80 Horizon 160	Fine Stitchers Ltd.	
G–AZRZ	Cessna U-206F	Army Parachute Association	
G–AZSA	Stampe SV-4B	J. K. Faulkner	
G–AZSC	AT-16 Harvard IIB (FT323)	D. W. Arnold	
G–AZSD	Slingsby T.29B Motor Tutor	R. G. Boynton	
G–AZSE	PA-28R Cherokee Arrow 200D	Amstrad Consumer Electronics Ltd.	
G–AZSF	PA-28R Cherokee Arrow 200D	P. Blamire	
G–AZSG	PA-28 Cherokee 180E	Scotia Safari Ltd.	
G–AZSH	PA-28R Cherokee Arrow 180	S. H. Hayward	
G–AZSK	Taylor JT.1 Monoplane	R. R. Lockwood	
G–AZSL	M.S.890B Rallye Commodore	F. J. Shevill	
G–AZSM	PA-28R Cherokee Arrow 180	Drumgate Ltd.	
G–AZSN	PA-28R Cherokee Arrow 200	Burch Aviation Ltd.	
G–AZSP	Cameron 0-84 balloon	Esso Petroleum Ltd. *Esso*	
G–AZSS	Jodel D.9 Bebe	M. W. Rice	
G–AZSU	H.S.748 Srs. 2A	Dan-Air Services Ltd.	
G–AZSW	Beagle 121 Pup 1	P. Evans & P. J. C. Graves	
G–AZSX	Beagle 121 Pup 1	P. W. Hunter	
G–AZSY	PA-24 Comanche 260	Christopher Foyle Aviation Leasing Co.	
G–AZSZ	PA-23 Aztec 250	Air Kilroe	
G–AZTA	MBB Bo 209 Monsun 150FF	R. S. Perks	
G–AZTD	PA-32 Cherokee Six 300	Presshouse Publications Ltd.	
G–AZTF	Cessna F.177RG	Carentals Ltd.	
G–AZTH	Bensen Autogyro	E. Henshaw	
G–AZTI	Bolkow Bo 105C	North Scottish Helicopters Ltd.	
G–AZTK	Cessna F.172F	C. C. Donald	
G–AZTM	Glos-Airtourer 115	I. J. Smith	
G–AZTN	Glos-Airtourer 115	Bernell Aviation Ltd.	
G–AZTO	PA-34-200 Seneca	C.S.E. Aviation Ltd.	
G–AZTR	SNCAN SV.4C	D. J. Shires	
G–AZTS	Cessna F.172L	Transgap Ltd.	
G–AZTT	PA-28R Cherokee Arrow 200	Rivermill Pyrford Ltd.	
G–AZTV	Stolp SA.500 Starlet	P. Russell	
G–AZTW	Cessna F.177RG	R. M. Clarke	
G–AZUG	AA-5 Traveler	Karen Peters Knitwear Ltd.	
G–AZUL	Stampe SV-4B	R. A. Seeley	
G–AZUM	Cessna F.172L	R. B. Lewis	
G–AZUO	Cessna F.177RG	Newbury Sand and Gravel Co. Ltd.	
G–AZUP	Cameron O-65 balloon	C. M. G. Ellis & ptnrs.	
G–AZUT	M.S.893A Rallye Commodore 180	Rallye Flying Group	

Notes	Reg.	Type	Owner or Operator
	G–AZUU	Fournier RF-4D	Gloster Aero Group
	G–AZUV	Cameron O-65 balloon	D. S. Bush *Icarus*
	G–AZUX	Western O-65 balloon	H. C. J. & Mrs. S. L. G. Williams *Slow Djinn*
	G–AZUY	Cessna E.310L	Euro Advertising Ltd.
	G–AZUZ	Cessna FRA.150L	D. J. Parker
	G–AZVA	MBB Bo 209 Monsun 150FF	K. H. Wallis
	G–AZVB	MBB Bo 209 Monsun 150FF	P. C. Logsdon
	G–AZVC	MBB Bo 209 Monsun 150FF	G. E. Horder & D. Cockcroft Ltd.
	G–AZVE	AA-5 Traveler	Mindon Engineering (Nottingham) Ltd.
	G–AZVF	M.S.894A Rallye Minerva	J. McCleary & T. S. Brown
	G–AZVG	AA-5 Traveler	R. B. Sandell & Co. Ltd.
	G–AZVH	M.S.894A Rallye Minerva	C. H. T. Trace
	G–AZVI	M.S.892A Rallye Commodore	Agricultural & Industrial Services (Wiltshire) Ltd. & J. F. Snook
	G–AZVJ	PA-34-200-2 Seneca	Business Air Travel Ltd.
	G–AZVL	Jodel D.119	C. Drinkwater
	G–AZVM	Hughes 369HS	Diagnostic Reagents Ltd.
	G–AZVP	Cessna F.177RG	R. W. Martin & R. G. Saunders
	G–AZVR	Cessna F.150L	E. P. Collier
	G–AZVS	H.S.125 Srs. 3B	Eastern Airways
	G–AZVT	Cameron O-84 balloon	Sky Soarer Ltd. *Jules Verne*
	G–AZVV	PA-28 Cherokee 180G	M. R. Woodgate
	G–AZVW	Bell 47G-5A	Helicopter Hire Ltd.
	G–AZVX	Bell 47G-5A	Helicopter Hire Ltd.
	G–AZVY	Cessna 310Q	Centreline Air Services Ltd.
	G–AZVZ	PA-28 Cherokee 140	Gordon King (Aviation) Ltd.
	G–AZWB	PA-28 Cherokee 140	Manchester School of Flying Ltd.
	G–AZWD	PA-28 Cherokee 140	Airways Aero Associations Ltd.
	G–AZWE	PA-28 Cherokee 140	Airways Aero Associations Ltd.
	G–AZWF	SAN Jodel DR.1050	S. R. Orwin & ptnrs.
	G–AZWS	PA-28R Cherokee Arrow 200D	Thames Estuary Flying Club
	G–AZWT	Westland Lysander III (V9441)	Strathallan Aircraft Collection
	G–AZWU	Cessna F.150L	D. W. Walton
	G–AZWW	PA-23 Aztec 250E	Merlix Air
	G–AZWY	PA-24 Comanche 260	Keymer Son & Co. Ltd.
	G–AZXA	Beechcraft 95-C55 Baron	Flight Refuelling Ltd.
	G–AZXB	Cameron O-65 balloon	London Balloon Club Ltd. *London Pride II*
	G–AZXC	Cessna F.150L	Brailsford Aviation Ltd.
	G–AZXD	Cessna F.172L	Dukdeed Ltd.
	G–AZXE	Jodel D.120A	Kestrel Flying Group
	G–AZXG	PA-23 Aztec 250	M. Priest & K. Lefevre
	G–AZXH	PA-34-200-2 Seneca	Tapehurst Ltd.
	G–AZXI	Hughes 269C	Litten Heating & Plumbing Co. Ltd.
	G–AZXM	H.S.121 Trident 2E	British Airways
	G–AZYA	GY-80 Horizon 160	T. Poole & G. L. Newbrook
	G–AZYB	Bell 47H-1	G. Watt
	G–AZYC	Cessna A.188B Agwagon	Mindacre Ltd.
	G–AZYD	M.S.893A Rallye Commodore	Deeside Gliding Club
	G–AZYF	PA-28 Cherokee 180	J. C. Glynn
	G–AZYG	PA-E23 Aztec 250	F. M. Barrett
	G–AZYI	Cessna E-310Q	Aircraft Mart
	G–AZYJ	PZL-104 Srs. 6 Wilga	Worcestershire Gliding Ltd.
	G–AZYK	Cessna 310Q	F. Adam
	G–AZYL	Portslade School free balloon	R. M. Glover *Mercury*
	G–AZYM	Cessna E-310Q	Summers Transport Ltd.
	G–AZYR	Cessna 340	Selflock Ltd.
	G–AZYS	CP.301C-1 Emeraude	J. R. Hughes
	G–AZYU	PA-E23 Aztec 250	Aviotec Ltd.
	G–AZYV	Burns O-77 balloon	B. F. G. Ribbans *Contrary Mary*
	G–AZYX	M.S.893A Rallye Commodore	Black Mountain Gliding Co. Ltd.
	G–AZYY	Slingsby T.61A Falke	J. A. Towers
	G–AZYZ	WA.51A Pacific	Tiger Club Ltd.
	G–AZZA	PA-E23 Aztec 250	Air Charter (Scotland) Ltd.
	G–AZZB	AB-206B JetRanger 2	Air Hanson Ltd.
	G–AZZC	Douglas DC-10-10	Laker Airways *Eastern Belle*
	G–AZZD	Douglas DC-10-10	Laker Airways *Western Belle*
	G–AZZE	Beech A.23-19 Musketeer	K. D. Price
	G–AZZF	M.S.880B Rallye Club	J. Meaden & ptnrs.
	G–AZZG	Cessna 188 Agwagon	W. P. Miller

Reg.	Type	Owner or Operator	Notes
G-AZZH	Practavia Pilot Sprite	K. G. Stewart	
G-AZZK	Cessna 414	Unifix Air Ltd.	
G-AZZO	PA-28 Cherokee 140	P. R. Dewing & J. Laidler	
G-AZZP	Cessna F.172H	R. C. Wilkinson Farms	
G-AZZR	Cessna F.150L	Herefordshire Aero Club Ltd.	
G-AZZT	PA-28 Cherokee 180	Stapleford Flying Club Ltd.	
G-AZZV	Cessna F.172L	Linskill Air Charter Ltd.	
G-AZZW	Fournier RF-5	Gloster Aero Group	
G-AZZX	Cessna FRA.150L	J. E. Uprichard & ptnrs.	
G-AZZZ	D.H.82A Tiger Moth	S. W. McKay	
G-BAAD	Evans Super VP-1	R. W. Husband	
G-BAAF	Manning-Flanders MF 1 replica	D. E. Bianchi	
G-BAAG	Beechcraft B.55 Baron	Mannin Aviation Ltd.	
G-BAAH	Coates SA.111 Swalesong	J. R. Coates	
G-BAAI	M.S.893A Rallye Commodore	A. F. Butcher	
G-BAAK	Cessna 207	Sunderland Parachute Centre Ltd.	
G-BAAL	Cessna 172A	V. H. Bellamy	
G-BAAP	PA-28R Cherokee Arrow 200	Shirley A. Shelley	
G-BAAR	PA-28R Cherokee Arrow 200	Diplomatic & Consular Yearbook Ltd.	
G-BAAT	Cessna 182P Skylane	S. J. Martin Ltd.	
G-BAAU	Enstrom F-28C	RHM Investments Ltd.	
G-BAAW	Jodel D.112	J. M. Alexander & R. J. Allan	
G-BAAX	Cameron O-84 balloon	The New Holker Estate Co. Ltd. *Holker Hall*	
G-BAAY	Valton Viima II	Shipping & Airlines Ltd.	
G-BAAZ	PA-28R Cherokee Arrow 200D	A. W. Rix	
G-BABA	D.H.82A Tiger Moth	S. W. McKay	
G-BABB	Cessna F.150L	George House Holdings Ltd.	
G-BABC	Cessna F.150L	E. P. Collier	
G-BABD	Cessna FRA.150L	Wycombe Air Centre Ltd.	
G-BABE	Taylor JT.2 Titch	J. Berry	
G-BABG	PA-28 Cherokee 180	M. J. McKenzie	
G-BABH	Cessna F.150L	N. F. O'Neill & E. J. Leathem	
G-BABK	PA-34-200 Seneca	D. F. J. & N. R. Flashman	
G-BABW	Beech E90 King Air	The Rank Organisation	
G-BABY	Taylor JT.2 Titch	J. R. D. Bygraves *Barnstormer Two*	
G-BACA	BAC Petrel	British Aircraft Corporation Ltd.	
G-BACB	PA-34-200 Seneca	Ernair Ltd.	
G-BACC	Cessna FRA.150L	N. F. Whistler & ptnrs.	
G-BACE	Fournier RF-5	R. W. K. Stead	
G-BACF	Cessna F.337F	Wilson Salt Co. Ltd.	
G-BACH	Enstrom F.28A	J. J. Woodhouse	
G-BACJ	Jodel D.120	Wearside Flying Association	
G-BACK	D.H.82A Tiger Moth (DF130)	G. R. French & ptnrs.	
G-BACL	Jodel D.150	G. R. French	
G-BACM	Cessna FRA.150L	Air Compton Ltd.	
G-BACN	Cessna FRA.150L	Regent Motors	
G-BACO	Cessna FRA.150L	Miss B. Kennett	
G-BACP	Cessna FRA.150L	Norfolk & Norwich Aero Club Ltd.	
G-BADC	Luton Beta	H. M. Mackenzie	
G-BADE	PA-23 Aztec 250	Thurston Aviation Ltd.	
G-BADF	PA-34-200-2 Seneca	Strata Surveys Ltd.	
G-BADH	Slingsby T.61A Falke	E. M. Andrew & ptnrs.	
G-BADI	PA-E23 Aztec 250	W. London Aero Services Ltd.	
G-BADJ	PA-E23 Aztec 250	J. G. Hogg	
G-BADK	BN-2A-8 Islander	Brymon Aviation Ltd.	
G-BADL	PA-34-200 Seneca	Apollo Aviation Ltd.	
G-BADM	D.62B Condor	Rollason Aircraft & Engines Ltd.	
G-BADO	PA-32 Cherokee Six 300	D. Russell	
G-BADP	Boeing 737-204	Britannia Airways Ltd. *Sir Arthur Whitten Brown*	
G-BADR	Boeing 737-204	Britannia Airways Ltd. *Capt. Robert Falconer Scott*	
G-BADT	Cessna 402B	British Aircraft Corp. Ltd.	
G-BADU	Cameron O-56 balloon	J. Philp *Dream Machine*	
G-BADV	Brochet MB-50	P. A. Cairns	
G-BADW	Pitts S-2A Special	Rothmans International Ltd.	
G-BADY	Pitts S-2A Special	J. McGachy	
G-BADZ	Pitts S-2A Special	A. L. Brown & ptnrs.	

Notes	Reg.	Type	Owner or Operator
	G–BAEB	Robin DR.400/160	Bracknell Refrigeration Services Ltd.
	G–BAEC	Robin HR.100/210	Checkpoint Travel (Rushden) Ltd.
	G–BAED	PA-E23 Aztec 250	Andair International Ltd.
	G–BAEE	Jodel DR.1050/M1	Joan H. Martin
	G–BAEF	Boeing 727-46	Dan-Air Services Ltd.
	G–BAEG	PA-31-300 Navajo	Aviation Beauport (Finance) Ltd.
	G–BAEJ	AA-5 Traveler	A. Clarke
	G–BAEM	Robin DR.400/125	F. J. Lingham
	G–BAEN	Robin DR.400/180	Trans Europe Air Charter Ltd.
	G–BAEP	Cessna FRA.150L	Airwork Services Ltd.
	G–BAER	Cosmic Wind	R. S. Voice
	G–BAES	Cessna 337A	Page & Moy Ltd. & High Voltage Applications Ltd.
	G–BAET	Piper L-4H Cub	C. M. G. Ellis
	G–BAEU	Cessna F.150L	Skyviews & General Ltd.
	G–BAEV	Cessna FRA.150L	South Midland Communications Ltd.
	G–BAEW	Cessna F.172M	Northamptonshire School of Flying Ltd
	G–BAEX	Cessna F.172M	D. H. Stephenson & ptnrs.
	G–BAEY	Cessna F.172M	R. Fursman
	G–BAEZ	Cessna FRA.150L	F. Butterfield
	G–BAFA	AA-5 Traveler	Lewis Flying Group Ltd.
	G–BAFD	MBB Bo 105D	British Caledonian Helicopters Ltd.
	G–BAFG	D.H.82A Tiger Moth	C. D. Cyster
	G–BAFH	Evans VP-1 Volksplane	R. H. W. Beath
	G–BAFI	Cessna F.177RG	WSM Aviation
	G–BAFL	Cessna 182P	Ingham Aviation Ltd.
	G–BAFM	AT-16 Harvard IIB	Hon. P. Lindsay
	G–BAFN	Bell 212	British Airways Helicopters Ltd.
	G–BAFP	Robin DR.400/160	Miss G. A. Habsey
	G–BAFS	PA-18 Super Cub 150	Doncaster & District Gliding Club
	G–BAFT	PA-18 Super Cub 150	Cambridge University Gliding Trust Ltd.
	G–BAFU	PA-28 Cherokee 140	Goshawk Aviation Ltd.
	G–BAFV	PA-18 Super Cub 95	P. Elliott
	G–BAFW	PA-28 Cherokee 140	B. J. Poulten
	G–BAFX	Robin DR.400/140	Copthorne Precision Products Ltd.
	G–BAFZ	Boeing 727-46	Dan-Air Services Ltd.
	G–BAGA	Cessna 182A Skylane	Peterborough Parachute Centre Ltd.
	G–BAGB	SIAI-Marchetti SF.260	British Midland Airways Ltd.
	G–BAGC	Robin DR.400/140	Hempalm Ltd.
	G–BAGE	Cessna T.210L	Worksop Aircraft Sales Partnership Ltd.
	G–BAGF	Jodel D.92 Bebe	G. R. French & J. D. Watt
	G–BAGG	PA-32 Cherokee Six 300E	J. S. Horne
	G–BAGI	Cameron O-31 balloon	Cameron Balloons Ltd. *Vital Spark*
	G–BAGL	SA.341G Gazelle Srs. 1	Westland Helicopters Ltd.
	G–BAGM	Wassmer WA.41	Alderney Flying Services Ltd.
	G–BAGN	Cessna F.177RG	M. L. Rhodes
	G–BAGO	Cessna 421B	Barline Aviation Ltd.
	G–BAGR	Robin DR.400/125	F. C. Aris & ptnrs.
	G–BAGS	Robin DR.400 2+2	Headcorn Flying School Ltd.
	G–BAGT	Helio H.295 Courier	B. J. C. Woodhall Ltd.
	G–BAGU	Luton L.A.5A Major	J. Gawley
	G–BAGV	Cessna U.206F	Corbett Farms Ltd.
	G–BAGW	Cessna F.150J	Sherburn Aero Club Ltd.
	G–BAGX	PA-28 Cherokee 140	R. A. J. Turnbull
	G–BAGY	Cameron O-84 balloon	P. G. Dunnington *Beatrice*
	G–BAHC	PA-23 Aztec 250	Berkeley Hotel Ltd.
	G–BAHD	Cessna 182P Skylane	S. Brunt (Silverdale Staffs) Ltd.
	G–BAHE	PA-28 Cherokee 140	Mooney Aviation Ltd.
	G–BAHF	PA-28 Cherokee 140	Mooney Aviation Ltd.
	G–BAHG	PA-24 Comanche 260	Friendly Aviation (Jersey) Ltd.
	G–BAHH	Wallis WA.121	K. H. Wallis
	G–BAHI	Cessna F.150H	Coventry Air Training School Ltd.
	G–BAHJ	PA-24 Comanche 250	Videovision
	G–BAHL	Robin DR.400/160	Norvett Electronics Ltd.
	G–BAHN	Beech 58 Baron	Research Consultants Ltd.
	G–BAHO	Beech C.23 Sundowner	Fairflight Ltd.
	G–BAHP	Volmer VJ.22 Sportsman	J. H. H. Turner
	G–BAHR	PA-28 Cherokee 140	N. R. Goodwin & ptnrs.
	G–BAHS	PA-28R Cherokee Arrow 200-II	A. A. Wild & ptnrs.
	G–BAHU	Enstrom F-28A	RHM Investments Ltd.

Reg.	Type	Owner or Operator	Notes
G-BAHW	Cessna 310Q	G. J. Metcalfe	
G-BAHX	Cessna 182P	Clifford Leasing Co.	
G-BAHZ	PA-28R Cherokee Arrow 200-II	J. Burgess	
G-BAIA	PA-32 Cherokee Six 300E	Langham International (Aircraft) Ltd.	
G-BAIB	Enstrom F-28A	J. Lloyd	
G-BAIF	Western O-65 balloon	B. M. Smith *Captain Starlight*	
G-BAIG	PA-34-200-2 Seneca	Derby Foods Ltd.	
G-BAIH	PA-28R Cherokee Arrow 200-II	J. Pemberton	
G-BAII	Cessna FRA.150L	Taurus Trading Ltd.	
G-BAIK	Cessna F.150L	Wickenby Aviation Ltd.	
G-BAIL	Cessna FR.172J	E. A. Black	
G-BAIM	Cessna 310Q	Airwork Services Ltd.	
G-BAIN	Cessna FRA.150L	Airwork Services Ltd.	
G-BAIO	Cessna F.150L	Gordon King (Aviation) Ltd.	
G-BAIP	Cessna F.150L	J. S. Jones	
G-BAIR	Thunder Ax7-77 balloon	P. A. & Mrs. M. Hutchins	
G-BAIS	Cessna F.177RG	Loughton Aviation Ltd.	
G-BAIU	Hiller UH-12E	Heliwork Ltd.	
G-BAIW	Cessna F.172M	R. D. Green & M. J. Dawkins	
G-BAIX	Cessna F.172M	John Cordery Aviation Ltd.	
G-BAIY	Cameron O-65 balloon	Budget Rent A Car (UK) Ltd. *Lady Budget*	
G-BAIZ	Slingsby T.61A Falke	W. L. C. O'Neill & ptnrs.	
G-BAJA	Cessna F.177RG	Don Ward Productions Ltd.	
G-BAJB	Cessna F.177RG	Citation Flying Services Ltd.	
G-BAJC	Evans VP-1	J. R. Clements	
G-BAJE	Cessna 177 Cardinal	Triavia Ltd.	
G-BAJN	AA-5 Traveler	Janacrew Ltd.	
G-BAJO	AA-5 Traveler	J. B. Cuddy	
G-BAJR	PA-28 Cherokee 180	K. F. Davison	
G-BAJT	PA-28R Cherokee Arrow 200-II	DMR Computor Ltd.	
G-BAJU	PA-23 Aztec 250	Mediterranean Caravan Sales Ltd.	
G-BAJV	SA.102.5 Cavalier	A. J. Starkey	
G-BAJW	Boeing 727-46	Dan-Air Services Ltd.	
G-BAJX	PA-E23 Aztec 250	A. J. Walgate & Son Ltd.	
G-BAJY	Robin DR.400/180	J. F. Ingledew	
G-BAJZ	Robin DR.400/125	J. F. Durcan	
G-BAKA	Sikorsky S-61N	Bristow Helicopters Ltd. *West Sole*	
G-BAKB	Sikorsky S-61N	Bristow Helicopters Ltd. *Montrose*	
G-BAKC	Sikorsky S-61N	Bristow Helicopters Ltd. *Forties*	
G-BAKD	PA-34-200-2 Seneca	NIC Instruments Ltd.	
G-BAKF	Bell 206B JetRanger 2	M. J. K. Belmont	
G-BAKG	Hughes 269C	W. R. Finance Ltd.	
G-BAKH	PA-28 Cherokee 140	Woodgate Aviation Ltd.	
G-BAKJ	PA-30 Twin Comanche 160	Overdraft Aviation	
G-BAKK	Cessna F.172H	P. J. Brown	
G-BAKL	F.27 Friendship 200	Air UK	
G-BAKM	Robin DR.400/140	F. Goodison	
G-BAKN	SNCAN SV-4C	M. Holloway	
G-BAKO	Cameron O-84 balloon	D. C. Dokk-Olsen *Pied Piper*	
G-BAKP	PA-E23 Aztec 250	Executair Ltd.	
G-BAKR	Jodel D.117	A. B. Bailey	
G-BAKS	A-B 206B JetRanger 2	G. M. H. Willis	
G-BAKT	A-B 206B JetRanger 2	Burnthills Plant Hire Ltd.	
G-BAKV	PA-18 Super Cub 150	Pounds Marine Shipping Ltd.	
G-BAKW	B.121 Pup 2	J. Trevor-Hicks Ltd.	
G-BAKY	Slingsby T.61C Falke	D. R. C. Reeves	
G-BAKZ	BN-2A Islander	Fairey Surveys Ltd.	
G-BALB	Air & Space Model 18A	Interflight Ltd.	
G-BALC	Bell 206B JetRanger 2	Dollar Air Services Ltd.	
G-BALE	Enstrom F.28A	C.S.E. Aviation Ltd.	
G-BALF	Robin DR.400/140	F. A. Spear	
G-BALG	Robin DR.400/180	R. Jones	
G-BALH	Robin DR.400/140	Pennine Leisure Ltd.	
G-BALI	Robin DR.400 2+2	E. F. Rowe	
G-BALJ	Robin DR.400/180	Barlodz Ltd.	
G-BALK	SNCAN SV-4C	J. C. Brierley	
G-BALL	Bede BD-5	J. P. Turner	
G-BALM	Cessna 340	Patgrove Ltd.	
G-BALN	Cessna T.310Q	Wadkin (Aviation) Ltd.	
G-BALP	PA-39 Twin Comanche 160 C/R	Maynards (Heels) Ltd.	
G-BALR	Wittman W.8 Tailwind	D. O. A. Elmer	

Notes	Reg.	Type	Owner or Operator
	G–BALS	Nipper T.66 Mk. 3	L. W. Shaw
	G–BALT	Enstrom F.28A	Franklin Aviation Ltd.
	G–BALU	PA-E23 Aztec 250C	Avio-Lec Ltd. (G–BADD)
	G–BALW	PA-28R Cherokee Arrow 200-II	H. R. Fenwick
	G–BALX	D.H.82A Tiger Moth (N6848)	C. P. B. Horsley & R. G. Annis
	G–BALY	Practavia Pilot Sprite 150	A. L. Young
	G–BAMB	Slingsby T.61C Falke	Universities of Glasgow & Strathclyde Gliding Club
	G–BAMC	Cessna F.150L	D. R. Calo & M. McDonald
	G–BAME	Volmer VJ-22 Sportsman	V. H. Bellamy
	G–BAMF	MBB Bo 105D	Management Aviation Ltd.
	G–BAMG	Avions Lobet Ganagobie	J. A. Brompton
	G–BAMI	Beech 95-B55 Baron	Ace Belmont International Ltd.
	G–BAMJ	Cessna 182P	Aylesbury Mushrooms Ltd.
	G–BAMK	Cameron D-96 hot-air airship	Cameron Balloons Ltd.
	G–BAML	Bell 206A JetRanger	Somerton-Rayner Helicopters Ltd.
	G–BAMM	PA-28 Cherokee 235	E. R. Walters
	G–BAMN	Cessna U.206C Super Skywagon	M. E. Robinson
	G–BAMR	PA-16 Clipper	H. Boyce
	G–BAMS	Robin DR.400/160	Alouette Leasing
	G–BAMU	Robin DR.400/160	Anvil Flying Group
	G–BAMV	Robin DR.400/180	Craven Aviation Ltd.
	G–BAMY	PA-28R Cherokee Arrow 200-II	B. Gittins & ptnrs.
	G–BAMZ	PA-34-200-2 Seneca	G. R. Air Services Ltd.
	G–BANA	Robin DR.221	G. T. Pryor
	G–BANB	Robin DR.400/180	Time Electronics Ltd.
	G–BANC	GY-201 Minicab	C. D. B. Trollope
	G–BAND	Cameron O-84 balloon	Mid-Bucks Farmers Balloon Group *Clover*
	G–BANE	Cessna FRA.150L	J. Turnbull
	G–BANF	Luton L.A.4A Minor	D. W. Bosworth
	G–BANG	Cameron O-84 balloon	R. Harrower *Salamander*
	G–BANK	PA-34-200-2 Seneca	Airde Ltd.
	G–BANL	BN-2A-8 Islander	Loganair Ltd.
	G–BANS	PA-34-200-2 Seneca	G. Knowles
	G–BANT	Cameron O-65 balloon	M. D. Tweedie & ptnrs. *Shades*
	G–BANU	Wassmer Jodel D.120	C. E. McKinney
	G–BANV	Phoenix Currie Wot	K. Knight
	G–BANW	CP.1330 Super Emeraude	J. D. McCracker & ptnrs.
	G–BANX	Cessna F.172M	A. J. Keen & ptnrs.
	G–BANY	Glos-AESL Airtourer 115	Bernell Aviation Ltd.
	G–BAOB	Cessna F.172M	Gordon King (Aviation) Ltd.
	G–BAOC	M.S.894E Rallye Minerva	P. V. & Mrs. E. M. Gilliar
	G–BAOD	M.S.880B Rallye Club	E. K. Chalke
	G–BAOF	M.S.880B Rallye Club	G. B. Instrument Panel Co. Ltd.
	G–BAOG	M.S.880B Rallye Club	P. L. M. Moss & S. W. Biroth
	G–BAOH	M.S.880B Rallye Club	S. P. Bryant & ptnrs.
	G–BAOJ	M.S.880B Rallye Club	D. W. Busby & J. L. Howard
	G–BAOM	M.S.880B Rallye Club	J. A. Aldridge & ptnrs.
	G–BAOP	Cessna FRA.150L	Renco Aviation Ltd.
	G–BAOS	Cessna F.172M	Allen Baker Photography Ltd.
	G–BAOT	M.S.880B Rallye Club	B. Findley
	G–BAOU	AA-5 Traveler	W. H. Ingram
	G–BAOV	AA-5 Traveler	Hornet Aviation Ltd.
	G–BAOW	Cameron O-65 balloon	P. A. White *Winslow Boy*
	G–BAOX	Cessna 310Q	Gempen Ltd.
	G–BAOY	Cameron S-31 balloon	Shell-Mex BP Ltd. *New Potato*
	G–BAPA	Fournier RF-5B Sperber	R. Pasold & D. Stuynor
	G–BAPB	DHC-1 Chipmunk 22	R. C. P. Brookhouse
	G–BAPC	Luton L.A.4A Minor	Midland Aircraft Preservation Soc.
	G–BAPF	V.814 Viscount	British Midland Airways Ltd.
	G–BAPH	Cessna FRA.150L	B. Flay & T. C. Hocking
	G–BAPI	Cessna FRA.150L	Industrial Supplies (Peterborough) Ltd.
	G–BAPJ	Cessna FRA.150L	M. D. Page
	G–BAPK	Cessna F.150L	Ulster Flying Club (1961) Ltd.
	G–BAPL	PA-23 Aztec 250E	Scottish Malt Distillers Ltd.
	G–BAPM	Fuji FA.200-160	M. J. Flanagan
	G–BAPN	PA-28 Cherokee 180	Capricorn Air Ltd.
	G–BAPP	Evans VP-1	N. Crow

Reg.	Type	Owner or Operator	Notes
G-BAPR	Jodel D.11	E. W. Osbourn & ptnrs.	
G-BAPS	*Campbell Cougar	British Rotorcraft Museum	
G-BAPT	Fuji FA.200-180	T. F. Turner	
G-BAPV	Robin DR.400/160	J. D. Millne & ptnrs.	
G-BAPW	PA-28R Cherokee Arrow 180	G. & R. Consultants Ltd.	
G-BAPX	Robin DR.400/160	R. R. Hall & R. H. Richards	
G-BAPY	Robin HR.100/210	Engineering Appliances Ltd.	
G-BARB	PA-34-200-2 Seneca	Maykind Ltd.	
G-BARC	Cessna FR.172J	C. Porter & ptnrs.	
G-BARD	Cessna 337C	Europa Aviation Ltd.	
G-BARF	Jodel D.112 Club	G. J. Anderson	
G-BARG	Cessna E.310Q	G. E. Platt	
G-BARH	Beech C.23 Sundowner	Hadley Green Garage Ltd.	
G-BARJ	Bell 212	B.E.A.S. Ltd.	
G-BARN	Taylor JT.2 Titch	R. G. W. Newton	
G-BARP	Bell 206B JetRanger 2	S.W. Electricity Board	
G-BARR	H.S.125 Srs. 600B	Rolls-Royce Ltd.	
G-BARS	D.H.C.1. Chipmunk 22	T. I. Sutton	
G-BART	H.S.125 Srs. 600B	Green Shield Trading Stamp Co. Ltd.	
G-BARV	Cessna 310Q	Old England Watches Ltd.	
G-BARW	Cessna 402B	C. Love & H. G. Woodsend	
G-BARX	Bell 206B JetRanger 2	W.R. Finance Ltd.	
G-BARY	CP.301A Emeraude	W. C. C. Meyer	
G-BARZ	Scheibe SF.28A	J. A. Fox & ptnrs.	
G-BASB	Enstrom F-28A	Travel Centre (Norwich) Ltd.	
G-BASD	B.121 Pup 2	G. W. Archer	
G-BASE	Bell 206B JetRanger 2	Air Hanson	
G-BASG	AA-5 Traveler	Argus Credit Management Services Ltd.	
G-BASH	AA-5 Traveler	A. Forrest-Gairaud	
G-BASI	PA-28 Cherokee 140	CFS Aircraft Ltd.	
G-BASJ	PA-28 Cherokee 180	Robinson-Wyllie Ltd.	
G-BASL	PA-28 Cherokee 140	Air Navigation & Trading Ltd.	
G-BASM	PA-34-200-2 Seneca	A. Jacobs *Joie de Vivre*	
G-BASN	Beech C.23 Sundowner	M. F. Fisher	
G-BASO	Lake LA-4 Amphibian	P. B. W. Spearing & H. W. A. Deacon	
G-BASP	B.121 Pup 1	Northamptonshire School of Flying Ltd.	
G-BASR	PA-25 Pawnee 235C	C. M. G. Ellis & M. M. James	
G-BASU	PA-31-350 Navajo Chieftain	Casair Aviation Services Ltd.	
G-BASV	Enstrom F-28A	Baldwin Leasing Ltd.	
G-BASX	PA-34-200-2 Seneca	P. J. Brown	
G-BASY	Jodel D.9 Bebe	R. L. Sambell	
G-BATA	H.S.125 Srs. 403B	Beecham-Imperial Aviation Ltd.	
G-BATC	MBB Bo 105D	Management Aviation Ltd.	
G-BATE	PA-23 Aztec 250E	Trehaven Trust Ltd.	
G-BATH	Cessna F.337G	S. White & L. H. Bedden	
G-BATJ	Jodel D.119	N. Shepherd & ptnrs.	
G-BATM	PA-32 Cherokee Six 300	Patgrove Ltd. & J. Wakeman & Co.	
G-BATN	PA-E23 Aztec 250	Marshall of Cambridge Ltd.	
G-BATR	PA-34-200-2 Seneca	Executive Aviation Ltd.	
G-BATS	Taylor JT.1 Monoplane	J. Jennings	
G-BATT	Hughes 269C	Farm Supply (Thirsk) Ltd.	
G-BATU	Enstrom F-28A	Trent Park Stables	
G-BATV	PA-28 Cherokee 180D	The Scoresby Flying Group	
G-BATW	PA-28 Cherokee 140	Mooney Aviation Ltd.	
G-BATX	PA-23 Aztec 250E	Tayside Aviation Ltd.	
G-BAUA	PA-E23 Aztec 250	David Parr & Associates Ltd.	
G-BAUC	PA-25 Pawnee 235C	P. M. Charles	
G-BAUD	Robin DR.400/160	Triton Air Travel Ltd.	
G-BAUE	Cessna 310Q	A. J. Dyer	
G-BAUF	Hughes 269C	Point to Point Helicopters Ltd.	
G-BAUH	Jodel D.112	I. G. Glenn & ptnrs.	
G-BAUI	PA-E23 Aztec 250	Simulated Flight Training Ltd.	
G-BAUJ	PA-E23 Aztec 250	Airde Ltd.	
G-BAUK	Hughes 269C	Curtis Engineering (Frome) Ltd.	
G-BAUM	Bell 206B JetRanger 2	PLM Helicopters Ltd.	
G-BAUN	Bell 206B JetRanger 2	Bristow Helicopters Ltd.	
G-BAUO	PA-E23 Aztec 250	Jones & Bailey (Contractors) Ltd.	
G-BAUR	F.27 Friendship Mk. 200	Air UK	
G-BAUV	Cessna F.150L	Skyviews & General Ltd.	
G-BAUW	PA-E23 Aztec 250	Myson Group Ltd.	
G-BAUX	Limba Lapwing	B. J. Jacobson & R. M. Fisher	

Notes	Reg.	Type	Owner or Operator
	G–BAUY	Cessna FRA.150L	Inverness Flying Services Ltd.
	G–BAUZ	Nord NC.854S	T. Atkinson
	G–BAVB	Cessna F.172M	Hudson Bell Aviation
	G–BAVC	Cessna F.150L	Elles Aviation
	G–BAVE	Beech A.100 King Air	Vernair Transport Services
	G–BAVF	Beech 58 Baron	Tarlworth Ltd.
	G–BAVG	Beech E.90 King Air	Allied Breweries (UK) Ltd.
	G–BAVH	D.H.C.1 Chipmunk 22	Portsmouth Naval Gliding Club
	G–BAVL	PA-E23 Aztec 250	Shipboard Maintenance Ltd.
	G–BAVM	PA-31-350 Navajo Chieftain	Air Commuter Ltd.
	G–BAVN	Boeing Stearman PT-17	London Publicity Services Ltd.
	G–BAVO	Boeing Stearman N2S	Keenair Services Ltd.
	G–BAVP	Beech A.23-24 Musketeer	Wearside Flying Group
	G–BAVR	AA-5 Traveler	Rabhart Ltd.
	G–BAVS	AA-5 Traveler	Crystal Heart Salad Co. Ltd.
	G–BAVU	Cameron A-105 balloon	J. D. Michaelis
	G–BAVW	PA-E23 Aztec 250	Baldock's of Wivelsfield Ltd.
	G–BAVX	HPR-7 Herald 214	British Air Ferries Ltd. *Timothy Keegan*
	G–BAVY	PA-E23 Aztec 250	Stellaris Ltd.
	G–BAVZ	PA-E23 Aztec 250	Dismore Aviation Ltd.
	G–BAWA	PA-28R-200-2 Cherokee Arrow	Airways Aero Associations Ltd.
	G–BAWB	PA-E23 Aztec 250	J. T. Tyer
	G–BAWG	PA-28R-200-2 Cherokee Arrow	Richard Flint & Co. Ltd.
	G–BAWI	Enstrom F-28A	Galaxy Aviation Ltd.
	G–BAWK	PA-28 Cherokee 140	Newcastle-Upon-Tyne Aero Club Ltd.
	G–BAWL	Airborne Industries gas airship	A. F. J. Smith *The Santos-Dumont*
	G–BAWN	PA-30C Twin Comanche 160	J. & Y. Plastics (Aviation) Ltd.
	G–BAWR	Robin HR.100/210	J. W. O'Sullivan
	G–BAWU	PA-30 Twin Comanche 160	C. P. Francis
	G–BAWV	PA-E23 Aztec 250	Golden Lion Plant Hire Ltd.
	G–BAWW	Thunder Ax7-77 balloon	Miss M. L. C. Hutchins *Taurus*
	G–BAWX	PA-28 Cherokee 180	Bawxair Ltd.
	G–BAWZ	Cessna 402B	E. M. Brain
	G–BAXD	BN-2A Mk. III Trislander	Loganair Ltd.
	G–BAXE	Hughes 269A	Reethorpe Engineering Ltd.
	G–BAXF	Cameron O-77 balloon	C. W. Aston *Granna*
	G–BAXH	Cessna 310Q	Mindglass Ltd.
	G–BAXI	PA-39 Twin Comanche C/R	C.S.E. (Aircraft Services) Ltd.
	G–BAXK	Thunder Ax7-77 balloon	Newbury Balloon Group *Jack O'Newbury*
	G–BAXL	H.S.125 Srs. 3B	Dennis Vanguard International (Switchgear) Ltd.
	G–BAXM	Beech B.24R Sierra	Strangford Flying Group
	G–BAXN	PA-34-200-2 Seneca	Ards Aviation
	G–BAXP	PA-E23 Aztec 250	Peregrine Air Services Ltd.
	G–BAXR	Beech B.55 Baron	Anglian Windows Double Glazing [Ltd.
	G–BAXS	Bell 47G-5	T. C. Barton
	G–BAXT	PA-28R-200 Cherokee Arrow	Williams & Griffin Ltd.
	G–BAXU	Cessna F.150L	W. Lancs Aero Club Ltd.
	G–BAXV	Cessna F.150L	Cardiff Aviation Ltd.
	G–BAXW	Cessna F.150L	Wycombe Air Centre Ltd.
	G–BAXY	Cessna F.172M	Risk Management Group (London) Ltd.
	G–BAXZ	PA-28 Cherokee 140	A. J. Smith & D. N. Sharpe
	G–BAYC	Cameron O-65 balloon	D. Whitlock & R. T. F. Mitchell
	G–BAYL	Nord 1203/111 Norecrin	D. M. Fincham
	G–BAYO	Cessna 150L	Cheshire Air Training School Ltd.
	G–BAYP	Cessna 150L	J. N. Collins
	G–BAYR	Robin HR.100/210	Gilbey Warren Co. Ltd.
	G–BAYT	H.S.125 Srs. 600B	Management Agency and Music Ltd.
	G–BAYX	Bell 47G-5	Helicopter Hire Ltd.
	G–BAYY	Cessna 310C	Allen Baker Aviation Ltd.
	G–BAYZ	Bellanca 7GC BC Citabria	Cambridge University Gliding Trust Ltd.
	G–BAZA	H.S.125 Srs. 403B	British Aerospace
	G–BAZB	H.S.125 Srs. 400B	Short Bros. Ltd.
	G–BAZC	Robin DR.400/160	Greataslot Ltd.

Reg.	Type	Owner or Operator	Notes
G–BAZF	AA-5 Traveler	Carpfinch Ltd.	
G–BAZG	Boeing 737-204	Britannia Airways Ltd. *Florence Nightingale*	
G–BAZH	Boeing 737-204	Britannia Airways Ltd. *Isambard Kingdom Brunel*	
G–BAZI	Boeing 737-204	Britannia Airways Ltd. *Sir Walter Rayleigh*	
G–BAZJ	HPR-7 Herald 209	Air UK	
G–BAZM	Jodel D.11	Bingley Flying Group	
G–BAZN	Bell 206B JetRanger	Somerton-Rayner Helicopters Ltd.	
G–BAZS	Cessna F.150L	Sherburn Aero Club Ltd.	
G–BAZT	Cessna F.172M	Murray Fraser (Aviation) Ltd.	
G–BAZU	PA-28R-200 Cherokee Arrow	Andytroe Ltd.	
G–BAZV	PA-E23 Aztec 250	Wings International	
G–BBAB	M.S.894A Rallye Minerva	Stenloss Ltd.	
G–BBAE	L.1011–385 TriStar	British Airways *The Stargazer Rose*	
G–BBAF	L.1011–385 TriStar	British Airways *The Coronation Gold Rose*	
G–BBAG	L.1011–385 TriStar	British Airways *The Caroline Davison Rose*	
G–BBAH	L.1011–385 TriStar	British Airways *The Sunsilk Rose*	
G–BBAI	L.1011–385 TriStar	British Airways *The Molly McGredy Rose*	
G–BBAJ	L.1011–385 TriStar	British Airways *The Elizabeth Harkness Rose*	
G–BBAK	M.S.894A Rallye Minerva	Scramble Flying Group	
G–BBAR	Jodel D.117	J. F. Wright	
G–BBAU	Enstrom F.28A	J. D. M. Roberts	
G–BBAV	PA-E23 Aztec 250	Aviation Enterprises	
G–BBAW	Robin HR.100/210	Scoba Ltd.	
G–BBAX	Robin DR.400/140	S. R. Young	
G–BBAY	Robin DR.400/140	G. A. Pentelow & D. B. Roadnight	
G–BBBB	Taylor JT.1 Monoplane	S. A. MacConnacher	
G–BBBC	Cessna F.150L	Norfolk & Norwich Aero Club Ltd.	
G–BBBD	PA-E23 Aztec 250	Moseley Group (PSV) Ltd.	
G–BBBI	AA-5 Traveler	The Bravo India Group	
G–BBBJ	PA-E23 Aztec 250	Worldmaster Ltd.	
G–BBBK	PA-28 Cherokee 140	Bolton Air Training School Ltd.	
G–BBBL	Cessna 337B	Alderney Air Charter Ltd.	
G–BBBM	Bell 206B JetRanger 2	Keluma Ltd.	
G–BBBN	PA-28 Cherokee 180	G. Nicholson	
G–BBBO	SIPA 903	M. C. Wroe	
G–BBBW	FRED Series 2	D. L. Webster	
G–BBBX	Cessna E310L	Air Atlantique Ltd.	
G–BBBY	PA-28 Cherokee 140	Channel Aviation Ltd.	
G–BBBZ	Enstrom F-28A	Spooner Aviation (Enstrom Helicopters) Ltd.	
G–BBCA	Bell 206B JetRanger 2	Barkers Plant Hire (Alsager) Ltd.	
G–BBCB	Western O-65 balloon	M. Westwood *Cee Bee*	
G–BBCC	PA-E23 Aztec 250	Goodridge (UK) Ltd.	
G–BBCD	Beech 95 B.55 Baron	Jackson Aviation Ltd.	
G–BBCF	Cessna FRA.150L	Union Beech Ltd.	
G–BBCG	Robin DR.400/2+2	Headcorn Flying School Ltd.	
G–BBCH	Robin DR.400/2+2	Headcorn Flying School Ltd.	
G–BBCI	Cessna 150H	D. Machin	
G–BBCJ	Cessna 150J	Mattick Aviation Ltd.	
G–BBCK	Cameron O-77 balloon	R. J. Leathart *The Mary Gloster*	
G–BBCL	H.S.125 Srs. 600B	British Aerospace (G–BJCB)	
G–BBCM	PA-E23 Aztec 250	Keenair Services Ltd.	
G–BBCN	Robin HR.100/210	B. D. Steel Holdings Ltd.	
G–BBCP	Thunder Ax6-56 balloon	P. H. Keene *Jack Frost*	
G–BBCS	Robin DR.400/140	Albert J. Parsons & Sons Ltd.	
G–BBCT	PA-31-350 Navajo Chieftain	McGirr Aviation Ltd.	
G–BBCW	PA-E23 Aztec 250	Deborah Services Ltd.	
G–BBCX	Airship (hot-air) radio-controlled	E. A. Wills & G. W. Moger *Dew Drop*	
G–BBCY	Luton L.A.4A. Minor	C. H. Difford	
G–BBCZ	AA-5 Traveler	Stronghill Flying Group	
G–BBDA	AA-5 Traveler	Headcorn Flying School Ltd.	
G–BBDB	PA-28 Cherokee 180	Eleuthero Ltd.	
G–BBDC	PA-28 Cherokee 140	Manchester School of Flying Ltd.	
G–BBDD	PA-28 Cherokee 140	FR Aviation	

Notes	Reg.	Type	Owner or Operator
	G–BBDE	PA-28R-200-2 Cherokee Arrow	Harding of Winsford
	G–BBDG	Concorde 100	British Aerospace
	G–BBDH	Cessna F.172M	G. Jones
	G–BBDI	PA-18-150 Super Cub	Scottish Gliding Union Ltd.
	G–BBDJ	Thunder Ax6-56 balloon	S. W. D. & H. B. Ashby *Jack Tar*
	G–BBDK	V.808F Viscount	—
	G–BBDL	AA-5 Traveler	A. Howard
	G–BBDM	AA-5 Traveler	E. M. Pettit Construction Ltd.
	G–BBDN	Taylor JT.1 Monoplane	D. A. Nice
	G–BBDO	PA-E23 Aztec 250	R. Long
	G–BBDP	Robin DR.400/160	Braddon & Sons (Haulage) Ltd.
	G–BBDS	PA-31 Navajo	Broad Oak Air Services
	G–BBDT	Cessna 150H	J. M. McCloy
	G–BBDU	PA-31 Navajo	ITT Components Ltd.
	G–BBDV	SIPA S.903	A. W. Webster
	G–BBEA	Luton L.A.4A Minor	D. J. Wells & ptnrs.
	G–BBEB	PA-28R-200-2 Cherokee Arrow	Atkinson Aviation Ltd.
	G–BBEC	PA-28 Cherokee 180	P. F. Blake
	G–BBED	M.S.894B Rallye Minerva	Trago Mills Ltd.
	G–BBEF	PA-28 Cherokee 140	Air Navigation & Trading Co. Ltd.
	G–BBEI	PA-31 Navajo	Survey Flights Ltd.
	G–BBEJ	PA-31-350 Navajo Chieftain	Aaronite Ltd.
	G–BBEL	PA-28R Cherokee Arrow 180	J. M. McRitchie
	G–BBEM	Beech B.55 Baron	M. J. Coburn & L. C. G. Hughes
	G–BBEN	Bellanca 7GCBC Citabria	Ulster Gliding Club Ltd.
	G–BBEO	Cessna FRA.150L	Granair Ltd.
	G–BBEU	Bell 206B JetRanger 2	International Messengers Ltd.
	G–BBEV	PA-28 Cherokee 140	Spooner Aviation Ltd.
	G–BBEW	PA-E23 Aztec 250	Lease Air Ltd.
	G–BBEX	Cessna 185A Skywagon	R. G. Brooks & D. E. Wilson
	G–BBEY	PA-E23 Aztec 250	Berrard Aviation
	G–BBFC	AA-1B Trainer	P. C. Blake
	G–BBFD	PA-28R-200-2 Cherokee Arrow	Peacock & Archer Ltd.
	G–BBFE	Bell 206 JetRanger	W. Holmes
	G–BBFL	GY-201 Minicab	D. H. Greenwood
	G–BBFS	Van Den Bemden gas balloon	A. J. F. Smith *Le Tomate*
	G–BBFT	Cessna A.188B AgTruck	Mindacre Ltd.
	G–BBFU	PA-E23 Aztec 250	Bees Flight Ltd.
	G–BBFV	PA-32 Cherokee Six 260	Southend Securities Ltd.
	G–BBFW	PA-E23 Aztec 250B	T. Bartlett
	G–BBFX	PA-34-200-2 Seneca	C. D. Weiswall
	G–BBFY	PA-31P Navajo	Peters Stores Ltd.
	G–BBFZ	PA-28R-200-2 Cherokee Arrow	Larkfield Garage (Chepstow) Ltd.
	G–BBGB	PA-E23 Aztec 250	K. J. F. Aircraft Ltd.
	G–BBGC	M.S.893E Rallye Commodore 180	D. D. Caines
	G–BBGE	PA-E23 Aztec 250	Dollar Air Services Ltd.
	G–BBGF	Cessna 340	Hillair Ltd. & Lawrence Wilson & Son Ltd.
	G–BBGG	AA-5 Traveler	H. Richards & Sons
	G–BBGH	AA-5 Traveler	P. D. Bingham & ptnrs.
	G–BBGI	Fuji FA.200–160	Sureway Security Ltd.
	G–BBGJ	Cessna 180	J. D. Maxwell-Fenning
	G–BBGK	Lake LA-4-200 Buccaneer	James Duncan (Plumbers) Ltd.
	G–BBGL	Baby Great Lakes	P. W. Thomas
	G–BBGO	Robin HR.100/210	Home & Overseas International Oil Co. Ltd.
	G–BBGR	Cameron O-65 balloon	Thames Valley Balloon Group
	G–BBGS	Sikorsky S-61N	Bristow Helicopters Ltd. *Indefatigable*
	G–BBGU	H.S.125 Srs. 403B	McAlpine Aviation Ltd.
	G–BBGX	Cessna 182P Skylane	Lockwoods Technical Services (Liverpool) Ltd.
	G–BBGZ	CHABA 42 balloon	J. Haigh & ptnrs. *Phlogiston*
	G–BBHB	PA-31-300 Navajo	Distance No Object Ltd.
	G–BBHC	Enstrom F-28A	Dumbrill Plant & Engineering Co. Ltd.
	G–BBHD	Enstrom F-28A	Hyde Industrial Holdings Ltd.
	G–BBHE	Enstrom F-28A	Complant Engineering Services Ltd.
	G–BBHF	PA-23 Aztec 250E	Bevan Lynch Aviation Ltd.
	G–BBHG	Cessna E-310Q	Airwork Services Ltd.

Reg.	Type	Owner or Operator	Notes
G-BBHI	Cessna 177RG	Hi-Lines Ltd.	
G-BBHJ	Piper J-3C-65 Cub	R. V. Miller & R. H. Heath	
G-BBHK	AT-16 Harvard IIB	Bob Warner Aviation	
G-BBHL	Sikorsky S-61N Mk II	Bristow Helicopters Ltd. *Glamis*	
G-BBHM	Sikorsky S-61N Mk II	Bristow Helicopters Ltd. *Braemar*	
G-BBHU	SA.341G Gazelle I	Solaria Investments Ltd.	
G-BBHW	SA.341G Gazelle I	McAlpine Aviation Ltd.	
G-BBHX	M.S.893E Rallye Commodore	P. J. Pitts	
G-BBHY	PA-28 Cherokee 180	Air Operations Ltd.	
G-BBIA	PA-28R-200 Cherokee Arrow	A. G. (Commodities) Ltd.	
G-BBIC	Cessna 310Q	ATA Grinding Processes	
G-BBID	PA-28 Cherokee 140	T. K. Aero Enterprises Ltd.	
G-BBIF	PA-E23 Aztec 250	Northern Executive Aviation Ltd.	
G-BBIH	Enstrom F-28A	Sir G. Grant-Suttie	
G-BBII	Fiat G-46-3B	The Hon. Patrick Lindsay	
G-BBIJ	Cessna 421B	Executive Express	
G-BBIL	PA-28 Cherokee 140	Leisure Lease	
G-BBIN	Enstrom F-28A	Churnbury Ltd.	
G-BBIO	Robin HR.100/210	R. A. King	
G-BBIT	Hughes 269B	Contract Development & Projects (Leeds) Ltd.	
G-BBIU	Hughes 269C	W. R. Finance Ltd.	
G-BBIV	Hughes 269C	W. R. Finance Ltd.	
G-BBIW	Hughes 269C	W. R. Finance Ltd.	
G-BBIX	PA-28 Cherokee 140	R. C. Harvey & W. G. Best	
G-BBJB	Thunder AX7-77 balloon	St. Crispin Balloon Group *Dick Darby*	
G-BBJD	Cessna 172M	A. F. Hall	
G-BBJH	Cessna A-188B Agtruck	W. R. C. Wauchope	
G-BBJI	Isaacs Spitfire	R. C. Tyler	
G-BBJT	Robin HR.200/100	M. J. McRobert	
G-BBJU	Robin DR.400/140	W. Hosie	
G-BBJV	Cessna F.177RG	Pilot Magazine	
G-BBJW	Cessna FRA.150L	Gordon King (Aviation) Ltd.	
G-BBJX	Cessna F.150L	Yorkshire Flying Services Ltd.	
G-BBJY	Cessna F.172M	J. Lucketti	
G-BBJZ	Cessna F.172M	Border Flying Group	
G-BBKA	Cessna F.150L	Executive Air	
G-BBKB	Cessna F.150L	Shoreham Flight Simulation Ltd.	
G-BBKC	Cessna F.172M	D. E. H. Designs Ltd.	
G-BBKE	Cessna F.150L	Wickenby Aviation Ltd.	
G-BBKF	Cessna FRA.150L	George House Holdings Ltd.	
G-BBKG	Cessna FR.172J	S. Richman & M. J. H. Raymont	
G-BBKI	Cessna F.172M	G. Crawford	
G-BBKJ	Cessna FT.337G	Carters Gold Medal Soft Drinks Ltd.	
G-BBKL	CP.301A Emeraude	W. J. Walker	
G-BBKO	Thunder Ax7-77 balloon	J. A. Clarke	
G-BBKP	Bell 47G-5A	Alan Mann Helicopters Ltd.	
G-BBKR	Scheibe SF.24A Motorspatz	D. W. Bridson	
G-BBKU	Cessna FRA.150L	Balgia Ltd.	
G-BBKV	Cessna FRA.150L	Laarbruch Powered Flying Club	
G-BBKX	PA-28 Cherokee 180	P. E. Eglington	
G-BBKY	Cessna F.150L	W. of Scotland Flying Club Ltd.	
G-BBKZ	Cessna 172M	Exeter Flying Club Ltd.	
G-BBLA	PA-28 Cherokee 140	Southport Aviation Co. Ltd.	
G-BBLC	Hiller UH-12E	Agricopters Ltd.	
G-BBLD	Hiller UH-12E	TK Aero Enterprises Ltd.	
G-BBLE	Hiller UH-12E	Agricopters Ltd.	
G-BBLG	Hiller UH-12E	Sloane Aviation Ltd.	
G-BBLH	Piper O-59A Grasshopper	P. A. Mann	
G-BBLJ	Cessna 402B	W. H. Crispe & Son Ltd.	
G-BBLL	Cameron O-84 balloon	University of East Anglia Hot-Air Ballooning Club *Boadicea*	
G-BBLM	MS.880 Rallye 100 Sport	R. J. Lewis & P. Walker	
G-BBLP	PA-E23 Aztec 250D	Elecwind (Clay Cross) Ltd.	
G-BBLS	AA-5 Traveler	Prestwick Flying Club	
G-BBLU	PA-34-200-2 Seneca	F. Tranter	
G-BBMB	Robin DR.400/180	R. B. Tyler	
G-BBME	BAC One-Eleven 401	British Airways (G-AZMI)	
G-BBMF	BAC One-Eleven 401	British Airways (G-ATVU)	
G-BBMG	BAC One-Eleven 408	British Airways (G-AWEJ)	
G-BBMH	E.A.A. Sports Biplane Model P.I.	K. Dawson	
G-BBMI	Dewoitine D.26 (282)	R. J. Willies	
G-BBMJ	PA-E23 Aztec 250	Northern Pig Development Co. Ltd.	

Notes	Reg.	Type	Owner or Operator
	G–BBMK	PA-31-300 Navajo	Steer Aviation Ltd.
	G–BBML	PA-31-300 Navajo	Cooper Merseyside Ltd.
	G–BBMN	D.H.C.1 Chipmunk 22	R. Steiner
	G–BBMO	D.H.C.1 Chipmunk 22	A. J. Hurst
	G–BBMR	*D.H.C.1 Chipmunk T.10 (WB763)	Southall Technical College
	G–BBMT	D.H.C.1 Chipmunk 22	A. T. Letts & ptnrs.
	G–BBMV	D.H.C.1 Chipmunk 22 (WG348)	M. D. Payne & M. J. Mead
	G–BBMW	D.H.C.1 Chipmunk 22	Andrew Edie Aviation
	G–BBMX	D.H.C.1 Chipmunk 22	B. R. C. Wild
	G–BBMZ	D.H.C.1 Chipmunk 22	A. J. Baggarley
	G–BBNA	D.H.C.1 Chipmunk 22	Coventry Gliding Club Ltd.
	G–BBNB	D.H.C.1 Chipmunk 22	A. J. Hurst
	G–BBNC	*D.H.C.1 Chipmunk T.10 (WP790)	Mosquito Aircraft Museum
	G–BBND	D.H.C.1 Chipmunk 22	West Johnson Property Holdings
	G–BBNG	Bell 206B JetRanger 2	Bristow Helicopters Ltd.
	G–BBNH	PA-34-200-2 Seneca	Lawrence Goodwin Machine Tools Ltd.
	G–BBNI	PA-34-200-2 Seneca	Colneray Ltd.
	G–BBNJ	Cessna F.150L	K. D. Wickenden
	G–BBNK	PA-23 Aztec 250E	George House (Holdings) Ltd.
	G–BBNM	PA-23 Aztec 250	Helmvole Ltd.
	G–BBNN	PA-E23 Aztec 250D	British Caledonian Airways Ltd.
	G–BBNO	PA-E23 Aztec 250E	Distance No Object Ltd.
	G–BBNR	Cessna 340	J. Lipton
	G–BBNS	Cessna 310Q	Hammond Pallets Ltd.
	G–BBNT	PA-31-350 Navajo Chieftain	Northern Executive Aviation Ltd.
	G–BBNV	Fuji FA.200-160	C.S.E. (Aircraft Services) Ltd.
	G–BBNX	Cessna FRA.150L	Three Counties Aero Club Ltd.
	G–BBNY	Cessna FRA.150L	Fairoaks Aviation Services Ltd.
	G–BBNZ	Cessna F.172M	Airwork Ltd.
	G–BBOA	Cessna F.172M	George House (Holdings) Ltd.
	G–BBOB	Cessna 421B	R. Gregory
	G–BBOC	Cameron O-77 balloon	A. G. Hopkins & ptnrs. *Bacchus*
	G–BBOD	Thunder 005 balloon	Thunder Balloons Ltd. *Eric the Lad*
	G–BBOE	Robin HR.200/100	Aberdeen Flying Group
	G–BBOH	Pitts S-1S Special	P. Meeson
	G–BBOI	Bede BD-5B	Heather V. B. Wheeler
	G–BBOK	PA-E23 Aztec 250E	NEI Clarke Chapman Ltd.
	G–BBOL	PA-18 150 Super Cub	Lakes Gliding Club Ltd.
	G–BBOM	PA-E23 Aztec 250E	T. S. Grimshaw Ltd.
	G–BBOO	Thunder Ax6-56 balloon	K. Meehan *Tigerjack*
	G–BBOR	Bell 206B JetRanger 2	Cabair Ltd.
	G–BBOX	Thunder Ax7-77 balloon	R. C. Weyda *Rocinante*
	G–BBOY	Thunder Ax6-56A balloon	N. C. Faithfull *Eric of Titchfield*
	G–BBPJ	Cessna F.172M	Simmette Ltd.
	G–BBPK	Evans VP-1	P. D. Kelsey
	G–BBPM	Enstrom F-28A	P. A. Masters
	G–BBPN	Enstrom F-28A	C.S.E. Aviation Ltd.
	G–BBPO	Enstrom F-28A	Flairair
	G–BBPP	PA-28 Cherokee 180	J. D. B. Hamilton
	G–BBPS	Jodel D.117	A. Appleby
	G–BBPU	Boeing 747-136	British Airways *Henry Hudson*
	G–BBPW	Robin HR.100–210	Parlway Ltd.
	G–BBPX	PA-34-200-2 Seneca	Richel Investments Ltd.
	G–BBPY	PA-28 Cherokee 180	Ulster Flying Club (1961) Ltd.
	G–BBPZ	PA-E23 Aztec 250D	Casair Aviation Services Ltd.
	G–BBRA	PA-E23 Aztec 250E	T. S. Grimshaw Ltd.
	G–BBRB	D.H.82A Tiger Moth (DF198)	R. Barham
	G–BBRC	Fuji FA.200-180	W. & L. Installations & Co. Ltd.
	G–BBRE	Fuji FA.200-160	M. R. Howse
	G–BBRG	Bell 47G-5A	N. H. Andrew
	G–BBRH	Bell 47G-5A	Heliwork Ltd.
	G–BBRI	Bell 47G-5A	Camlet Helicopters Ltd.
	G–BBRJ	PA-E23 Aztec 250E	Park Management Ltd.
	G–BBRN	Procter Kittiwake	RNGSA
	G–BBRO	H.S.125 Srs. 600B	McAlpine Aviation Ltd.
	G–BBRP	BN-2A-9 Islander	Army Parachute Association
	G–BBRS	Enstrom F-28A	A. G. Chrismas Ltd.
	G–BBRV	D.H.C.1 Chipmunk 22	HSA (Chester) Sports & Social Club
	G–BBRW	PA-28 Cherokee 140	J. Martin

Reg.	Type	Owner or Operator	Notes
G–BBRX	SIAI-Marchetti S.205-18F	W. Chrystal	
G–BBRY	Cessna 210	R. Q. & A. S. Bond	
G–BBRZ	AA-5 Traveler	Galaxy Publications Ltd.	
G–BBSA	AA-5 Traveler	Watsons Anodising Ltd.	
G–BBSB	Beech C23 Sundowner	Sundowner Group	
G–BBSC	Beech B24R Sierra	V. G. Orme	
G–BBSD	Beech 58 Baron	A.B.I. Caravans Ltd.	
G–BBSE	D.H.C.1 Chipmunk 22	Hartley House Investments Ltd.	
G–BBSF	Cessna 310Q	R. B. W. Enterprises Ltd.	
G–BBSM	PA-32 Cherokee Six 300	All Seasons Aviation Co. Ltd.	
G–BBSN	PA-E23 Aztec 250	Burnthills (Contractors) Ltd.	
G–BBSO	PA-28 Cherokee 140	K. Edwards	
G–BBSR	PA-E23 Aztec 250D	LDL Enterprises	
G–BBSS	D.H.C.1A Chipmunk 22	Northumbria Gliding Club Ltd.	
G–BBST	PA-E23 Aztec 250	Thurston Aviation Ltd.	
G–BBSU	Cessna 421B	Westward Television Ltd.	
G–BBSV	Cessna 421B	Owledge Ltd.	
G–BBSW	Pietenpol Air Camper	J. K. S. Wills	
G–BBSZ	Douglas DC-10-10	Laker Airways *Canterbury Belle*	
G–BBTB	Cessna FRA.150L	George House (Holdings) Ltd.	
G–BBTG	Cessna F.172M	P. Murphy & R. Loveday	
G–BBTH	Cessna F.172M	H. J. C. Townley & ptnrs.	
G–BBTJ	PA-E23 Aztec 250E	Milford Haven Dry Dock Ltd.	
G–BBTK	Cessna FRA.150L	Airwork Services Ltd.	
G–BBTL	PA-E23 Aztec 250C	Air Navigation & Trading Co. Ltd.	
G–BBTS	Beech V35B Bonanza	Golden Sands Estates Ltd.	
G–BBTU	ST-10 Diplomate	P. Campion	
G–BBTW	PA-31P Navajo	M. G. Tyrell & Co. Ltd.	
G–BBTX	Beech C23 Sundowner	J. P. Danton	
G–BBTY	Beech C23 Sundowner	Torlid Ltd.	
G–BBTZ	Cessna F.150L	Ulster Aviation Ltd.	
G–BBUD	Sikorsky S-61N Mk. II	British Airways Helicopters Ltd.	
G–BBUE	AA-5 Traveler	C. R. Arnold	
G–BBUF	AA-5 Traveler	Eglinton Flying Club	
G–BBUG	PA-16 Clipper	F. Beardsell	
G–BBUJ	Cessna 421B	The Automobile Association	
G–BBUK	Bell 47G-2	Bristow Helicopters Ltd.	
G–BBUL	Mitchell-Procter Kittiwake I	R. Bull	
G–BBUO	Cessna 150L	Exeter Flying Club Ltd.	
G–BBUP	B.121 Pup I	C. C. Brown	
G–BBUT	Western O-65 balloon	Wg. Cdr. G. F. Turnbull & Mrs. K. Turnbull *Christabelle II*	
G–BBUU	Piper L-4B Cub	Cooper Bros.	
G–BBUW	SA.102.5 Cavalier	P. C. H. Clarke	
G–BBUY	Bell 206B JetRanger 2	N. Denes Aerodrome Ltd.	
G–BBVA	Sikorsky S-61N Mk. II	Bristow Helicopters Ltd. *Vega*	
G–BBVB	Sikorsky S-61N Mk. II	Bristow Helicopters Ltd. *Aboyne*	
G–BBVC	Slingsby T-59D Kestrel	Slingsby Engineering Ltd.	
G–BBVE	Cessna 340	R. M. Cox Ltd.	
G–BBVF	SA Twin Pioneer III	Flight One Ltd.	
G–BBVG	PA-E23 Aztec 250D	R. F. Wanbon & P. G. Warmerdan	
G–BBVH	V.807 Viscount	GB Airways Ltd.	
G–BBVI	Enstrom F-28A	C.S.E. Aviation Ltd.	
G–BBVJ	Beech B24R Sierra	Oldfield Music Ltd.	
G–BBVM	Beech A.100 King Air	Laura Ashley Ltd.	
G–BBVO	Isaacs Fury II	D. B. Wilson	
G–BBVP	Westland-Bell 47G-3B1	Freemans of Bewdley (Aviation) Ltd.	
G–BBWM	PA-E23 Aztec 250E	Alidair Ltd.	
G–BBWN	D.H.C.1 Chipmunk 22	G. R. Tait & J. Ripley	
G–BBWZ	AA-1B Trainer	Warner Aviation	
G–BBXA	Beech B.55 Baron	La Villette Court Apartments Ltd.	
G–BBXB	Cessna FRA.150L	Cambridge Technical Developments Ltd.	
G–BBXG	PA-34-200-2 Seneca	Tabair	
G–BBXH	Cessna FR.172F	Vale Hire & Contracting Co. Ltd.	
G–BBXI	HPR-7 Herald 203	Air UK	
G–BBXK	PA-34-200-2 Seneca	J. A. Galt	
G–BBXL	Cessna E310Q	Kearns-Barker Associates Ltd.	
G–BBXO	Enstrom F-28A	C.S.E. Aviation Ltd.	
G–BBXR	PA-31-350 Navajo Chieftain	Rolls-Royce Ltd.	
G–BBXS	Piper J-3C-65 Cub	N. Simpson (G–ALMA)	
G–BBXT	Cessna F.172M	I. B. Wilkens	
G–BBXU	Beech B24R Sierra	Picador Leasing Ltd.	

Notes	Reg.	Type	Owner or Operator
	G–BBXV	PA-28-151 Warrior	Beechwood Marine Ltd.
	G–BBXW	PA-28-151 Warrior	Hypozone Ltd.
	G–BBXX	PA-31-350 Navajo Chieftain	Rolls-Royce Ltd.
	G–BBXY	Bellanca 7GC BC Citabria	Cambridge University Gliding Trust Ltd
	G–BBXZ	Evans VP-1	J. A. Naughton
	G–BBYB	PA-18 Super Cub 95	E. C. Lee & ptnrs.
	G–BBYE	Cessna 195	Wilrow Products Ltd.
	G–BBYH	Cessna 182P	Sanderson (Forklifts) Ltd.
	G–BBYK	PA-E23 Aztec 250	T. S. Grimshaw Ltd.
	G–BBYL	Cameron O-77 balloon	Buckingham Balloon Club *Jammy*
	G–BBYM	H.P.137 Jetstream 200	Glos-Air (Sales) Ltd. (G–AYWR)
	G–BBYN	PA-30 Twin Comanche 160	Express Aviation Services Ltd.
	G–BBYO	BN-2A Mk. III Trislander	Aurigny Air Services (G–BBWR)
	G–BBYP	PA-28 Cherokee 140	Channel Aviation Ltd.
	G–BBYR	Cameron O-65 balloon	D. M. Winder *Phoenix*
	G–BBYS	Cessna 182P Skylane	Forth Engineering Ltd.
	G–BBYT	Cessna 414	Cheshire Scaffold Co. Ltd.
	G–BBYU	Cameron O-56 balloon	C. J. T. Davey *Chieftain*
	G–BBYW	PA-28 Cherokee 140	C.S.E. (Aircraft Services) Ltd.
	G–BBZE	PA-28 Cherokee 140	C.S.E. (Aircraft Services) Ltd.
	G–BBZF	PA-28 Cherokee 140	R. F. Grute & M. K. Taylor
	G–BBZH	PA-28R-200 Cherokee Arrow	George House (Holdings) Ltd.
	G–BBZI	PA-31-310 Navajo	Depike Engineers Ltd.
	G–BBZJ	PA-34-200-2 Seneca	C.S.E. (Aircraft Services) Ltd.
	G–BBZL	Westland-Bell 47G-3B1	Dollar Air Services Ltd.
	G–BBZN	Fuji FA.200-180	Jasper Carrott Ltd.
	G–BBZO	Fuji FA.200-160	D. G. Foreman
	G–BBZR	Enstrom F-28A	C.S.E. Aviation Ltd.
	G–BBZS	Enstrom F-28A	Spooner Aviation (Enstrom Helicopters) Ltd
	G–BBZV	PA-28R-200-2 Cherokee Arrow	Unicol Engineering
	G–BCAB	M.S.894A Rallye Minerva 220	P. V. & Mrs. E. M. Gilliar
	G–BCAC	M.S.894A Rallye Minerva 220	D. Hollinshead
	G–BCAD	M.S.894A Rallye Minerva 220	RMA Aviation
	G–BCAH	D.H.C.1 Chipmunk 22 (WG316)	P. D. Southerington
	G–BCAN	Thunder Ax7-77 balloon	D. & L. Cole *Billboard*
	G–BCAP	Cameron O-56 balloon	A. T. Willmer *Honey Child*
	G–BCAR	Thunder Ax7-77 balloon	T. J. Woodbridge *Marie Antoinette*
	G–BCAS	Thunder Ax7-77 balloon	Zebedee Balloon Service *Drifter*
	G–BCAT	PA-31-310 Turbo Navajo	Falaise Investments Ltd.
	G–BCAY	R. Commander 685	Glos-Air (Sales) Ltd.
	G–BCAZ	PA-12 Super Cruiser	Dr. C. J. Pennycuick
	G–BCBD	Bede BD-5	Brockmore-Bede Aircraft (UK) Ltd
	G–BCBG	PA-E23 Aztec 250	D. V. Beadon
	G–BCBI	Cessna 402B	Systime Ltd.
	G–BCBJ	PA-25 Pawnee 235	Westwick Distributors Ltd.
	G–BCBK	Cessna 421B	Lloyds & Scottish Development Ltd
	G–BCBL	Fairchild 24R-46A Argus III (HB751)	Battle of Britain Prints International Ltd
	G–BCBN	Scheibe SF.27M-Ci	D. B. James
	G–BCBP	M.S.880B Rallye 100S Sport	B. W. J. Pring
	G–BCBR	AJEP/Wittman Tailwind	G. McMillan
	G–BCBV	PA-25 Pawnee 235	Farmair Ltd.
	G–BCBW	Cessna 182P	Rogers Aviation Sales Ltd.
	G–BCBX	Cessna F.150L	Ulster Aviation Ltd.
	G–BCBY	Cessna F.150L	R. A. Ranscombe
	G–BCBZ	Cessna 337C	H. Tempest Ltd.
	G–BCCA	Cessna A.188B Agwagon	A. H. Scott-Riddell
	G–BCCB	Robin HR.200/100	Goodwood Terrena Ltd.
	G–BCCC	Cessna F.150L	Crescent Leasing Ltd.
	G–BCCD	Cessna F.172M	Clyde Developments Ltd.
	G–BCCE	PA-E23 Aztec 250	Barry Sheen Racing Ltd.
	G–BCCF	PA-28 Cherokee 180	J. T. Friskney Ltd.
	G–BCCG	Thunder Ax7-65 balloon	D. W. Claridge *Emitape*
	G–BCCH	Thunder Ax6-56A balloon	Wrangler, Bluebell Apparel Ltd. *Wrangler*
	G–BCCJ	AA-5 Traveler	T. Needham & ptnrs.
	G–BCCK	AA-5 Traveler	Donald Healey Motor Co. Ltd.
	G–BCCL	HS.125 Srs. 600B	McAlpine Aviation Ltd.

Reg.	Type	Owner or Operator	Notes
G–BCCN	Robin HR.200/100	June Hall Services Ltd.	
G–BCCP	Robin HR.200/100	Northampton School of Flying Ltd.	
G–BCCR	CP.301A Emeraude	A. B. Fisher	
G–BCCX	D.H.C.1 Chipmunk 22 (Lycoming)	RAFGSA	
G–BCCY	Robin HR.200/100	N. D. Anderson	
G–BCDA	Boeing 727–46	Dan-Air Services Ltd.	
G–BCDB	PA-34-200-2 Seneca	C.S.E. (Aircraft Services) Ltd.	
G–BCDC	PA-18 Super Cub 95	Aly Aviation Ltd.	
G–BCDH	MBB Bo 105D	North-Scottish Helicopters Ltd.	
G–BCDI	Cessna T.310Q-11	Leygatecourt Ltd.	
G–BCDJ	PA-28 Cherokee 140	Andrewsfield Flying Club Ltd.	
G–BCDK	Partenavia P.68B	H. E. Peacock	
G–BCDL	Cameron O-42 balloon	D. P. & Mrs. B. O. Turner *Chums*	
G–BCDN	F.27 Friendship Mk. 200	Air UK	
G–BCDO	F.27 Friendship Mk. 200	Air UK	
G–BCDR	Thunder Ax7-77 balloon	W. G. Johnston & ptnrs. *Obelix*	
G–BCDS	PA-E23 Aztec 250	Hamilton Aviation Ltd.	
G–BCDY	Cessna FRA. 150L	P. R. Pykett	
G–BCDZ	H.S.748 Srs. 2A	British Aerospace	
G–BCEA	Sikorsky S-61N Mk. II	British Airways Helicopters Ltd.	
G–BCEB	Sikorsky S-61N Mk. II	British Airways Helicopters Ltd.	
G–BCEC	Cessna F.172M	David Scott-Moncrieff & Son Ltd.	
G–BCEE	AA-5 Traveler	Echo Echo Ltd.	
G–BCEF	AA-5 Traveler	Echo Fox Ltd.	
G–BCEN	BN-2A Islander	Alderney Air Ferries (Holdings) Ltd.	
G–BCEO	AA-5 Traveler	Amex Automobile Exports	
G–BCEP	AA-5 Traveler	Nottingham Industrial Cleaners Ltd.	
G–BCER	GY-201 Minicab	T. Worrall	
G–BCEU	Cameron O-42 balloon	Entertainment Services Ltd. *Harlequin*	
G–BCEX	PA-E23 Aztec 250	Weekes Bros. (Welling) Ltd.	
G–BCEY	D.H.C.1 Chipmunk 22	D. O. Wallis	
G–BCEZ	Cameron O-84 balloon	Anglia Aeronauts Ascension Association *Stars and Bars*	
G–BCFB	Cameron O-77 balloon	J. J. Harris & P. Pryce-Jones *Teutonic Turkey*	
G–BCFC	Cameron O-65 balloon	B. H. Mead *Candy Twist*	
G–BCFD	West balloon	E. D. & J. N. West *Hellfire*	
G–BCFE	Odyssey 4000 balloon	R. M. Glover *Odyssey*	
G–BCFF	Fuji FA-200-160	C.S.E. (Aircraft Services) Ltd.	
G–BCFN	Cameron O-65 balloon	T. Aston & ptnrs. *Fireball*	
G–BCFO	PA-18-150 Super Cub	Bristol & Gloucestershire Gliding Club (Pty) Ltd.	
G–BCFP	Enstrom F-28A	RHM Investments Ltd.	
G–BCFR	Cessna FRA.150L	Chipglow Ltd.	
G–BCFV	Saab 91D Safir	J. Townhill	
G–BCFW	Saab 91D Safir	D. R. Williams	
G–BCFY	Luton L.A.4A Minor	M. G. E. Hutton	
G–BCFZ	Cameron A-500 balloon	C. J. T. Davy & ptnrs. *Le Geant*	
G–BCGB	Bensen B.8	A. Melody	
G–BCGC	D.H.C.1 Chipmunk 22	Culdrose Gliding Club	
G–BCGG	Jodel DR.250 Srs. 160	C. G. Gray (G–ATZL)	
G–BCGH	Nord NC.854S	G. T. Roberts & A. W. Hughes	
G–BCGI	PA-28 Cherokee 140	C.S.E.(Aircraft Services) Ltd.	
G–BCGJ	PA-28 Cherokee 140	P. L. Naylor	
G–BCGK	PA-28 Cherokee 140	C.S.E.(Aircraft Services) Ltd.	
G–BCGL	Jodel D.112	G. Hughes & R. E. Jones	
G–BCGM	Jodel D.120	I. E. Fisher	
G–BCGN	PA-28 Cherokee 140	C.S.E.(Aircraft Services) Ltd.	
G–BCGP	Gazebo Ax6 balloon	A. R. Wilson *Aries*	
G–BCGS	PA-28R-200 Cherokee Arrow	D. W. Hermiston-Hooper	
G–BCGT	PA-28 Cherokee 140	I. M. Fieldsend	
G–BCGU	HP.137 Jetstream 200	Terravia (Aircraft) Ltd. (G–AXRI)	
G–BCGW	Jodel D.11	G. H. & M. D. Chittenden	
G–BCGX	Bede BD-5A/B	R. Hodgson	
G–BCHJ	Cessna F.172H	Toon Ghose Aviation Ltd.	
G–BCHK	Cessna F.172H	Skyview & General Ltd.	
G–BCHL	D.H.C.1 Chipmunk 22A	B. D. Bate & R. Rutherford	
G–BCHM	SA.341G Gazelle	Westland Helicopters Ltd.	
G–BCHT	Schleicher ASK.16	K. M. Barton & ptnrs.	
G–BCHU	Dawes VP-2	G. Dawes	
G–BCHV	D.H.C.1 Chipmunk 22	N. F. Charles	

Notes	Reg.	Type	Owner or Operator
	G–BCHX	SF.23A Sperling	T. Davies & P. H. Chamberlain
	G–BCID	PA-34-200-2 Seneca	Comanche Air Services Ltd.
	G–BCIE	PA-28-151 Warrior	W. Legg
	G–BCIF	PA-28 Cherokee 140	Fyer-Robins Aviation Ltd.
	G–BCIH	D.H.C.1 Chipmunk 22	J. M. Hosey & R. A. Schofield
	G–BCII	Cessna 500 Citation	IDS Aircraft Ltd.
	G–BCIJ	AA-5 Traveler	C. Wilkinson & ptnrs.
	G–BCIK	AA-5 Traveler	W. Nutt & Son Ltd.
	G–BCIM	AA-1B Trainer	Special Alloys (Northern) Ltd.
	G–BCIN	Thunder Ax7-77 balloon	P. G. & R. A. Vale
	G–BCIO	PA-39 Twin Comanche C/R	C.S.E. (Aircraft Services) Ltd.
	G–BCIP	PA-39 Twin Comanche C/R	C.S.E. (Aircraft Services) Ltd.
	G–BCIR	PA-28-151 Warrior	W. M. F. Taylor
	G–BCIS	Beagle B.206 Srs. 1	R. Nathan
	G–BCIT	CIT/AI Srs. 1	Cranfield Institute of Technology
	G–BCIU	Beagle B.206 Srs. 1	Bradburn & Wedge (Garages) Ltd.
	G–BCIV	Beagle B.206 Srs. 1	Bradburn & Wedge Ltd.
	G–BCIW	D.H.C.1 Chipmunk 22 (WZ868)	R. K. J. Hadlow & ptnrs.
	G–BCJF	Beagle B.206 Srs. 1	A. A. Mattacks
	G–BCJH	Mooney M.20F	V. J. Cousin
	G–BCJL	PA-28 Cherokee 140	C.S.E. (Aircraft Services) Ltd.
	G–BCJM	PA-28 Cherokee 140	MB Aviation Ltd.
	G–BCJN	PA-28 Cherokee 140	A. J. Steed
	G–BCJO	PA-28R-200 Cherokee Arrow	G. I. Cooper
	G–BCJP	PA-28 Cherokee 140	C. C. Smith
	G–BCJR	PA-E23 Aztec 250	Alisons Automated Machinery Ltd.
	G–BCJS	PA-E23 Aztec 250	Woodgate Aviation Ltd.
	G–BCJU	HS.125 Srs. 600B	McAlpine Aviation Ltd.
	G–BCKB	R. Thrush Commander 600	G. W. Peek
	G–BCKC	R. Thrush Commander 600	ADS (Aerial) Ltd.
	G–BCKD	PA-28R-200-2 Cherokee Arrow	A. B. Plant (Aviation) Ltd.
	G–BCKF	SA.102.5 Cavalier	K. Fairness
	G–BCKJ	PA-23 Aztec 250	Renco (Aviation) Ltd.
	G–BCKM	Cessna 500 Citation	IDS Fanjets Ltd.
	G–BCKN	D.H.C.1A Chipmunk 22	RAFGSA
	G–BCKO	PA-E23 Aztec 250	NPD Aviation Ltd.
	G–BCKP	Luton L.A.5A Major	J. R. Callow
	G–BCKS	Fuji FA.200-180	J. T. Hicks
	G–BCKT	Fuji FA.200-180	Littlewick Green Service Station Ltd.
	G–BCKU	Cessna FRA.150L	Airwork Services Ltd.
	G–BCKV	Cessna FRA.150L	Airwork Services Ltd.
	G–BCLA	Sikorsky S-61N	Bristow Helicopters Ltd.
	G–BCLC	Sikorsky S-61N	Bristow Helicopters Ltd. *Craigievar*
	G–BCLD	Sikorsky S-61N	Bristow Helicopters Ltd. *Slains*
	G–BCLI	AA-5 Traveler	C. G. Whittaker (Commercials) Ltd.
	G–BCLJ	AA-5 Traveler	C. P. W. Villa & ptnrs.
	G–BCLL	PA-28 Cherokee 180	Agricultural & Industrial Services (Yorkshire) Ltd.
	G–BCLM	GY-201 Minicab	J. H. Hill
	G–BCLS	Cessna 170B	C. W. Proffitt-White
	G–BCLU	Jodel D.117	N. A. Wallace
	G–BCLV	Bede BD-5A	R. A. Gardiner
	G–BCLW	AA-1B Trainer	Whisky Flying Group Ltd.
	G–BCMB	Partenavia P.68B	DK Aviation Ltd.
	G–BCMC	Bell 212	Bristow Helicopters Ltd.
	G–BCMD	PA-19 Super Cub 95	R. W. M. & B. N. C. Mogg
	G–BCMF	Levi Go-Plane RL-6 Srs. 1	R. Levi
	G–BCMJ	SA.102.5 Cavalier	M. Johnson
	G–BCMR	Robin HR.100/285	D. Betts
	G–BCMT	Isaacs Fury II	M. H. Turner
	G–BCNC	GY.201 Minicab	J. R. Wraight
	G–BCNO	BN-2A Mk. III-I Trislander	Aurigny Air Services
	G–BCNP	Cameron O-77 balloon	J. E. H. Quilliam & ptnrs.
	G–BCNR	Thunder Ax7-77A balloon	S. J. Milliken & ptnrs. *Howdy*
	G–BCNS	Cameron O-77 balloon	Cathay Pacific Airways Ltd. *Cathay I*
	G–BCNT	Partenavia P.68B	Josef D. J. Jons & Co. Ltd.
	G–BCNX	Piper J-3C-65 Cub	N. J. R. Empson & ptnrs.
	G–BCNZ	Fuji FA.200-160	Condor Flying Club
	G–BCOA	Cameron O-65 balloon	E. E. J. & Mrs. E. Sutton *Bunny*
	G–BCOB	Piper J-3C-65 Cub	R. W. & Mrs. J. W. Marjoram
	G–BCOE	HS.748 Srs. 2A	British Airways
	G–BCOF	HS.748 Srs. 2A	British Airways

Reg.	Type	Owner or Operator	Notes
G–BCOG	Jodel D.112	B. A. Bower & ptnrs.	
G–BCOH	Avro 683 Lancaster 10	Strathallan Aircraft Collection	
G–BCOI	D.H.C.1 Chipmunk 22	D. S. McGregor & A. T. Letham	
G–BCOJ	Cameron O-56 balloon	Anglian Industrial Gases Ltd. *Vulcan*	
G–BCOL	Cessna F.172M	J. Birkett	
G–BCOM	Piper J-3C-65 Cub	M. J. Miles	
G–BCOO	D.H.C.1 Chipmunk 22	T. G. Fielding & M. S. Morton	
G–BCOP	PA-28R-200 Cherokee Arrow	J. H. Parker & ptnrs.	
G–BCOR	SOCATA Rallye 100ST	H. J. Pincombe	
G–BCOT	Enstrom F-28C-UK	Spooner Aviation Ltd.	
G–BCOU	D.H.C.1 Chipmunk 22	P. J. Loweth	
G–BCOV	Hawker Sea Fury TT.20 (VX302)	D. W. Arnold	
G–BCOY	D.H.C.1 Chipmunk 22	Coventry Gliding Club Ltd.	
G–BCPA	SA.315B Alouette II	Dollar Air Services Ltd.	
G–BCPB	Howes radio-controlled model free balloon	R. B. & Mrs. C. Howes *Posbee I*	
G–BCPD	GY.201 Minicab	A. H. K. Denniss	
G–BCPE	Cessna F.150M	Channel Islands Aero Holdings Ltd.	
G–BCPF	PA-23 Aztec 250	Keenair Services Ltd.	
G–BCPG	PA-28R-200 Cherokee Arrow	Yates Burgess Finance Ltd.	
G–BCPH	Piper J-3C-65 Cub	M. B. C. Ballard	
G–BCPI	PA-18 Super Cub 135	Kent Gliding Club Ltd.	
G–BCPJ	Piper J-3C-65 Cub	M. C. Barraclough & T. M. Storey	
G–BCPK	Cessna F.172M	Skegness Air Taxi Services Ltd.	
G–BCPN	AA-5 Traveler	B.W. Agricultural Equipments Ltd.	
G–BCPO	Partenavia P.68B	Astra Aviation Ltd.	
G–BCPU	D.H.C.1 Chipmunk T.10	P. Waller	
G–BCPX	SZEP HFC.125	A Szep	
G–BCPZ	R.500S Shrike Commander	Cleanacres Ltd.	
G–BCRA	Cessna F.150M	Three Counties Aero Club	
G–BCRB	Cessna F.172M	F. D. & M. D. Forbes	
G–BCRC	D.31 Turbulent	R. A. G. Chapman	
G–BCRE	Cameron O-77 balloon	A. R. Langston & R. J. Fuller *Snapdragon*	
G–BCRF	PA-23 Aztec 250	Linco Poultry Machinery Co. Ltd.	
G–BCRH	Alaparma Baldo B.75	A. L. Scadding	
G–BCRI	Cameron O-65 balloon	V. J. Thorne *Joseph*	
G–BCRJ	Taylor JT.1 Monoplane	Canary Flying Group	
G–BCRK	SA.102.5 Cavalier	R. E. Ward & D. A. Thorpe	
G–BCRL	PA-28-151 Warrior	Nordec (Builders) Ltd.	
G–BCRN	Cessna FRA.150L	Airwork Services Ltd.	
G–BCRP	PA-E23 Aztec 250	LEC Refrigeration Ltd.	
G–BCRR	AA-5B Tiger	Travelworth Ltd.	
G–BCRT	Cessna F.150M	Executive Air	
G–BCRX	D.H.C.1 Chipmunk 22	J. P. V. Hunt & P. G. H. Tory	
G–BCSA	D.H.C.1 Chipmunk 22	RAFGSA	
G–BCSB	D.H.C.1 Chipmunk 22	RAFGSA	
G–BCSL	D.H.C.1 Chipmunk 22	Jalawain Ltd.	
G–BCSM	Bellanca 8GC BC Scout	Hendon Air Services Ltd.	
G–BCSR	Bellanca 7ECA Citabria	Hendon Air Services Ltd.	
G–BCSS	M.S.892A Rallye Commodore 150	R. E. Ward & ptnrs.	
G–BCST	M.S.893A Rallye Commodore 180	P. J. Wilcox	
G–BCSV	Cessna 421B	Northair Aviation Ltd.	
G–BCSX	Thunder Ax7-77 balloon	A. T. Wood *Whoopski*	
G–BCSY	Taylor JT.2 Titch	P. L. Mines	
G–BCSZ	PA-28R-200 Cherokee Arrow	Marlow Chemical Co. Ltd.	
G–BCTA	PA-28-151 Warrior	G. Capes	
G–BCTF	PA-28-151 Warrior	Seasons & Systems Ltd.	
G–BCTH	PA-28 Cherokee 140	John V. White Ltd.	
G–BCTI	Schleicher ASK.16	R. J. Steward	
G–BCTJ	Cessna 310Q	Airwork Services Ltd.	
G–BCTK	Cessna FR.172J	Shobdon Aviation Co. Ltd.	
G–BCTR	Taylor JT.2 Titch	T. Reagan	
G–BCTT	Evans VP-2	B. J. Boughton	
G–BCTU	Cessna FRA.150M	Mona Aviation Ltd.	
G–BCTV	Cessna F.150M	Smith School of Flying Ltd.	
G–BCTW	Cessna F.150M	Ulster Aviation Ltd.	
G–BCUB	Piper J-3C-65 Cub	M. J. Mead	
G–BCUF	Cessna F.172M	G. H. C. B. Kirke	
G–BCUH	Cessna F.150M	Gordon King (Aviation) Ltd.	

Notes	Reg.	Type	Owner or Operator
	G–BCUI	Cessna F.172M	Hillhouse Estates Ltd.
	G–BCUJ	Cessna F.150M	S.B. International Aviation Ltd.
	G–BCUK	Cessna F.172M	Q. Quitmann Ltd.
	G–BCUL	SOCATA Rallye 100ST	F. A. Tanks
	G–BCUM	Stinson HW-75	P. J. Sellar
	G–BCUW	Cessna F.177RG Cardinal	A. J. Wheeler
	G–BCUY	Cessna FRA.150M	Vectair Aviation Ltd.
	G–BCUZ	Beech A200 Super King Air	Allied Breweries (UK) Ltd.
	G–BCVA	Cameron O-65 balloon	J. C. Bass & ptnrs. *Crepe Suzette*
	G–BCVB	PA-17 Vagabond	A. T. Nowak & B. Holland
	G–BCVC	SOCATA Rallye 100ST	Brettshire Ltd.
	G–BCVE	Evans VP-2	G. A. Bentley
	G–BCVF	Practavia Pilot Sprite	G. B. Castle
	G–BCVG	Cessna FRA.150L	Westair Flying Services Ltd.
	G–BCVH	Cessna FRA.150L	W. Lancs Aero Club Ltd.
	G–BCVI	Cessna FR.172J	B. D. & Mrs. W. Phillips
	G–BCVJ	Cessna F.172M	J. Males
	G–BCVV	PA-28-151 Warrior	Channel Islands Aero Holdings Ltd.
	G–BCVW	GY-80 Horizon 180	P. M. A. Parrett
	G–BCVY	PA-34-200T Seneca	Galaxy Aviation Ltd.
	G–BCWA	BAC One-Eleven 518	Dan-Air Services Ltd. (G–AXMK)
	G–BCWB	Cessna 182P	Gaymark Trustees Ltd.
	G–BCWD	Sikorsky S-58T	Management Aviation Ltd.
	G–BCWE	HPR-7 Herald 206	British Air Ferries *Caroline Frost*
	G–BCWF	Scottish Aviation Twin Pioneer	Flight One Ltd. (G–APRS)
	G–BCWH	Practavia Pilot Sprite	K. B. Parkinson & R. Tasker
	G–BCWI	Bensen B.8M	C. J. Blundell
	G–BCWK	Alpavia Fournier RF-3	D. I. Nickolls & ptnrs.
	G–BCWL	Westland Lysander III (V9281)	D. W. Arnold
	G–BCWM	AB-206B JetRanger 2	N. R. Foster
	G–BCWU	BN-2A-27 Islander	Marchwiel Plant & Engineering Co. Ltd.
	G–BCWW	HP.137 Jetstream 200	The Distillers Co. Ltd. (G–AXUN)
	G–BCXB	SOCATA Rallye 100ST	Inchberry Ltd.
	G–BCXD	Pitts S-2A Special	Atalona Ltd.
	G–BCXE	Robin DR.400/2+2	Headcorn Flying School Ltd.
	G–BCXF	H.S.125 Srs. 600B	Montagius Ltd.
	G–BCXH	PA-28 Cherokee 140F	Airborne Electronics Ltd.
	G–BCXI	G-164A Ag-Cat	W. P. Miller
	G–BCXJ	Piper J-3C-65 Cub	W. F. Stockdale
	G–BCXN	D.H.C.1 Chipmunk 22	J. D. Scott
	G–BCXO	MBB Bo 105D	Management Aviation Ltd.
	G–BCXR	BAC One-Eleven 517	Monarch Airlines Ltd. (G–BCCV)
	G–BCXT	Cessna F.150M	Peterborough Aero Club Ltd.
	G–BCXZ	Cameron O-56 balloon	Olives from Spain Ltd. *Olives from Spain*
	G–BCYC	BN-2A Mk. III-2 Trislander	Loganair Ltd.
	G–BCYE	D.H.C.1 Chipmunk 22	L. T. Mersh
	G–BCYF	Dassault Mystère 20	Nidiva Services (UK) Ltd.
	G–BCYI	Schleicher ASK-16	D. J. Pearce & ptnrs.
	G–BCYJ	D.H.C.1 Chipmunk 22 (WG307)	G. A. Warner
	G–BCYK	Avro CF.100 Mk. 4 Canuck (18393)	—
	G–BCYL	D.H.C.1 Chipmunk 22	P. C. Henry
	G–BCYM	D.H.C.1 Chipmunk 22	C. R. R. Eagleton
	G–BCYP	AB-206B JetRanger 2	Alan Mann Helicopters Ltd.
	G–BCYR	Cessna F.172M	Laxtonbridge Ltd.
	G–BCYS	Piper J-3C-65 Cub	R. P. M. Dawson
	G–BCYY	Westland-Bell 47G-3B1	Minster Helicopters Ltd.
	G–BCYZ	Westland-Bell 47G-3B1	Helicrops Ltd.
	G–BCZF	PA-28 Cherokee 180	Proto Kote Ltd.
	G–BCZG	HPR-7 Herald 202	Air UK
	G–BCZH	D.H.C.1 Chipmunk 22	The London Gliding Club (Pty.) Ltd.
	G–BCZI	Thunder Ax7-77 balloon	Motor Tyres & Accessories *Motorway for Tyres*
	G–BCZL	Westland-Bell 47G-3B1	British Executive Air Services Ltd.
	G–BCZM	Cessna F.172M	D. Morgan & C. Gurley
	G–BCZN	Cessna F.150M	T. M. G. Hanley
	G–BCZO	Cameron O-77 balloon	W. O. T. Holmes *Leo*
	G–BCZP	Cessna T.210L	K. & F.L. Equipment & Aircraft Hiring Ltd.
	G–BDAB	SA.102.5 Cavalier	A. H. Brown

Reg.	Type	Owner or Operator	Notes
G–BDAC	Cameron O-77 balloon	D. Fowler & J. Goody *Chocolate Ripple*	
G–BDAD	Taylor JT.1 Monoplane	J. F. Bakewell	
G–BDAE	BAC One-Eleven 518	Dan-Air Services Ltd. (G–AXMI)	
G–BDAG	Taylor JT.1 Monoplane	R. S. Basinger	
G–BDAH	Evans VP-1	R. W. Lowe	
G–BDAI	Cessna FRA.150M	Arjellees Ltd.	
G–BDAJ	R. Commander 112A	Ortonpost Ltd.	
G–BDAK	R. Commander 112A	Chalke Valley Constructing Ltd.	
G–BDAL	R. 500S Shrike Commander	Micro Consultants Ltd.	
G–BDAM	AT-16 Harvard IIB (FE992)	M. V. Gauntlett	
G–BDAO	SIPA 91	A. L. Rose	
G–BDAP	AJEP Tailwind	J. Whiting	
G–BDAR	Evans VP-1	S. C. Foggin & M. T. Dugmore	
G–BDAS	BAC One-Eleven 518	Dan-Air Services Ltd. (G–AXMH)	
G–BDAT	BAC One-Eleven 518	Dan-Air Services Ltd. (G–AYOR)	
G–BDAU	Cessna FRA.150M	Airwork Services Ltd.	
G–BDAV	PA-23 Aztec 250	Sound Powered Telephone Co. Ltd.	
G–BDAW	Enstrom F-28A	Usoland Ltd.	
G–BDAX	PA-E23 Aztec 250	A. M. Collins	
G–BDAY	Thunder Ax5-42A balloon	Thunder Balloons Ltd. *Tom's Balloon*	
G–BDBA	Thunder Ax7-77 balloon	P. Porati *Briaza*	
G–BDBB	Cessna F.150M	Northamptonshire School of Flying Ltd.	
G–BDBD	Wittman W.8 Tailwind	H. Best-Devereux	
G–BDBE	Thunder Ax7-77A balloon	St. Ivel Yoghurt *Prize Guy*	
G–BDBF	FRED Srs. 2	W. T. Morrell	
G–BDBH	Bellanca 7GCBC Citabria	Mastergear Co. Ltd.	
G–BDBI	Cameron O-77 balloon	Royston Cooper Design Consultants *Sunny Money*	
G–BDBJ	Cessna 182P	Cosworth Engineering Ltd.	
G–BDBK	Cameron O-56 balloon	J. G. Green *Square Baby*	
G–BDBL	D.H.C.1 Chipmunk 22	Wycombe Gliding School Syndicate	
G–BDBM	Cameron O-56 balloon	A. H. K. Olphin *Scorpio*	
G–BDBP	D.H.C.1 Chipmunk 22	Sherwood Flying Club Ltd.	
G–BDBR	AB-206B JetRanger 2	P.L.M. Helicopters Ltd.	
G–BDBS	Short SD3-30	Short Bros. Ltd.	
G–BDBU	Cessna F.150M	Channel Islands Aero Holdings Ltd.	
G–BDBV	Jodel D.11A	J. P. de Hevingham	
G–BDBW	Heintz Zenith 100 A18	D. B. Winstanley	
G–BDBX	Evans VP-1	Montgomeryshire Ultra-Light Flying Club	
G–BDBZ	WS.55 Whirlwind Srs. 2	Autair Ltd.	
G–BDCA	SOCATA Rallye 150ST	B. W. J. Pring & ptnrs.	
G–BDCB	D.H.C.1 Chipmunk 22	D. R. Hodgson	
G–BDCC	D.H.C.1 Chipmunk 22	Coventry Gliding Club Ltd.	
G–BDCD	Piper J-3C-65 Cub (480308)	Suzanne C. Brooks	
G–BDCE	Cessna F.172H	H. A. Baillie	
G–BDCI	CP.301A Emeraude	H. A. R. Horesign	
G–BDCK	AA-5 Traveler	Sequoia Air Ltd.	
G–BDCM	Cessna F.177RG	Park Plant Ltd.	
G–BDCO	B.121 Pup 1	Dr. R. D. H. & Mrs. K. N. Maxwell	
G–BDCP	AB-206B JetRanger 2	Ben Turner & Sons (Helicopters) Ltd.	
G–BDCS	Cessna 421B	Marchwiel Aviation Ltd.	
G–BDCT	PA-25 Pawnee 235C	Sprayfields (Scothern) Ltd.	
G–BDCU	Cameron O-77 balloon	Swanflight Balloon Group	
G–BDDA	Sikorsky S-61N Mk. II	British Airways Helicopters Ltd.	
G–BDDD	D.H.C.1 Chipmunk 22	RAE Aero Club Ltd.	
G–BDDE	Douglas DC-8-54F	—	
G–BDDF	Jodel D.120	Sywell Skyriders Flying Group	
G–BDDG	Jodel D.112	A. C. Watt & T. C. Greig	
G–BDDH	F.27 Friendship Mk. 200	Air UK	
G–BDDJ	Luton LA.4A Minor	D. D. Johnson	
G–BDDS	PA-25 Pawnee 235	Farm Air (Shropshire) Ltd.	
G–BDDT	PA-25 Pawnee 235	Farm Aviation Services Ltd.	
G–BDDV	BN-2A-26 Islander	Loganair Ltd.	
G–BDDX	*Whittaker MW.2B Excalibur	Cornwall Aero Park, Helston	
G–BDDZ	CP.301A Emeraude	M. Jones	
G–BDEA	Boeing 707-338C	British Caledonian Airways *Loch Thom*	
G–BDEB	SOCATA Rallye 100ST	W. G. Dunn & ptnrs.	
G–BDEC	SOCATA Rallye 100ST	Cambridge Chemical Co. Ltd.	

Notes	Reg.	Type	Owner or Operator
	G-BDEF	PA-34-200T-2 Seneca	European Paper Sales Ltd.
	G-BDEH	Jodel D.120A	J. Brown
	G-BDEI	Jodel D.9 Bebe	P. J. Griggs
	G-BDEJ	R. Commander 112	Brigstone Ltd.
	G-BDEN	SIAI-Marchetti SF.260	Micro Consultants Ltd.
	G-BDER	Auster AOP.9 (WZ672)	F. & H. (Aircraft) Ltd.
	G-BDES	Sikorsky S-61N Mk. II	British Airways Helicopters Ltd.
	G-BDET	D.H.C.1 Chipmunk 22	A. J. C. Plowman & ptnrs.
	G-BDEU	D. H. C.1 Chipmunk 22 (WP808)	A. Taylor
	G-BDEV	Taylor JT.1 Monoplane	P. J. Houston
	G-BDEW	Cessna FRA.150M	George House (Holdings) Ltd.
	G-BDEX	Cessna FRA.150M	George House (Holdings) Ltd.
	G-BDEY	Piper J-3C-65 Cub	W. J. & J. Morecraft
	G-BDEZ	Piper J-3C-65 Cub	P. Elliott
	G-BDFB	Currie Wot	D. F. Faulkner-Bryant
	G-BDFC	R. Commander 112A	Zaar International Cinema & TV Programmes Ltd.
	G-BDFE	HPR-7 Herald 206	British Air Ferries Ltd. *Rory Keegan*
	G-BDFF	Supermarine S.5 Replica (N220)	Leisure Sport Ltd.
	G-BDFG	Cameron O-65 balloon	N. A. Robertson *Golly II*
	G-BDFH	Auster AOP.9 (XR240)	R. O. Holden
	G-BDFI	Cessna F.150M	Coventry Civil Aviation Ltd.
	G-BDFJ	Cessna F.150M	American Airspeed Inc. Ltd.
	G-BDFL	PA-28R-200-2 Cherokee Arrow	Standard Aviation Ltd.
	G-BDFM	Caudron C.270 Luciole	G. V. Gower
	G-BDFO	Hiller UH-12E	Heliscot Ltd. & F. F. Chamberlain
	G-BDFP	Hughes 369HS	John E. Clarke & Co. (Bournemouth) Ltd.
	G-BDFR	Fuji FA.200-160	C.S.E. (Aircraft Services) Ltd.
	G-BDFS	Fuji FA.200-160	D. J. Carding
	G-BDFT	V.668 Varsity T.1 (WJ897)	D. W. Mickleburgh
	G-BDFU	*Dragonfly MPA Mk. 1	Museum of Flight, E. Fortune
	G-BDFW	R. Commander 112A	Ace Aviation
	G-BDFX	Auster 5	I. A. Haddon & M. J. Kirk
	G-BDFY	AA-5 Traveler	E. O. Liebert
	G-BDFZ	Cessna F.150M	Skyviews & General Ltd.
	G-BDGA	Bushby-Long Midget Mustang	J. R. Owen
	G-BDGB	GY-20 Minicab	D. G. Burden
	G-BDGH	Thunder Ax6-56 balloon	The London Balloon Club Ltd. *London Pride III*
	G-BDGJ	Cameron O-56 balloon	Cathay Pacific Airways Ltd. *Cathy II*
	G-BDGK	Beechcraft D.17S	Customline Ltd.
	G-BDGL	Cessna U.206 Super Skywagon	C. Wren
	G-BDGM	PA-28-151 Warrior	Channel Islands Aero Services Ltd.
	G-BDGN	AA-5B Tiger	Hamblin & Glover Oil Field (Services) Ltd.
	G-BDGO	Thunder Ax7-77 balloon	International Distillers & Vintners Ltd. *J. & B. Rare*
	G-BDGP	Cameron V-65 balloon	P. G. Dunnington
	G-BDGY	PA-28 Cherokee 140	R. E. Woolridge
	G-BDHA	Douglas DC-8-54F	*For Sale*
	G-BDHB	Isaacs Fury II	D. H. Berry
	G-BDHC	D.H.C.6 Twin Otter 310	Chubb Security Services Ltd.
	G-BDHJ	Pazmany PL.1	H. James
	G-BDHK	Piper J-3C-65 Cub (329417)	H. Knight
	G-BDHL	PA-E23 Aztec 250E	Granpack Ltd.
	G-BDHM	SA.102.5 Cavalier	D. H. Mitchell
	G-BDIC	D.H.C.1 Chipmunk 22	T. Bibby & D. Halliwell
	G-BDID	D.H.C.1 Chipmunk 22	Coventry Gliding Club Ltd.
	G-BDIE	R. Commander 112A	T. Saveker Ltd.
	G-BDIG	Cessna 182P	Gunton Electronics Ltd.
	G-BDIH	Jodel D.117	J. Chisholm
	G-BDII	Sikorsky S-61N Mk. II	Bristow Helicopters Ltd. *Drum*
	G-BDIJ	Sikorsky S-61N Mk. II	Bristow Helicopters Ltd. *Crathes*
	G-BDIK	V.708 Viscount	*Derelict at E. Midlands*
	G-BDIL	Bell 212	Bristow Helicopters Ltd.
	G-BDIM	D.H.C.1 Chipmunk 22	L. W. Gruber & ptnrs.
	G-BDIT	*D.H.106 Comet 4C	D. W. Arnold, Blackbushe
	G-BDIV	D.H.106 Comet 4C	Dan-Air Preservation Group

Reg.	Type	Owner or Operator	Notes
G-BDIW	*D.H.106 Comet 4C	Air Classik, Dusseldorf	
G-BDIX	*D.H.106 Comet 4C	Museum of Flight, E. Fortune	
G-BDIY	Luton L.A.4A Minor	M. A. Musselwhite	
G-BDJB	Taylor JT.1 Monoplane	J. F. Barber	
G-BDJC	AJEP Tailwind	A. Whiting	
G-BDJD	Jodel D.112	J. V. Derrick	
G-BDJE	H.S.125 Srs. 600B	McAlpine Aviation Ltd.	
G-BDJF	Bensen B.8MV	R. P. White	
G-BDJG	Luton L.A.4A Minor	D. J. Gaskin	
G-BDJN	Robin HR.200/100	Northampton School of Flying Co. Ltd.	
G-BDJP	Piper J-3C-65 Cub	Mrs. J. M. Pothecary	
G-BDJR	Nord NC.858	V. Panteli & J. Spanton	
G-BDJY	BN-2A-27 Islander	J. Cunningham	
G-BDKB	SOCATA Rallye 150ST	KK Aviation	
G-BDKC	Cessna A185F	Bridge of Tilt Co. Ltd.	
G-BDKD	Enstrom F-28A	Renco (Aviation) Ltd.	
G-BDKG	Beech 65-A80 Queen Air	I. R. Martin	
G-BDKH	CP.301A Emeraude	R. F. Bridge	
G-BDKI	Sikorsky S-61N Mk. II	British Airways Helicopters Ltd.	
G-BDKJ	SA.102.5 Cavalier	H. B. Yardley	
G-BDKK	Bede BD-5B	A. W. Odell	
G-BDKL	Hughes 369HM	A & B Cars (Distributors) Ltd.	
G-BDKM	SIPA 903	S. W. Markham	
G-BDKR	BN-2A Mk. III-2 Trislander	Loganair Ltd.	
G-BDKS	Pitts S-2A Special	Airmiles Ltd.	
G-BDKT	Cameron O-84 balloon	A. R. M. Fraser & R. Fermor-Hesketh	
G-BDKU	Taylor JT.1 Monoplane	A. C. Dove	
G-BDKV	PA-28R-200-2 Cherokee Arrow	Patricia Swanson	
G-BDKW	R. Commander 112A	Link Systems Ltd.	
G-BDLM	Boeing 707-338C	British Caledonian Airways *Loch Fyne*	
G-BDLO	AA-5A Cheetah	Chisnally Ltd.	
G-BDLR	AA-5B Tiger	Coldmix Ltd.	
G-BDLS	AA-1B Trainer	Warner Aviation	
G-BDLT	R. Commander 112A	Wintergrain Ltd.	
G-BDLV	Chilton DW.1A	R. E. Nerou	
G-BDLY	SA.102.5 Cavalier	B. S. Reeve	
G-BDLZ	B.175 Britannia 253F	Air Faisel *Al Mubarak*	
G-BDMB	Robin HR.100/210	R. J. Hitchman & Son	
G-BDMC	MBB Bo 105D	Management Aviation Ltd.	
G-BDMD	PA-31-350 Navajo Chieftain	Berrard Aviation	
G-BDME	Robin DR.400/140B	Miss S. A. Pound	
G-BDMM	Jodel D.11	D. M. Metcalf	
G-BDMO	Thunder Ax7-77A balloon	H. G. Twilley Ltd. *Flash Harry*	
G-BDMS	Piper J-3C-65 Cub	A. T. H. Martin & K. G. Harris	
G-BDMW	Jodel DR.100	J. T. Nixon	
G-BDNC	Taylor JT.1 Monoplane	N. J. Cole	
G-BDNE	Harker Hawk	D. Harker	
G-BDNF	Bensen B.8M	W. F. O'Brien	
G-BDNG	Taylor JT.1 Monoplane	D. J. Phillips	
G-BDNO	Taylor JT.1 Monoplane	R. A. Bragger	
G-BDNP	BN-2A Islander	Jersey European Airways	
G-BDNR	Cessna FRA.150M	W. D. & P. M. Jasper	
G-BDNS	Saro Skeeter 12 (XM529)	M. Eastman & M. Ridley	
G-BDNT	Jodel D.92	D. Beaumont	
G-BDNU	Cessna F.172M	Skyline Aviation Ltd.	
G-BDNW	AA-IB Trainer	J. G. Hill	
G-BDNX	AA-IB Trainer	J. E. M. Evans	
G-BDNY	AA-IB Trainer	M. R. Langford	
G-BDNZ	Cameron O-77 balloon	N. R. Page *Winston Churchill*	
G-BDOA	H.S.125 Srs. 600B	McAlpine Aviation Ltd.	
G-BDOC	Sikorsky S-61N Mk. II	Bristow Helicopters Ltd. *Tolquhoun*	
G-BDOD	Cessna F.150M	Latharp Ltd.	
G-BDOE	Cessna FR.172J	P.A.V.H. (International) Ltd.	
G-BDOF	Cameron O-56 balloon	New Holker Estates Co. *Fred Cavendish*	
G-BDOG	SA Bullfinch Srs. 2100	Dukeries Aviation Ltd.	
G-BDOH	Hiller UH-12E	Bridge Helicopters Ltd.	
G-BDOI	Hiller UH-12E	Heli-Spray Ltd.	
G-BDOL	Piper J-3C-65 Cub	Allman & Son (Redhill) Ltd.	
G-BDOM	BN-2A Mk. III-2 Trislander	Loganair Ltd.	

Notes	Reg.	Type	Owner or Operator
	G–BDON	Thunder Ax7-77A balloon	York Hot-Air Balloon Club *Fred*
	G–BDOO	Thunder Ax7-77 balloon	Thunder Balloons Ltd.
	G–BDOR	Thunder Ax6-56A balloon	Outspan Sales Ltd. *Peeler*
	G–BDOS	BN-2A Mk. III-2 Trislander	Loganair Ltd.
	G–BDOU	Cessna FRA.150M	T. W. R. Case
	G–BDOW	Cessna FRA.150M	Hartmann Ltd.
	G–BDOY	Hughes 369HS	Cosworth Engineering Ltd.
	G–BDOZ	Sportavia-Fournier RF-5	A. J. Noble & ptnrs.
	G–BDPA	PA-28-151 Warrior	Noblair Ltd.
	G–BDPB	Falconar F-11-3	A. E. Pritchard
	G–BDPC	Bede BD-5A	P. R. Cremer
	G–BDPF	Cessna F.172M	Westair Flying Services Ltd.
	G–BDPH	Cessna F.172M	Astro Data Ltd.
	G–BDPI	PA-25 Pawnee 235B	Farmair Ltd.
	G–BDPK	Cameron O-56 balloon	Manpower Ltd. *Manpower*
	G–BDPL	Falconar F-11	P. J. Shone & D. L. Scott
	G–BDPR	BN-2A Islander	BN (Bembridge) Ltd.
	G–BDPV	Boeing 747-136	British Airways *City of Aberdeen*
	G–BDRB	AA-5B Tiger	Northern Strip Mining Ltd.
	G–BDRC	V.724 Viscount	Guernsey Airlines Ltd. *Sarnia II*
	G–BDRD	Cessna FRA.150M	Airwork Services Ltd.
	G–BDRE	AA-1B Trainer	KAAL Electrics Ltd.
	G–BDRF	Taylor JT.1 Monoplane	R. A. Codling
	G–BDRG	Taylor JT.2 Titch	D. R. Gray
	G–BDRI	PA-34-200T-2 Seneca	Aylesbury Mushrooms Ltd.
	G–BDRJ	D.H.C.1 Chipmunk 22 (WP857)	J. K. Avis
	G–BDRK	Cameron O-65 balloon	D. L. Smith *Smirk*
	G–BDRL	Stitts SA-3 Playboy	D. L. MacLean
	G–BDRY	Hiller UH-12E	G. & S. G. Neal (Helicopters) Ltd.
	G–BDSB	PA-28-181 Archer II	C. H. C. Coombe
	G–BDSC	Cessna F.150M	Lyndair (Aviation) Ltd.
	G–BDSD	Evans VP-1	J. E. Worthington
	G–BDSE	Cameron O-77 balloon	British Airways *Concorde*
	G–BDSF	Cameron O-56 balloon	J. R. Joiner *Itzuma*
	G–BDSH	PA-28 Cherokee 140	Bamberhurst Ltd.
	G–BDSK	Cameron O-65 balloon	Southern Balloon Group *Carousel II*
	G–BDSL	Cessna F.150M	Cleveland Flying School Ltd.
	G–BDSM	Slingsby/Kirby Cadet Mk. 3	D. W. Savage
	G–BDSN	Wassmer WA.52 Europa	Tiger Club Ltd. (G–BADN)
	G–BDSO	Cameron O-31 balloon	Budget Rent-a-Car *Baby Budget*
	G–BDSP	Cessna U.206F Stationair	F. M. Usher-Smith
	G–BDSR	PA-25 Pawnee 235	A. G. Edwards
	G–BDSZ	BN-2A Islander	Fletcher & Stewart Ltd.
	G–BDTB	Evans VP-1	T. E. Boyes
	G–BDTL	Evans VP-1	A. K. Lang
	G–BDTT	Bede BD-5	The TT Group
	G–BDTU	Omega III gas balloon	Mrs. K. E. Turnbull *Omega III*
	G–BDTV	Mooney M.20F	J. A. Holgate Ltd.
	G–BDTW	Cassutt Racer	B. E. Smith & C. S. Thompson
	G–BDTX	Cessna F.150M	A. A. & R. N. Croxford
	G–BDUI	Cameron V-56 balloon	G. H. Dorrell *Pig Bucket*
	G–BDUJ	PA-31-310 Navajo	Vickers Shipbuilding Group Ltd.
	G–BDUK	R. Commander 685	Air Service Training Ltd.
	G–BDUL	Evans VP-1	C. Goodman
	G–BDUM	Cessna F.150M	Lattey Aviation Ltd.
	G–BDUN	PA-34-200T-2 Seneca	Coriolis Ltd.
	G–BDUO	Cessna F.150M	Fishbourne Garage Ltd.
	G–BDUP	B.175 Britannia 253	Afrek Ltd.
	G–BDUR	B.175 Britannia 253	Afrek Ltd.
	G–BDUX	Slingsby T.31B motor glider	R. H. Hearn
	G–BDUY	Robin DR.400/140B	Waveney Flying Group
	G–BDUZ	Cameron V-56 balloon	Balloon Stable Ltd. *Hot Lips*
	G–BDVA	PA-17 Vagabond	Mrs. H. S. & I. M. Callier
	G–BDVB	PA-15 (PA-17) Vagabond	S. Lowe
	G–BDVC	PA-17 Vagabond	A. T. Christian
	G–BDVG	Thunder Ax6-56A balloon	R. F. Pollard *Argonaut*
	G–BDVH	H.S.748 Srs. 2A	British Aerospace
	G–BDVI	D.H.82A Tiger Moth	K. D. J. Ecclestone
	G–BDVJ	Westland-Bell 47G-3B1	Hawkspare Ltd.
	G–BDVS	F.27 Friendship 200	Air UK
	G–BDVT	F.27 Friendship 200	Air UK
	G–BDVU	Mooney M.20F	Freecomp Ltd.
	G–BDVW	BN-2A Islander	Loganair Ltd.

Reg.	Type	Owner or Operator	Notes
G–BDWA	SOCATA Rallye 150ST	C. B. Roberts	
G–BDWB	SOCATA Rallye 150ST	M. Hazzeldine	
G–BDWE	Flaglor Scooter	D. W. Evernden	
G–BDWG	BN-2A Islander	Euroair Transport Ltd.	
G–BDWH	SOCATA Rallye 150ST	O. G. Owen	
G–BDWI	PA-34-200T-2 Seneca	Rentons Garages Ltd.	
G–BDWJ	SE-5A Replica (F8010)	M. L. Beach	
G–BDWK	Beech 95-B58 Baron	David Huggett Motor Factors Ltd.	
G–BDWL	PA-25 Pawnee 235	Summerhouse Farm Ltd.	
G–BDWM	Mustang replica	D. C. Bonsall	
G–BDWN	SA.318C Alouette II	AB Aviation Holdings Ltd.	
G–BDWO	Howes Ax6 balloon	R. B. & Mrs. C. Howes *Griffin*	
G–BDWP	PA-32R-300 Cherokee Lance	Starling (Sales Ideas) Ltd.	
G–BDWU	BN-2A Mk.III-2 Trislander	Anglo-Thai Corporation Ltd.	
G–BDWW	Cameron O-77 balloon	Dan-Air Hot-Air Balloon Group *Dan-Air*	
G–BDWX	Jodel D.120A	J. P. Lassey	
G–BDWY	PA-28 Cherokee 140	Tees-side Aviation Ltd.	
G–BDXA	Boeing 747-236B	British Airways *City of Cardiff*	
G–BDXB	Boeing 747-236B	British Airways	
G–BDXC	Boeing 747-236B	British Airways	
G–BDXD	Boeing 747-236B	British Airways *City of Plymouth*	
G–BDXE	Boeing 747-236B	British Airways *City of Glasgow*	
G–BDXF	Boeing 747-236B	British Airways *City of York*	
G–BDXG	Boeing 747-236B	British Airways *City of Oxford*	
G–BDXH	Boeing 747-236B	British Airways	
G–BDXI	Boeing 747-236B	British Airways	
G–BDXJ	Boeing 747-236B	British Airways *City of Birmingham*	
G–BDXW	PA-28R-200 Cherokee Arrow	Camlet Helicopters Ltd.	
G–BDXX	Nord NC.858S	A. F. Cashin	
G–BDXY	Auster AOP.9 (XR269)	Tyre & Tyne Transport Ltd.	
G–BDXZ	Pitts S-1S Special	P. Meeson	
G–BDYB	AA-5B Tiger	M. D. Joy	
G–BDYC	AA-1B Trainer	Cabair Ltd.	
G–BDYD	R. Commander 114	Conforth Ltd.	
G–BDYF	Cessna 421C	Mining Supplies Ltd.	
G–BDYG	Percival Provost T.1 (WV493)	Museum of Flight, E. Fortune	
G–BDYH	Cameron V-56 balloon	Aeolus Balloon Company *Novacastrian*	
G–BDYL	Beech C23 Sundowner	J. C. & Mrs. V. Hale	
G–BDYM	Skysales S-31 balloon	Miss A. I. Smith & M. J. Moore *Cheeky Devil*	
G–BDYY	Hiller UH-12E	AGN Helicopters Ltd.	
G–BDYZ	MBB Bo 105D	Management Aviation Ltd.	
G–BDZA	Scheibe SF.25E Super Falke	Norfolk Gliding Club Ltd.	
G–BDZB	Cameron S-31 balloon	Kenning Motor Group Ltd. *Kenning*	
G–BDZC	Cessna F.150M	Air Tows Ltd.	
G–BDZD	Cessna F.172M	Three Counties Aero Club Ltd.	
G–BDZF	G.164 Ag-Cat B	Miller Aerial Spraying Ltd.	
G–BDZS	Scheibe SF.25E Super Falke	A. D. Gubbay	
G–BDZU	Cessna 421C	European Ferries Ltd.	
G–BDZW	PA-28 Cherokee 140	Oldbus Ltd.	
G–BDZX	PA-28-151 Warrior	Data Precision Equipment Ltd.	
G–BDZY	Phoenix LA-4A Minor	P. J. Dalby	
G–BEAA	Taylor JT.1 Monoplane	R. C. Hobbs	
G–BEAB	Jodel DR.1051	Laxtonbridge Ltd.	
G–BEAC	PA-28 Cherokee 140	Eileen R. Purfield	
G–BEAD	Westland WG.13 Lynx	Westland Helicopters Ltd.	
G–BEAE	PA-25 Pawnee 235	Farm Aviation Services Ltd.	
G–BEAG	PA-34-200T-2 Seneca	R. P. Yeoward	
G–BEAH	J/2 Arrow	W. J. & Mrs. M. D. Horler	
G–BEAK	L-1011-385 TriStar	British Airways *The Northern Lights Rose*	
G–BEAL	L-1011-385 TriStar	British Airways *The Red Devil Rose*	
G–BEAM	L-1011-385 TriStar	British Airways *The Silver Jubilee Rose*	
G–BEAR	Viscount V.5 balloon	B. Hargraves & B. King	
G–BEAU	Pazmany PL.4A	B. H. R. Smith	
G–BEBA	H.S.748 Srs. 2	Dan-Air Services Ltd.	
G–BEBB	HPR-7 Herald 214	British Air Ferries	
G–BEBC	*WS.55 Whirlwind 3 (XP355)	Museum, Norwich Airport	
G–BEBE	AA-5A Cheetah	Special Alloys (Northern) Ltd.	
G–BEBF	Auster AOP.9	M. D. N. & Mrs. A. C. Fisher	

Notes	Reg.	Type	Owner or Operator
	G–BEBG	WSK-PZL SDZ-45A Ogar	Anglo-Polish Sailplanes Ltd.
	G–BEBH	Cessna FR.172J	D. L. B. Upjohn
	G–BEBI	Cessna F.172M	Calder Equipment Ltd.
	G–BEBJ	PA-E23 Aztec 250	Burnthills Aviation Ltd.
	G–BEBK	PA-31-300 Turbo Navajo	Dubilier Ltd.
	G–BEBL	Douglas DC-10-30	British Caledonian Airways *Sir Alexander Fleming—The Scottish Challenger*
	G–BEBM	Douglas DC-10-30	British Caledonian Airways *Robert Burns—The Scottish Bard*
	G–BEBN	Cessna 177B	J. Parry
	G–BEBO	Phoenix Currie Wot	C. Turner
	G–BEBR	GY-201 Minicab	A. S. Jones & D. R. Upton
	G–BEBS	Andreasson BA-4B	D. M. Fenton
	G–BEBT	Andreasson BA-4B	D. M. Fenton
	G–BEBU	R. Commander 112A	Judd Studios Ltd.
	G–BEBZ	PA-28-151 Warrior	Goodwood Terrena Ltd.
	G–BECA	SOCATA Rallye 100ST	Air Manchester Ltd.
	G–BECB	SOCATA Rallye 100ST	J. D. Brook & B. K. Arthur
	G–BECC	SOCATA Rallye 150ST	Air Touring Services Ltd.
	G–BECD	SOCATA Rallye 150ST	J. B. Roberts
	G–BECF	Scheibe SF.25A Falke	D. A. Wilson & ptnrs.
	G–BECG	Boeing 737-204ADV	Britannia Airways Ltd. *Amy Johnson*
	G–BECH	Boeing 737-204ADV	Britannia Airways Ltd. *Viscount Montgomery of Alamein*
	G–BECJ	Partenavia P.68B	Seven Seas Finance Ltd.
	G–BECK	Cameron V-56 balloon	Miss D. M. Bragg *Noddy*
	G–BECL	C.A.S.A. C.352L (N9+AA)	War Birds of Great Britain Ltd.
	G–BECM	Aerotec Pitts S-2A Special	Airmiles Ltd.
	G–BECN	Piper J-3C-65 Cub	E. J. Cottle & Sons
	G–BECO	Beech A.36 Bonanza	Thorney Machinery Co. Ltd.
	G–BECP	PA-31-310 Turbo Navajo	Kenton Utilities & Developments Ltd.
	G–BECR	Cessna 182B	Midland Parachute Centre
	G–BECT	C.A.S.A. 1.131 Jungmann	Reidermere Ltd.
	G–BECW	C.A.S.A. 1.131 Jungmann	N. C. Jensen
	G–BECX	C.A.S.A. 1.131 Jungmann	R. A. Seeley
	G–BECZ	CAARP CAP.10B	Aerobatic Associates Ltd.
	G–BEDA	C.A.S.A. 1.131 Jungmann	M. G. Kates & D. J. Berry
	G–BEDB	Nord 1203 Norecrin	B. F. G. Lister
	G–BEDD	Jodel D.117A	B. R. Vickers
	G–BEDE	Bede BD-5A	Biggin Hill BD5 Syndicate
	G–BEDF	Boeing B-17G-105-VE (485784)	M. H. Campbell
	G–BEDG	R. Commander 112A	N. J. Orr
	G–BEDH	R. Commander 114	Ian Anthony (Sales) Ltd.
	G–BEDI	Sikorsky S-61N Mk. II	British Airways Helicopters Ltd.
	G–BEDJ	Piper J-3C-65 Cub	D. J. Elliott & A. E. Molton
	G–BEDK	Hiller UH-12E	T. C. Jay
	G–BEDL	Cessna T337D	R. M. Cox Ltd.
	G–BEDO	BN-2A Mk. III-2 Trislander	BN (Bembridge) Ltd.
	G–BEDS	Thunder Ax7-77A balloon	Thunder Balloons Ltd.
	G–BEDU	Scheibe SF.23C Sperling	Doncaster & District Gliding Club Ltd.
	G–BEDV	V.668 Varsity T.1 (WJ945)	D. S. Selway & A. G. Battersby
	G–BEDZ	BN-2A Islander	Loganair Ltd.
	G–BEEA	SOCATA Rallye 235GT	Air Sinclair
	G–BEEE	Thunder Ax6-56A balloon	I. R. M. Jacobs *Avia*
	G–BEEF	Thunder Ax6-56A balloon	R. Mirray *Beefeater*
	G–BEEG	BN-2A Islander	Loganair Ltd.
	G–BEEH	Cameron V-56 balloon	J. M. Langley *Kaleidoscope*
	G–BEEI	Cameron N-77 balloon	D. W. A. Legg *Master McGrath*
	G–BEEJ	Cameron O-77 balloon	DAL (Builders Merchants) Ltd. *Dal's Pal*
	G–BEEK	Enstrom F-280C Shark	Uniweld Ltd.
	G–BEEL	Enstrom F-280C-UK Shark	Slea Aviation Ltd.
	G–BEEN	Cameron O-56 balloon	Swire Bottlers Ltd. *Coke*
	G–BEEP	Thunder Ax5-42 balloon	Mrs. B. C. Faithful *Also Kenneth*
	G–BEER	Isaacs Fury II	M. J. Clark
	G–BEEU	PA-28 Cherokee 140E	Berkshire Aviation Services Ltd.
	G–BEEV	PA-28 Cherokee 140E	K. W. Watts
	G–BEEW	Taylor JT.1 Monoplane	K. Wigglesworth
	G–BEFA	PA-28-151 Warrior	A. J. Blois-Brooke
	G–BEFC	AA-5B Tiger	John Holt & Sons

Reg.	Type	Owner or Operator	Notes
G–BEFF	PA-28 Cherokee 140	Sherwood Flying Club Ltd.	
G–BEFH	—	—	
G–BEFR	Fokker DR.I Repilca (FI425)	Leisure Sport Ltd.	
G–BEFT	Cessna 421C	Lucas Industries Ltd.	
G–BEFU	Sturgeonair MJ.7 Mustang	W. E. Wilks	
G–BEFV	Evans VP-2	Yeadon Aeroplane Group	
G–BEFW	PA-39 Twin Comanche C/R	Warnell Motors Ltd. (G–AYZP)	
G–BEFX	Hiller UH-12E	Agricopters Ltd.	
G–BEFY	Hiller UH-12E	Farm Supply (Thirsk) Ltd.	
G–BEFZ	H.S.125 Srs. 700B	McAlpine Aviation Ltd.	
G–BEGA	Westland Bell 47G-3B1	Harold Poupart Ltd.	
G–BEGG	Scheibe SF.25E Super Falke	RAFGSA	
G–BEGV	PA-23 Aztec 250F	Stoitwell Ltd.	
G–BEGZ	Boeing 727-193	Dan-Air Services Ltd.	
G–BEHC	BN-2A Mk. III-2 Trislander	BN (Bembridge) Ltd.	
G–BEHD	BN-2A Mk. III-2 Trislander	BN (Bembridge) Ltd.	
G–BEHE	BN-2A Mk. III-2 Trislander	BN (Bembridge) Ltd.	
G–BEHF	BN-2A Mk. III-2 Trislander	BN (Bembridge) Ltd.	
G–BEHG	AB-206B JetRanger 2	Willowbrook International Ltd.	
G–BEHH	PA-32R-300 Cherokee Lance	SMK Engineering Ltd.	
G–BEHJ	Evans VP-1	K. Heath	
G–BEHK	Agusta-Bell 47G-3B1 (Soloy)	Dollar Air Services Ltd.	
G–BEHL	Agusta-Bell 47G-3B1	Dollar Air Services Ltd.	
G–BEHM	Taylor JT.1 Monoplane	H. McGovern	
G–BEHR	Beech A200 Super King Air	Eagle Aircraft Services Ltd.	
G–BEHS	PA-25 Pawnee 260C	Farm Aviation Services Ltd.	
G–BEHT	PA-25 Pawnee 260C	Farm Aviation Services Ltd.	
G–BEHU	PA-34-200T-2 Seneca	Nottingham Self-Fly & Aircraft Hire	
G–BEHV	Cessna F.172N	L. Mitchell	
G–BEHW	Cessna F.150M	S. R. Taylor	
G–BEHX	Evans VP-2	G. S. Adams	
G–BEHY	PA-28-181 Archer II	Q. P. Cope	
G–BEIA	Cessna FRA.150M	Airwork Services Ltd.	
G–BEIB	Cessna F.172N	R. L. Orsborn & Son Ltd.	
G–BEIC	Sikorsky S-61N	British Airways Helicopters Ltd.	
G–BEID	Sikorsky S-61N	British Airways Helicopters Ltd.	
G–BEIE	Evans VP-2	F. G. Morris	
G–BEIF	Cameron O-65 balloon	C. Vening & S. W. C. Hall	
G–BEIG	Cessna F.150M	Gordon King (Aviation) Ltd.	
G–BEIH	PA-25 Pawnee 235D	Farmair Ltd.	
G–BEII	PA-25 Pawnee 235D	Miller Aerial Spraying Ltd.	
G–BEIJ	G.164B Ag-Cat	Miller Aircraft Hire Ltd.	
G–BEIK	Beech A.36 Bonanza	Scot-Stock Ltd.	
G–BEIL	SOCATA Rallye 150T	D. H. Tonkin	
G–BEIP	PA-28-181 Archer II	S. Webb	
G–BEIS	Evans VP-1	D. J. Park	
G–BEIZ	Cessna 500 Citation	Casair Ltd.	
G–BEJA	Thunder Ax6-56A balloon	P. A. Hutchins *Jackson*	
G–BEJB	Thunder Ax6-56A balloon	International Distillers & Vintners Ltd. *Baby J. & B.*	
G–BEJD	H.S.748 Srs. 1	Dan-Air Services Ltd.	
G–BEJE	H.S.748 Srs. 1	Dan-Air Services Ltd.	
G–BEJK	Cameron S-31 balloon	Esso Petroleum Ltd.	
G–BEJL	Sikorsky S-61N	British Airways Helicopters Ltd.	
G–BEJM	BAC One-Eleven 423	Ford Motor Co. Ltd.	
G–BEJO	Saffery S.250 free balloon	Cupro Sapphire Ltd. *Firefly*	
G–BEJP	D.H.C.-6 Twin Otter 310	Loganair Ltd.	
G–BEJT	PA-23 Aztec 250F	Worldair Sales Ltd.	
G–BEJV	PA-34-200T-2 Seneca	C.S.E. Aviation Ltd.	
G–BEJW	BAC One-Eleven 423	Ford Motor Co. Ltd.	
G–BEJX	Partenavia P.68B	DK Aviation Ltd.	
G–BEJY	Hughes 369D	Dollar Air Services Ltd.	
G–BEJZ	Aerostar 601PE	Mann Aviation Sales Ltd.	
G–BEKA	BAC One-Eleven 520	Dan-Air Services Ltd.	
G–BEKB	PA-23 Aztec 250	Alidair Ltd.	
G–BEKC	H.S.748 Srs. 1	Dan-Air Services Ltd.	
G–BEKD	H.S.748 Srs. 1	Dan-Air Services Ltd.	
G–BEKE	H.S.748 Srs. 1	Dan-Air Services Ltd.	
G–BEKG	H.S.748 Srs. 1	Dan-Air Services Ltd.	
G–BEKH	AB-206B JetRanger 2	W.R. Finance Ltd.	
G–BEKL	Bede BD-4E	G. A. Hodges	
G–BEKM	Evans VP-1	G. J. McDill	
G–BEKN	Cessna FRA.150M	Balgin Ltd.	

Notes	Reg.	Type	Owner or Operator
	G–BEKO	Cessna F.182Q	C. J. Leonard
	G–BEKR	Rand KR-2	K. B. Raven
	G–BEKS	PA-25 Pawnee 235D	Farmair Ltd.
	G–BEKT	PA-25 Pawnee 235D	Peter Charles (Air Farmers) Ltd.
	G–BELK	BN-2A Islander	Brown & Root (UK) Ltd.
	G–BELO	Douglas DC-10-10	Laker Airways *Southern Belle*
	G–BELP	PA-28-151 Warrior	Channel Aviation Ltd.
	G–BELR	PA-28 Cherokee 140	Woodgate Aviation Ltd.
	G–BELS	D.H.C.-6 Twin Otter 310	Loganair Ltd.
	G–BELT	Cessna F.150J	Yorkshire Light Aircraft Ltd. (G–AWUV)
	G–BELV	Cessna 404 Titan	Executive Express Ltd.
	G–BELX	Cameron V-56 balloon	W. H. & Mrs. J. P. Morgan *Topsy*
	G–BEMA	Cessna 310R II	Air Charter & Travel Ltd.
	G–BEMB	Cessna F.172M	Northair Aviation Ltd.
	G–BEMC	Cessna F.172M	Stevair
	G–BEMD	Beech 95-B55 Baron	Vaux (Aviation) Ltd.
	G–BEMF	Taylor JT.1 Monoplane	J. Simpson
	G–BEMI	Thunder Ax6-56A balloon	Thunder Balloons Ltd.
	G–BEMM	Slingsby T.31B Motor Cadet	M. N. Martin
	G–BEMT	Bede BD-5G	G. Smith
	G–BEMU	Thunder Ax5-42 balloon	P. J. Langford
	G–BEMV	PA-28 Cherokee 140	Worldair Sales Ltd.
	G–BEMW	PA-28-181 Archer II	Encee Services
	G–BEMX	Cessna 404 Titan	Executive Express
	G–BEMY	Cessna FRA.150M	TE-ELF Ltd.
	G–BEMZ	B.175 Britannia 253F	Redcoat Air Cargo
	G–BENC	Cessna 402B	Medburn Air Services Ltd.
	G–BEND	Cameron V-56 balloon	Dante Balloon Group *Le Billet*
	G–BENE	Cessna 402B	Greenline Refrigerated Transport Ltd.
	G–BENF	Cessna T.210L	Laughton Aviation Ltd.
	G–BENG	Westland Bell 47G-3B1	GSM Helicopters Ltd.
	G–BENH	Phoenix LA.5A Major	C. D. Macartney
	G–BENI	Bell 47G-4A	R. J. A. Brown
	G–BENJ	R. Commander 112B	F. T. Arnold
	G–BENK	Cessna F.172M	Capeston Aviation Ltd.
	G–BENL	PA-25 Pawnee 235D	Farm Aviation Services Ltd.
	G–BENM	PA-31-325 Navajo	Gull Air Ltd.
	G–BENN	Cameron V-56 balloon	S. L. G. & H. C. J. Williams
	G–BENO	Enstrom F-280C Shark	Jack Tatties Ltd.
	G–BENP	D.31A Turbulent	N. H. Ponsford
	G–BENS	Saffery S.330 balloon	D. Whitlock *Hot Plastic*
	G–BENT	Cameron N-77 balloon	N. Tasker
	G–BEOC	BN-2A Islander	Alderney Air Ferries Ltd.
	G–BEOD	Cessna 180	Farm Supply (Thirsk) Ltd.
	G–BEOE	Cessna FRA.150M	The Aerobat Flying Group Ltd.
	G–BEOH	PA-28R-201T Arrow III	Flockvale Ltd.
	G–BEOI	PA-18-150 Super Cub	S. Down Gliding Club Ltd.
	G–BEOK	Cessna F.150M	Gordon King (Aviation) Ltd.
	G–BEON	Sikorsky S-61N Mk. II	British Airways Helicopters Ltd.
	G–BEOO	Sikorsky S-61N Mk. II	British Airways Helicopters Ltd.
	G–BEOT	PA-25 Pawnee 235D	C.S.E. Aviation Ltd.
	G–BEOV	PA-28 Cherokee 140	Aerogulf Services Co. Ltd.
	G–BEOW	PA-28 Cherokee 140	Aerogulf Services Co. Ltd.
	G–BEOX	*L-414 Hudson IIIA (A16-199)	RAF Museum
	G–BEOY	Cessna FRA.150L	Wycombe Air Centre Ltd.
	G–BEOZ	A.W.650 Argosy 101	Air Bridge Carriers Ltd.
	G–BEPB	Pereira Osprey II	J. J. & A. J. C. Zwetsloot
	G–BEPC	SNCAN SV-4C	M. Harbron
	G–BEPD	SA.102.5 Cavalier	P. & Mrs. E. A. Donaldson
	G–BEPE	SC.5 Belfast	HeavyLift Cargo Airlines Ltd. (G–ASKE)
	G–BEPF	SNCAN SV-4A	M. Stelfox & R. G. Harrington
	G–BEPI	BN-2A Mk. III-2 Trislander	BN (Bembridge) Ltd.
	G–BEPJ	BN-2A Mk. III-2 Trislander	BN (Bembridge) Ltd.
	G–BEPK	BN-2A Mk. III-2 Trislander	BN (Bembridge) Ltd.
	G–BEPO	Cameron N-77 balloon	Sungas Ltd. *Sungas*
	G–BEPP	AB-206B JetRanger 2	T. W. Walker Ltd.
	G–BEPS	SC.5 Belfast	HeavyLift Cargo Airlines Ltd.
	G–BEPV	Fokker S.11-1 Instructor	Strathallan Aircraft Collection
	G–BEPY	R. Commander 112B	D. S. Thomas
	G–BEPZ	Cameron D-96 hot-air airship	IAZ International (UK) Ltd.

Reg.	Type	Owner or Operator	Notes
G-BERA	SOCATA Rallye 150ST	B. J. Durrant-Peatfield	
G-BERB	SOCATA Rallye 150ST	Martin Ltd.	
G-BERC	SOCATA Rallye 150ST	Velcourt (East) Ltd.	
G-BERD	Thunder Ax6-56A balloon	Thunder Balloons Ltd. *Goldfinger*	
G-BERE	Thunder Ax6-56A balloon	Fisons Ltd. *Fisons*	
G-BERG	SA 330J Puma	Bristow Helicopters Ltd.	
G-BERH	SA 330J Puma	Bristow Helicopters Ltd.	
G-BERI	R. Commander 114	Rockville Motors Ltd.	
G-BERJ	Bell 47G-4A	Hon. G. M. H. Wills	
G-BERK	Osprey Mk. 6 balloon	A. P. Chown & ptnrs.	
G-BERL	AA-5B Tiger	Scotia Safari Ltd.	
G-BERM	AA-5A Cheetah	John Farbon & Co. Ltd.	
G-BERN	Saffery S-330 balloon	B. Martin *Beeze*	
G-BERR	Thunder Ax7-77A balloon	M. Kendrick *Woolworths*	
G-BERT	Cameron V-56 balloon	Southern Balloon Group *Bert*	
G-BERW	R. Commander 114	Allison (Contractors) Ltd.	
G-BERY	AA-1B Trainer	E. G. Whelan & D. I. Hall	
G-BESA	Beech 95-58 Baron	Oldham Aviation	
G-BESG	BN-2A Islander	BN (Bembridge) Ltd.	
G-BESO	BN-2A Islander	Jersey European Airways	
G-BESR	BN-2A Islander	T. E. Oxley	
G-BESS	Hughes 369D	Rassler Aero Services	
G-BESW	BN-2A Islander	Alderney Air Ferries Ltd.	
G-BETA	Rollason B.2A Beta	J. L. Kinch	
G-BETC	Cameron V-56 balloon	P. G. Dunnington *Etcetera*	
G-BETD	Robin HR.200/100	D. W. Jackson	
G-BETE	Rollason B.2 Beta	T. M. Jones	
G-BETF	Cameron 'Champion' balloon	Balloon Stable Ltd. *Champion*	
G-BETG	Cessna 180K Skywagon	E. H. S. Warner	
G-BETH	Thunder Ax6-56A balloon	Debenhams Ltd. *Debenhams I*	
G-BETI	Pitts S-1D Special	E. B. Bray	
G-BETJ	Douglas DC-8-33	*Derelict at Stansted*	
G-BETL	PA-25 Pawnee 235D	Lincs Aerial Spraying Co.	
G-BETM	PA-25 Pawnee 235D	Skegness Air Taxi Services Ltd.	
G-BETO	MS.885 Super Rallye	A. Somerville	
G-BETP	Cameron O-65 balloon	J. R. Rix & Sons Ltd.	
G-BETS	Cessna A.188B AgTruck	Shoreham Flight Simulation Ltd.	
G-BETT	PA-34-200-2 Seneca	Andrews Professional Colour Laboratories Ltd.	
G-BETU	Piper J-3C-65 Cub	M. J. Curran & R. G. Fitton	
G-BETV	HS.125 Srs. 600B	Tenneco Aviation Ltd.	
G-BETW	Rand KR-2	T. A. Wiffen	
G-BEUA	PA-18-150 Super Cub	London Gliding Club (Pty.) Ltd.	
G-BEUC	PA-28-181 Archer II	P. C. T. Warner	
G-BEUD	Robin HR.100/285R	E. A. & L. M. C. Payton	
G-BEUE	Bell 47G-2	Bristow Helicopters Ltd.	
G-BEUF	Bell 47G-2	Bristow Helicopters Ltd.	
G-BEUG	Bell 47G-2	Bristow Helicopters Ltd.	
G-BEUH	Bell 47G-2	Bristow Helicopters Ltd.	
G-BEUI	Piper J-3C-65 Cub	B. P. Gardner	
G-BEUK	Fuji FA-200-160	C.S.E. Aviation Ltd.	
G-BEUL	Beech 95-58 Baron	Basic Metal Co. Ltd.	
G-BEUM	Taylor JT.1 Monoplane	M. T. Taylor	
G-BEUN	Cassutt Racer 111m	R. S. Voice	
G-BEUP	Robin DR.400/180	Steelfields Ltd.	
G-BEUR	Cessna F.172M	G. A. Locke & Sons Ltd.	
G-BEUS	SNCAN SV-4C	G. V. Gower	
G-BEUU	PA-19 Super Cub 95	E. D. Burke & M. R. Smith	
G-BEUV	Thunder Ax6-56A balloon	Thunder Balloons Ltd. *Debenhams II*	
G-BEUW	AA-5A Cheetah	Cabair Ltd.	
G-BEUX	Cessna F.172N	Light Planes (Lancashire) Ltd.	
G-BEUY	Cameron N-31 balloon	Southern Balloon Group	
G-BEUZ	Beech A200 Super King Air	Anglian Double Glazing Ltd.	
G-BEVA	SOCATA Rallye 150ST	J. G. Wilson	
G-BEVB	SOCATA Rallye 150ST	R. & Mrs. S. McClelland	
G-BEVC	SOCATA Rallye 150ST	B. W. Walpole	
G-BEVE	Thunder Ax7-77A	Thunder Balloons Ltd.	
G-BEVG	PA-34-200T-2 Seneca	Skycabs	
G-BEVH	Holland D.700 balloon	D. I. Holland *Sally*	
G-BEVI	Thunder Ax7-77A balloon	The Painted Clouds Balloon Co. Ltd.	
G-BEVJ	Cessna U-206F Stationair	Twinguard Leasing Ltd.	
G-BEVL	Cessna 421C	Vickers Ltd.	
G-BEVO	Sportavia-Pützer RF-5	D. Lister & R. F. Bradshaw	

Notes	Reg.	Type	Owner or Operator
	G–BEVP	Evans VP-2	I. S. Walsh
	G–BEVR	BN-2A Mk. III-2 Trislander	BN (Bembridge) Ltd.
	G–BEVS	Taylor JT.I Monoplane	D. Hunter
	G–BEVT	BN-2A Mk. III-2 Trislander	BN (Bembridge) Ltd.
	G–BEVV	BN-2A Mk. III-2 Trislander	BN (Bembridge) Ltd.
	G–BEVW	SOCATA Rallye 150ST	G. V. Wallis
	G–BEVY	BN-2A Mk. III-2 Trislander	BN (Bembridge) Ltd.
	G–BEWJ	Westland-Bell 47G-3B-I	G. Tanner & P. O. Wicks Ltd.
	G–BEWL	Sikorsky S-61N Mk. II	British Airways Helicopters Ltd.
	G–BEWM	Sikorsky S-61N Mk. II	British Airways Helicopters Ltd.
	G–BEWN	D.H.82A Tiger Moth	H. D. Labouchere
	G–BEWO	Zlin 326 Trener Master	H. D. Labouchere
	G–BEWP	Cessna F.150M	Pegasus Aviation Ltd.
	G–BEWR	Cessna F.172N	Mersey Flying Club Ltd.
	G–BEWS	Cameron 56 'Lamp Bulb' hot-air balloon	Osram (GEC) Ltd. *Osram*
	G–BEWW	HS.125 Srs. 600B	Truzone Ltd. (G–AZUF)
	G–BEWX	PA-28R-201 Arrow III	Gidney & Kirby Holdings Ltd.
	G–BEWY	Bell 206B JetRanger 2	Bristow Helicopters Ltd.
	G–BEWZ	PA-32-300 Cherokee Six C	Sherwood Car Hire (Mansfield) Ltd.
	G–BEXK	PA-25 Pawnee 235D	Howard Avis (Aviation) Ltd.
	G–BEXL	PA-25 Pawnee 235D	Plains Aerial Spraying Ltd.
	G–BEXN	AA-IC Lynx	Scotia Safari Ltd.
	G–BEXO	PA-23 Apache 160	B. Burton
	G–BEXR	Mudry/CAARP CAP-10B	R. P. Lewis
	G–BEXS	Cessna F.150M	Hartmann Ltd.
	G–BEXV	Cameron O-56 balloon	Revlon (Hong Kong) Ltd. *Charlie*
	G–BEXW	PA-28-181 Archer II	J. Coyle
	G–BEXX	Cameron V-56 balloon	A. Tyler & ptnrs. *Rupert of Rutland*
	G–BEXY	PA-28 Cherokee 140	D. H. L. Wigan
	G–BEXZ	Cameron N-56 balloon	M. J. Moroney *Valor*
	G–BEYA	Enstrom F-280C Shark	Guy Morton & Sons Ltd.
	G–BEYB	Fairey Flycatcher (replica) (S1287)	John S. Fairey
	G–BEYD	HPR-7 Herald 401	British Air Ferries
	G–BEYE	HPR-7 Herald 401	British Air Ferries *Rupert Keegan*
	G–BEYF	HPR-7 Herald 401	British Air Ferries *Sheila Scott*
	G–BEYJ	HPR-7 Herald 401	British Air Ferries
	G–BEYK	HPR-7 Herald 401	Air UK
	G–BEYL	PA-28 Cherokee 180	B. G. & G. Airlines Ltd.
	G–BEYM	Cessna F.150M	Anglian Flight Training Ltd.
	G–BEYN	Evans VP-2	C. D. Denham
	G–BEYO	PA-28 Cherokee 140	Rimmer Scaffolding
	G–BEYP	Fuji FA-200-180AO	A. C. Pritchard
	G–BEYS	PA-28-181 Archer II	Scotia Safari Ltd.
	G–BEYU	Fuji FA-200-160	C.S.E. Aviation Ltd.
	G–BEYV	Cessna T.210L	Valley Motors
	G–BEYW	Taylor JT.I Monoplane	R. J. Smythe
	G–BEYX	PA-31P Navajo	Warwick Air Services Ltd.
	G–BEYY	PA-31-310 Turbo Navajo	Trans Europe Air Charters Ltd.
	G–BEYZ	Jodel DR.1051/MI	M. J. McCarthy & S. Aarons
	G–BEZA	Zlin 226T Trener	L. Bezak
	G–BEZB	HPR-7 Herald 209	Express Air Services Ltd.
	G–BEZC	AA-5 Traveler	R. M. Gosling & P. J. Schwind
	G–BEZE	Rutan VariEze	M. F. Sharples & ptnrs.
	G–BEZF	AA-5 Traveler	W. Wheatley
	G–BEZG	AA-5 Traveler	R. J. Stevens
	G–BEZH	AA-5 Traveler	R. J. Herbert
	G–BEZI	AA-5 Traveler	Fergusons (Blyth) Ltd.
	G–BEZJ	MBB Bo 105D	Management Aviation Ltd.
	G–BEZK	Cessna F.172H	R. J. Lock & ptnrs.
	G–BEZL	PA-31-310 Navajo	Barratt Developments (Northern) Ltd.
	G–BEZM	Cessna F.182Q	Northair Aviation Ltd.
	G–BEZO	Cessna F.172M	Rogers Aviation Ltd.
	G–BEZP	PA-32-300D Cherokee Six	Advanced Marketing Management Ltd.
	G–BEZR	Cessna F.172M	Kirmington Aviation Ltd.
	G–BEZU	PA-31-350 Navajo Chieftain	C.S.E. Aviation Ltd.
	G–BEZV	Cessna F.172M	S. G. Brady
	G–BEZW	Practavia Pilot Sprite	T. S. Wilkins & ptnrs.
	G–BEZY	Rutan VariEze	R. J. Jones
	G–BEZZ	Jodel D.112	A. J. Stevens & ptnrs.

Reg.	Type	Owner or Operator	Notes
G–BFAA	GY-80 Horizon 160	Dustalong Ltd.	
G–BFAB	Cameron N-56 balloon	Phonogram Ltd. *Phonogram*	
G–BFAC	Cessna F.177 RG	Brendan Butler (London) Ltd.	
G–BFAD	PA-28-161 Warrior II	J. G. Hogg & Broad Oak Chartering Ltd.	
G–BFAF	Aeronca 7BCM	D. C. W. Harper	
G–BFAH	Phoenix Currie Wot	T. G. Thomson	
G–BFAI	R. Commander 114	Biograft Ltd.	
G–BFAK	MS.892A Rallye Commodore 150	R. Jennings & ptnrs.	
G–BFAM	PA-31P Navajo	International Air Charter Ltd.	
G–BFAN	H.S.125 Srs. 600F	British Aerospace (G–AZHS)	
G–BFAO	PA-20 Pacer 135	C. A. G. Schofield	
G–BFAP	SIAI-Marchetti S.205-20R	Miss M. A. Eccles	
G–BFAR	Cessna 500-1 Citation	Yewlands Executive Transport Ltd.	
G–BFAS	Evans VP-1	A. I. Sutherland	
G–BFAV	Orion model free balloon	D. C. Boxall	
G–BFAW	D.H.C.-1 Chipmunk 22	G. Jones & ptnrs.	
G–BFAX	D.H.C.-1 Chipmunk 22	P. G. D. Bell	
G–BFBA	Jodel DR.100A	Wasp Flying Group	
G–BFBB	PA-23 Aztec 250E	Falaise Investments Ltd.	
G–BFBC	Taylor JT.1 Monoplane	A. Brooks	
G–BFBD	Partenavia P.68B	Pegasus Aviation Ltd.	
G–BFBE	Robin HR 200/100	Charles Major Ltd.	
G–BFBF	PA-28 Cherokee 140	P. H. de Havilland	
G–BFBH	PA-31-325 Turbo Navajo	Mont Arthur Finance Ltd.	
G–BFBJ	PA-34-200T-2 Seneca	Shorgard Ltd.	
G–BFBK	Agusta-Bell 47G-3B1	Dollar Air Services Ltd.	
G–BFBM	Saffery S.330 balloon	B. Martin *Beeze II*	
G–BFBN	PA-25 Pawnee 235D	J. E. F. Aviation Ltd.	
G–BFBR	PA-28-161 Warrior II	Linvic Ltd.	
G–BFBS	Boeing 707-351B	Laker Airways Ltd.	
G–BFBU	Partenavia P-68B	Clyde Developments Ltd.	
G–BFBV	Brügger Colibri M.B.2	J. D. Hutton & ptnrs.	
G–BFBW	PA-25 Pawnee 235D	Lawrence Agricultural Aviation Services Ltd.	
G–BFBX	PA-25 Pawnee 235D	Bowker Aircraft Services Ltd.	
G–BFBY	Piper J-3C-65 Cub	D. S. Quirk	
G–BFCA	L-1011-385 TriStar 500	British Airways *Princess Margaret Rose*	
G–BFCB	L-1011-385 TriStar 500	British Airways *The Harry Wheatcroft Rose*	
G–BFCC	L-1011-385 TriStar 500	British Airways *The English Mist Rose*	
G–BFCD	L-1011-385 TriStar 500	British Airways *The Astral Rose*	
G–BFCE	L-1011-385 TriStar 500	British Airways *The Gay Gordons Rose*	
G–BFCF	L-1011-385 TriStar 500	British Airways *The Elizabeth of Glamis Rose*	
G–BFCT	Cessna TU-206F	Cecil Aviation Ltd.	
G–BFCX	BN-2A Islander	Loganair Ltd.	
G–BFCY	AB-206B JetRanger 2	Window Boxes Ltd.	
G–BFCZ	Sopwith Camel (B7270)	Leisure Sport Ltd.	
G–BFDA	PA-31-350 Navajo Chieftain	Air Charter (Scotland) Ltd	
G–BFDB	PA-31-350 Navajo Chieftain	Fairflight Ltd.	
G–BFDC	D.H.C.-1 Chipmunk 22	N. F. O'Neill	
G–BFDD	BN-2A Mk III-2 Trislander	BN (Bembridge) Ltd.	
G–BFDE	Sopwith Tabloid (replica) (168)	D. M. Cashmore	
G–BFDF	SOCATA Rallye 235GT	S. G. Lawrence	
G–BFDG	PA-28R-201T Turbo-Arrow III	The Electric Rainbow Co. Ltd.	
G–BFDI	PA-28-181 Archer II	A. P. Grant & J. R. Sunderland	
G–BFDK	PA-28-161 Warrior II	C.S.E. Aviation Ltd.	
G–BFDL	Piper L-4J Cub	R. Windley	
G–BFDM	Jodel D.120	Worcestershire Gliding Ltd.	
G–BFDN	PA-31-350 Navajo Chieftain	Lease Air Ltd.	
G–BFDO	PA-28R-201T Turbo-Arrow III	Grangewood Press Ltd.	
G–BFDP	CP.301A Emeraude	J. P. McGrath & J. Robson	
G–BFDR	Beech 60 Duke	Penarth Commercial Properties Ltd.	
G–BFDZ	Taylor JT.1 Monoplane	D. C. Barber	
G–BFEB	Jodel D.150	D. Aldersea & ptnrs.	
G–BFEC	PA-23 Aztec 250F	Midas Airways	
G–BFED	PA-31-350 Navajo Chieftain	Woodgate Aviation Ltd.	
G–BFEE	Beech 95-E55 Baron	J. F. Pugh	

Notes	Reg.	Type	Owner or Operator
	G–BFEF	Agusta-Bell 47G-3B1	Trent Helicopters Ltd.
	G–BFEG	Westland-Bell 47G-3B1	Trent Helicopters Ltd.
	G–BFEH	Jodel D.117A	G. W. Talbot
	G–BFEI	Westland-Bell 47G-3B1	Trent Helicopters Ltd.
	G–BFEJ	Agusta-Bell 47G-3B1	Copley Farms Ltd.
	G–BFEK	Cessna F.152	Staverton Flying Services Ltd.
	G–BFEL	Cessna F.150M	A. L. Strongman
	G–BFEM	Cessna 421C	Brush Electrical Machines Ltd.
	G–BFEO	Boeing 707-323C	Tradewinds Airways Ltd.
	G–BFER	Bell 212	Bristow Helicopters Ltd.
	G–BFET	Partenavia P.68B	DK Aviation Ltd.
	G–BFEU	SA.330J Puma	Bristow Helicopters Ltd.
	G–BFEV	PA-25 Pawnee 235	Bowker Aircraft Services Ltd.
	G–BFEW	PA-25 Pawnee 235	Agricola Aerial Work Ltd.
	G–BFEX	PA-25 Pawnee 235	Harvest Air Ltd.
	G–BFEY	PA-25 Pawnee 235	Harvest Air Ltd.
	G–BFEZ	Beech B60 Duke	Turner Engineering Co. Ltd.
	G–BFFB	Evans VP-2	D. Bradley
	G–BFFC	Cessna F.152-II	Yorkshire Flying Services Ltd.
	G–BFFD	Cessna F.152-II	Skegness Air Taxi Service Co. Ltd.
	G–BFFE	Cessna F.152-II	Veetstor Ltd.
	G–BFFF	Cessna 188B Ag Truck	Shoreham Flight Simulation
	G–BFFG	Beech 95-B55 Baron	Basic Metal Co. Ltd.
	G–BFFJ	Sikorsky S-61N Mk. II	British Airways Helicopters Ltd.
	G–BFFK	Sikorsky S-61N Mk. II	British Airways Helicopters Ltd.
	G–BFFN	Enstrom F.28A	Grays of Cambridge Ltd.
	G–BFFP	PA-18-150 Super Cub	Airways Aero Associations Ltd.
	G–BFFR	PA-31-350 Navajo Chieftain	Vickers Shipbuilding Group
	G–BFFT	Cameron V-56 balloon	R. I. M. Kerr & D. C. Boxall
	G–BFFV	Agusta Bell 47G-3B1	Dollar Air Services Ltd.
	G–BFFW	Cessna F.152	W. A. Hanton
	G–BFFY	Cessna F.150M	Northair Aviation Ltd.
	G–BFFZ	Cessna FR.172 Hawk XP	Trent Helicopters Ltd.
	G–BFGA	MS.880B Rallye 150ST	Selles Dispensing Chemists Ltd.
	G–BFGC	Cessna F.150M	Northair Aviation Ltd.
	G–BFGD	Cessna F.172N-II	Clark Masts Ltd.
	G–BFGE	Cessna F.172N-II	Scottish Aero Club Ltd.
	G–BFGF	Cessna F.177RG	Landlink Two Ltd.
	G–BFGG	Cessna FRA.150M	Airwork Services Ltd.
	G–BFGH	Cessna F.337G	Barfax Distributing Co. Ltd.
	G–BFGI	Douglas DC-10-30	British Caledonian Airways *David Livingstone— The Scottish Explorer*
	G–BFGJ	Cameron D-96 hot-air airship	Cameron Balloons Ltd. *Sultan*
	G–BFGK	Jodel D.117	R. G. M. Bissett
	G–BFGL	Cessna FA.152	Yorkshire Flying Services Ltd.
	G–BFGM	Boeing 727-095	Dan-Air Services Ltd.
	G–BFGO	Fuji FA.200-160	Hillcrest Garage (Heseldon) Ltd.
	G–BFGP	D.H.C.-6 Twin Otter 310	Aurigny Air Services
	G–BFGR	R. Commander 685	Hamdani Investments Ltd.
	G–BFGS	SOCATA Rallye 180GT	P. A. Cairns
	G–BFGW	Cessna F.150H	South Humberside Flying Group Ltd.
	G–BFGX	Cessna FRA.150M	Airwork Services Ltd.
	G–BFGY	Cessna F.182Q	Oxford Controls Ltd.
	G–BFGZ	Cessna FRA.150M	Airwork Services Ltd.
	G–BFHD	C.A.S.A. C.352L (NB+AA)	Warbirds of Great Britain Ltd.
	G–BFHF	C.A.S.A. C.352L	Warbirds of Great Britain Ltd.
	G–BFHG	C.A.S.A. C.352L	Warbirds of Great Britain Ltd.
	G–BFHH	D.H.82A Tiger Moth	P. Harrison & M. J. Gambrell
	G–BFHI	Piper J-3C-65 Cub	J. M. Robinson
	G–BFHJ	Cessna FRA.150M	Airwork Services Ltd.
	G–BFHK	Cessna F.177RG-II	A. F. Hall
	G–BFHL	Cessna FRA.150M	Van Dusen Aircraft Supplies Co.
	G–BFHM	Steen Skybolt	W. R. Penaluna
	G–BFHN	Scheibe SF.25E Super Falke	W. J. Dyer & R. D. C. Hart
	G–BFHO	PA-31 Navajo	Willow Vale Electronics Ltd.
	G–BFHP	Champion 7GCAA Citabria	Buckminster Gliding Club Ltd.
	G–BFHR	Jodel DR.220/2+2	J. B. Keith & J. Bugg
	G–BFHS	AA-5B Tiger	J. E. Fazackerley
	G–BFHT	Cessna F.152	Riger Ltd.
	G–BFHU	Cessna F.152	Air Continental Securities Ltd.
	G–BFHV	Cessna F.152	P. J. Brown
	G–BFHX	Evans VP-1	D. F. Gibson

Reg.	Type	Owner or Operator	Notes
G-BFIB	PA-31-310 Turbo Navajo	Mann Aviation Ltd.	
G-BFIC	H.S.125 Srs. 600B	Cougar Aviation Ltd.	
G-BFID	Taylor JT.2 Titch Mk. III	W. F. Adams	
G-BFIE	Cessna FRA.150M	Rural Flying Club	
G-BFIF	Cessna FR.172XP	Air Navigation & Trading Co. Ltd.	
G-BFIG	Cessna FR.172XP	D. M. Balfour	
G-BFII	PA-23 Aztec 250E	European Ferries Ltd.	
G-BFIJ	AA-5A Cheetah	Goddard Aviation Ltd.	
G-BFIK	AA-5A Cheetah	Cabair Ltd.	
G-BFIL	AA-5A Cheetah	Goddard Aviation Ltd.	
G-BFIM	AA-5A Cheetah	D. E. Nixon & C. Heathcote	
G-BFIN	AA-5A Cheetah	Lupton Leasing	
G-BFIP	Wallbro Monoplane 1909 Replica	K. H. Wallis	
G-BFIR	Avro 652A Anson 21 (WD413)	G. M. K. Fraser	
G-BFIT	Thunder Ax6-56Z balloon	R. J. Nicholson & C. MacKinnon	
G-BFIU	Cessna FR.172XP	Fletcher Bakeries Ltd.	
G-BFIV	Cessna F.177RG	Murr International Constructions Ltd.	
G-BFIX	Thunder Ax7-77A balloon	E. Sawden Ltd.	
G-BFIY	Cessna F.150M	R. J. Ford-Sagers	
G-BFIZ	Cessna FT.337GP	Brotway Ltd.	
G-BFJA	AA-5B Tiger	G. W. Hind	
G-BFJB	Bell 212	British Airways Helicopters Ltd.	
G-BFJC	—	—	
G-BFJD	—	—	
G-BFJE	—	—	
G-BFJF	—	—	
G-BFJH	SA.102-5 Cavalier	B. F. J. Hope	
G-BFJI	Robin HR.100/250	H. Deville & C. J. Lear	
G-BFJJ	Evans VP-1	P. R. Pykett & B. J. Dyke	
G-BFJK	PA-23 Aztec 250E	CGH Managements Ltd.	
G-BFJM	Cessna F.152	Pegasus Aviation Ltd.	
G-BFJN	Westland-Bell 47G-3B1	A & B Agricultural Aviation Ltd.	
G-BFJO	G.164B Ag-Cat 450	Miller Aerial Spraying Ltd.	
G-BFJR	Cessna F.337G	John Roberts Services Ltd.	
G-BFJT	Westland-Bell 47G-3B1	GSM Helicopters Ltd.	
G-BFJU	Westland-Bell 47G-3B1	GSM Managements Ltd.	
G-BFJV	Cessna F.172H	H. C. Wilson	
G-BFJW	AB-206B JetRanger 2	Multi-Ownership Hotels Ltd.	
G-BFJZ	Robin DR.400/140B	Forge House Restaurant Ltd.	
G-BFKA	Cessna F.172N	C. Blackburn	
G-BFKB	Cessna F.172N	N. Denes Aerodrome Ltd.	
G-BFKC	Rand KR.2	K. K. Cutt	
G-BFKD	R. Commander 114B	K. W. Ford & S. R. Whitehead	
G-BFKF	Cessna FA.152	Klingair Ltd.	
G-BFKG	Cessna F.152	Western Air Training Ltd.	
G-BFKH	Cessna F.152	Western Air Training Ltd.	
G-BFKJ	PA-31-310 Navajo	Air Messenger Ltd.	
G-BFKL	Cameron N-56 balloon	Merrythought Toys Ltd. *Merrythought*	
G-BFKM	Westland-Bell 47G-3B1	A. & B. Agricultural Aviation Ltd.	
G-BFKN	PA-23 Aztec 250F	Air Envoy Ltd.	
G-BFKP	Partenavia P.68B	Twin Flight Ltd.	
G-BFKR	PA-24 Comanche 250	Conforex Ltd.	
G-BFKS	Cessna F.172L	Wycombe Air Centre Ltd.	
G-BFKT	Cessna F.172M	Wycombe Air Centre Ltd.	
G-BFKU	PA-25 Pawnee 235D	Farmair Ltd.	
G-BFKV	PA-25 Pawnee 235D	Moonraker Aviation Co. Ltd.	
G-BFKY	PA-34-200 Seneca	S.L.H. Construction Ltd.	
G-BFKZ	SA.330J Puma	Bristow Helicopters Ltd.	
G-BFLC	Cessna 210L	Jaserve Ltd.	
G-BFLD	Boeing 707-338C	British Midland Airways Ltd.	
G-BFLE	Boeing 707-338C	British Midland Airways Ltd.	
G-BFLH	PA-34-200T-2 Seneca	C.S.E. (Aircraft Services) Ltd.	
G-BFLI	PA-28R-201T Turbo Arrow III	Rite-Vent Ltd.	
G-BFLK	Cessna F.152	Gordon King (Aviation) Ltd.	
G-BFLL	H.S.748 Srs. 2A	Dan-Air Services Ltd.	
G-BFLM	Cessna 150M	Cornwall Flying Club Ltd.	
G-BFLN	Cessna 150M	J. G. Theobald	
G-BFLO	Cessna F.172M	W. A. Cook & K. Dando	
G-BFLP	Amethyst Ax6 balloon	K. J. Hendry *Amethyst*	
G-BFLR	Hiller UH-12E	Heliscot Ltd. & Major F. F. Chamberlain	

Notes	Reg.	Type	Owner or Operator
	G–BFLS	Westland-Bell 47G-3B1	Hon. M. H. Wills
	G–BFLT	—	—
	G–BFLU	Cessna F.152	Inverness Flying Services Ltd.
	G–BFLV	Cessna F.172N	Sir J. E. M. Conart
	G–BFLW	PA-39 Twin Comanche 160CR	Harris Aviation Services Ltd.
	G–BFLX	AA-5A Cheetah	Sheffield Auto Hire
	G–BFLZ	Beech 95-A55 Baron	K. K. Demel Ltd.
	G–BFMC	BAC One-Eleven 414	Ford Motor Co. Ltd.
	G–BFME	Cameron V-56 balloon	Warwick Balloons Ltd.
	G–BFMF	Cassutt Racer Mk. III	P. H. Lewis
	G–BFMG	PA-28-161 Warrior II	M. Abdel-Khalek
	G–BFMH	Cessna 177B	Span Aviation
	G–BFMJ	AA-5B Tiger	John Farbon & Co. Ltd.
	G–BFMK	Cessna FA.152	Denham Flying Training School Ltd.
	G–BFMM	PA-28-181 Archer II	Bristol & Wessex Aeroplane Club Ltd.
	G–BFMN	PA-28R-201T Turbo Arrow III	Adrian Alan Ltd.
	G–BFMR	PA-20 Pacer 135	S. B. Reay & ptnrs.
	G–BFMS	MS.893E Rallye 180GT	Mills Marketing Services Ltd.
	G–BFMT	Robin HR.200/100	Eagle Forms (GB) Ltd.
	G–BFMU	AA-5A Cheetah	Tentergate Trading
	G–BFMV	PA-28 Cherokee 160C	Overlantic Ltd.
	G–BFMW	V.735 Viscount	Inter City Airlines
	G–BFMX	Cessna F.172N	Bletchley Motor (Rentals) Ltd.
	G–BFMY	Sikorsky S-61N	Bristow Helicopters Ltd.
	G–BFMZ	Payne Ax6 balloon	G. F. Payne
	G–BFNB	PA-25 Pawnee 235D	A. G. Edwards
	G–BFNC	AS.350B Ecureuil	P. Pilkington & K. M. Armitage
	G–BFNE	C.A.S.A. I.131 Jungmann	J. Lipton
	G–BFNG	Jodel D.112	K. D. Bass
	G–BFNH	Cameron V-77 balloon	P. O. Atkins & R. Emms *Red Pepper*
	G–BFNI	PA-28-161 Warrior II	C.S.E. (Aircraft Services) Ltd.
	G–BFNJ	PA-28-161 Warrior II	C.S.E. (Aircraft Services) Ltd.
	G–BFNK	PA-28-161 Warrior II	C.S.E. (Aircraft Services) Ltd.
	G–BFNM	Globe GC.1 Swift	Nottingham Flying Group
	G–BFNP	Saffery S.330 balloon	N. H. Ponsford *Sidney*
	G–BFNU	BN-2A Islander	Brian G. Kelly Ltd.
	G–BFNV	BN-2A Islander	Loganair Ltd.
	G–BFOD	Cessna F.182Q	Staverton Flying School Ltd.
	G–BFOE	Cessna F.152	Northair Aviation Ltd.
	G–BFOF	Cessna F.152	Staverton Flying School Ltd.
	G–BFOG	Cessna 150M	Zincraft Ltd.
	G–BFOH	Westland-Bell 47G-3B1	Helicopter Hire Ltd.
	G–BFOI	Westland-Bell 47G-3B1	Helicopter Hire Ltd.
	G–BFOJ	AA-1 Yankee	P. A. MacCarty Ltd.
	G–BFOL	Beech A200 Super King Air	Bristow Helicopters Ltd.
	G–BFOM	PA-31-325 Navajo	Brew Bros. Ltd.
	G–BFON	PA-31-310 Navajo	Air Kilroe
	G–BFOP	Jodel D.120	H. Cope
	G–BFOR	Thunder Ax6-56Z balloon	Thunder Balloons Ltd.
	G–BFOS	Thunder Ax6-56A balloon	Thunder Balloons Ltd. *Milton Keynes*
	G–BFOT	Thunder Ax6-56A balloon	Thunder Balloons Ltd.
	G–BFOU	Taylor JT.1 Monoplane	G. Bee
	G–BFOV	Cessna F.172N	Gooda Walker Ltd.
	G–BFOW	Cessna F.172N	Lobby Ticketing Ltd.
	G–BFOX	D.H.83 Fox Moth	R. K. J. Hadlow
	G–BFOY	Cessna A.188B AgTruck	Plains Aerial Spraying Ltd.
	G–BFOZ	Thunder Ax6-56 balloon	Motorway Tyres *Motorway II*
	G–BFPA	Scheibe SF.25B Super Falke	Yorkshire Gliding Club (Pty) Ltd.
	G–BFPB	AA-5B Tiger	Seatoller Ltd.
	G–BFPC	AA-5B Tiger	K.K. Aviation
	G–BFPD	AA-5A Cheetah	Scotia Safari Ltd.
	G–BFPE	PA-28 Cherokee 140	Clacton Light Aviation Co.
	G–BFPF	Sikorsky S-61N	British Airways Helicopters Ltd.
	G–BFPH	Cessna F.172K	R. W. Dorch
	G–BFPI	H.S.125 Srs. 700B	McAlpine Aviation Ltd.
	G–BFPJ	Procter Petrel	S. G. Craggs
	G–BFPK	Beech A.23 Musketeer	Five Octa Flying Group
	G–BFPM	Cessna F.172M	Abbey Windows Ltd.
	G–BFPO	R. Commander 112B	Tillet Shipping Ltd.
	G–BFPP	Bell 47J-2	Avon Electrics (Wholesale) Ltd.
	G–BFPS	PA-25 Pawnee 235D	Bowker Aircraft Services Ltd.
	G–BFPX	Taylor JT.1 Monoplane	E. A. Taylor
	G–BFPY	PA-32 Cherokee Six 260	Air Service Training Ltd.

Reg.	Type	Owner or Operator	Notes
G–BFPZ	Cessna F.177RG	Partlease Ltd.	
G–BFRA	R. Commander 114	Glos-Air (Sales) Ltd.	
G–BFRB	Cessna F.152	W. Shyvers Ltd.	
G–BFRC	AA-5A Cheetah	Northgleam Ltd.	
G–BFRD	Bowers Flybaby 1A	F. R. Donaldson	
G–BFRE	AA-5B Tiger	Harlands of Hull Ltd.	
G–BFRF	Taylor JT.1 Monoplane	E. R. Bailey	
G–BFRI	Sikorsky S-61N	Bristow Helicopters Ltd.	
G–BFRJ	HPR-7 Herald 209	Express Air Services Ltd.	
G–BFRK	HPR-7 Herald 209	Express Air Services Ltd.	
G–BFRL	Cessna F.152	Westair Flying Services Ltd.	
G–BFRM	Cessna 550 Citation II	Marshall of Cambridge (Engineering) Ltd.	
G–BFRN	Cessna F.152	Rogers Aviation Sales Ltd.	
G–BFRO	Cessna F.150M	Skyviews & General Ltd.	
G–BFRP	Cessna F.150M	Transgap Ltd.	
G–BFRR	Cessna FRA.150M	Pegasus Aviation Ltd.	
G–BFRS	Cessna F.172N	Poplar Toys Ltd.	
G–BFRT	Cessna FR.172XP	English Country Magazines Ltd.	
G–BFRU	Cessna F.152	A. G. Chrismas Ltd.	
G–BFRV	Cessna FA.152	Rogers Aviation Ltd.	
G–BFRW	Agusta-Bell 47G-3B1	Autair Ltd.	
G–BFRX	PA-25 Pawnee 260	W. R. C. Wauchope	
G–BFRY	PA-25 Pawnee 260	Aero Cropcare Ltd.	
G–BFSA	Cessna F.182Q	Clark Masts Ltd.	
G–BFSB	Cessna F.152	R. M. Clarke	
G–BFSC	PA-25 Pawnee 235D	Farm Aviation Services Ltd.	
G–BFSD	PA-25 Pawnee 235D	Farm Aviation Services Ltd.	
G–BFSJ	Westland-Bell 47G-3B1	R.K.B. Leasing Services	
G–BFSK	*PA-23 Apache 160	Oxford Air Training School	
G–BFSL	Cessna U.206F Stationair	Range Air	
G–BFSN	Agusta-Bell 47G-2	Minster Helicopters Ltd.	
G–BFSO	H.S.125 Srs. 700B	Dravidian Air Service Ltd.	
G–BFSP	H.S.125 Srs. 700B	Dravidian Air Service Ltd.	
G–BFSR	Cessna F.150J	Aero Group 78	
G–BFSS	Cessna FR.172G	Minerva Services	
G–BFST	Partenavia P.68B	Air Oxford	
G–BFSY	PA-28-181 Archer II	K. F. Davison	
G–BFTA	PA-28-161 Warrior II	N. Clayton	
G–BFTC	PA-28R-201T Turbo Arrow III	R. I. S. A. Herbert	
G–BFTE	AA-5A Cheetah	B. Refson	
G–BFTF	AA-5B Tiger	F. C. Burrow Ltd.	
G–BFTG	AA-5B Tiger	Speedair Ltd.	
G–BFTH	Cessna F.172N	The Hamper People Ltd.	
G–BFTO	Rotorway Scorpion 133	Rotorway Helicopters Ltd.	
G–BFTR	Bell 206L Long Ranger	Air Hanson Ltd.	
G–BFTT	Cessna 421C	Comben Group Services Ltd.	
G–BFTU	Cessna FA.152	J. H. Wilmshurst Ltd.	
G–BFTW	PA-23 Aztec 250F	Southampton Airport Ltd.	
G–BFTX	Cessna F.172N	Westair Flying Services Ltd.	
G–BFTY	Cameron V-77 balloon	Regal Motors (Bilston) Ltd. *Regal Motors*	
G–BFTZ	MS.880B Rallye Club	R. & B. Legge Ltd.	
G–BFUB	PA-32RT-300 Turbo Lance II	Bumbles Ltd.	
G–BFUD	Scheibe SF.25E Super Falke	R. C. Bull	
G–BFUG	Cameron N-77 balloon	R. Sisterson *Blue Boar*	
G–BFUN	Hughes 269C	H. Downs & Sons (Huddersfield) Ltd.	
G–BFUO	PA-23 Aztec 250F	Portland Motor Supplies	
G–BFUS	Cessna 404 Titan	Euroair Transport Ltd.	
G–BFUZ	Cameron V-77 balloon	Skysales Ltd.	
G–BFVA	Boeing 737-204ADV	Britannia Airways Ltd. *Sir John Alcock*	
G–BFVB	Boeing 737-204ADV	Britannia Airways Ltd. *Sir Thomas Sopwith*	
G–BFVF	PA-38-112 Tomahawk	C.S.E. (Aircraft Services) Ltd.	
G–BFVG	PA-28-181 Archer II	J. M. Henderson	
G–BFVH	D.H.2 Replica (5964)	Leisure Sport Ltd.	
G–BFVI	H.S.125 Srs. 700B	Bristow Helicopters Ltd.	
G–BFVM	Westland-Bell 47G-3B1	Heliwork Ltd.	
G–BFVO	Partenavia P.68B	Anvil Aviation Ltd.	
G–BFVP	PA-23 Aztec 250	Dukes Transport (Craigavon) Ltd.	
G–BFVS	AA-5B Tiger	J. J. Mudford	
G–BFVU	Cessna 150L	IIP Associates	
G–BFVV	SA.365 Dauphin 2	Management Aviation Ltd.	

Notes	Reg.	Type	Owner or Operator
	G–BFVW	SA.365 Dauphin 2	Management Aviation Ltd.
	G–BFVX	Beech C90 King Air	Vernair Transport Services
	G–BFVY	Beech C90 King Air	Vernair Transport Services
	G–BFVZ	Beech A200 Super King Air	Executive Express
	G–BFWB	PA-28-161 Warrior II	C.S.E. (Aircraft Services) Ltd.
	G–BFWD	Currie Wot	F. E. Nuthall
	G–BFWE	PA-23 Aztec 250	Air Navigation & Trading Co. Ltd.
	G–BFWF	Cessna 421B	Alcon Oil Ltd.
	G–BFWG	R. Commander 112A	Alcon Oil Ltd.
	G–BFWK	PA-28-161 Warrior II	Woodgate Aviation Ltd.
	G–BFWL	Cessna F.150L	Eglinton Flying Club
	G–BFWM	Hiller UH-12D	Management Aviation Ltd.
	G–BFWW	Robin HR.100/210	Willingair Ltd.
	G–BFXC	Mooney M.20C	E. W. Passmore
	G–BFXD	PA-28-161 Warrior II	C.S.E. (Aircraft Services) Ltd.
	G–BFXE	PA-28-161 Warrior II	C.S.E. (Aircraft Services) Ltd.
	G–BFXF	Andreasson BA.4B	A. Brown
	G–BFXG	D.31 Turbulent	S. Griffin
	G–BFXH	Cessna F.152	Dayton Aerohire
	G–BFXI	Cessna F.172M	D. J. Stockley & P. S. Nicholson
	G–BFXK	PA-28 Cherokee 140	G. S. & Mrs. M. T. Pritchard
	G–BFXL	Albatross D.5A	Leisure Sport Ltd.
	G–BFXM	Jurca MJ.5 Sirocco	D. I. & W. A. Barker
	G–BFXN	PA-36-375 Brave	Farm Aviation Services Ltd.
	G–BFXO	Taylor JT.1 Monoplane	A. S. Nixon
	G–BFXR	Jodel D.112	P. E. Walker & M. Riddin
	G–BFXS	R. Commander 114	European Steel Sheets Ltd.
	G–BFXT	H.S.125 Srs. 700B	Coca Cola Export Corporation
	G–BFXU	American Beta Z Airship	G. Turnbull
	G–BFXV	Air Tractor AT-302	ADS Aerial Ltd.
	G–BFXW	AA-5B Tiger	Terry Giles Ltd.
	G–BFXX	AA-5B Tiger	Jones & Brooks Ltd.
	G–BFXY	AA-5A Cheetah	Cabair Ltd.
	G–BFXZ	PA-28-181 Archer II	Cavalair Ltd.
	G–BFYA	MBB Bo 105D	British Caledonian Helicopters Ltd.
	G–BFYB	PA-28-161 Warrior II	C.S.E. (Aircraft Services) Ltd.
	G–BFYC	PA-32RT-300 Lance II	Avingate Ltd.
	G–BFYD	PA-32RT-300T Turbo Lance II	Wells Timbercraft
	G–BFYE	Robin HR.100/285	Tollbridge Machine Tool Co. Ltd.
	G–BFYF	Westland-Bell 47G-3B1	H. J. Hofman
	G–BFYI	Westland-Bell 47G-3B1	Dollar Air Services Ltd.
	G–BFYJ	Hughes 369HE	Wilford Aviation Ltd.
	G–BFYL	Evans VP.2	A. G. Wilford
	G–BFYM	PA-28-161 Warrior II	C.S.E. (Aircraft Services) Ltd.
	G–BFYN	Cessna FA.152	Denham Flying Training School Ltd.
	G–BFYO	Spad XIII (replica)	Leisure Sport Ltd.
	G–BFYP	Benson B.7	A. J. Philpotts
	G–BFYU	SC.5 Belfast	HeavyLift Cargo Airlines Ltd.
	G–BFZA	Alpavia Fournier RF-3	T. J. Hartwell & D. R. Wilkinson
	G–BFZB	Piper J-3C-65 Cub	P. F. Ansell & ptnrs.
	G–BFZC	Sikorsky S-61N	Management Aviation Ltd.
	G–BFZD	Cessna FR.182Q	J. & R. Mac (Machine Tools) Ltd.
	G–BFZE	AS.350B Ecureuil	TBF Transport Ltd.
	G–BFZF	Boeing 707-321C	Scimitar Airlines Ltd.
	G–BFZG	PA-28-161 Warrior II	C.S.E. (Aircraft Services) Ltd.
	G–BFZH	PA-28R-200 Cherokee Arrow	R. R. Henderson
	G–BFZK	EMB-110P2 Bandeirante	Fairflight Ltd./Air Ecosse
	G–BFZL	V.836 Viscount	British Midland Airways Ltd.
	G–BFZM	R. Commander 112TC	Glos-Air Ltd.
	G–BFZN	Cessna FA.152	Leicestershire Aero Club Ltd.
	G–BFZO	AA-5A Cheetah	Northern Executive Aviation Ltd.
	G–BFZP	AA-5B Tiger	Scotia Safari Ltd.
	G–BFZS	Cessna F.152	R. T. Haddow
	G–BFZT	Cessna FA.152	Shirlstar Container Transport Ltd.
	G–BFZU	Cessna FA.152	W. D. & P. M. Jasper
	G–BFZV	Cessna F.172M	S. M. Rourke
	G–BGAA	Cessna 152 II	Solent Flight Centre
	G–BGAB	Cessna F.152 II	Humberside Aero Club
	G–BGAD	Cessna F.152 II	R. F. Howard
	G–BGAE	Cessna F.152 II	Tayside Aviation Ltd.
	G–BGAF	Cessna FA.152	E. P. Collier
	G–BGAG	Cessna F.172N	Medway Flying Group Ltd.

Reg.	Type	Owner or Operator	Notes
G–BGAH	FRED Srs. 2	G. A. Harris	
G–BGAI	Westland-Bell 47G-3B1	Dollar Air Services Ltd.	
G–BGAJ	Cessna F.182Q II	Channel Islands Aero Holdings Ltd.	
G–BGAK	Cessna F.182Q II	Northair Aviation Ltd.	
G–BGAL	Saffery S.330 balloon	A. M. Lindsay *English Lady*	
G–BGAP	Cessna F.182RG	B. D. Harris	
G–BGAS	Colting Ax8-105A balloon	British Gas Corporation	
G–BGAT	Douglas DC-10-30	British Caledonian Airways *James Watt – The Scottish Engineer*	
G–BGAU	Rearwin 9000L	Shipping & Airlines Ltd.	
G–BGAV	Rearwin 8135T	Shipping & Airlines Ltd.	
G–BGAW	Rotorway Scorpion 133	Rotorway Helicopters Ltd.	
G–BGAX	PA-28 Cherokee 140	D. E. Nicholl	
G–BGAY	Cameron O-77 balloon	Dante Balloon Group *Antonia*	
G–BGAZ	Cameron V-77 balloon	Cameron Balloons Ltd. *Silicon Chip*	
G–BGBA	Robin R.2100A	D. Faulkner	
G–BGBB	L.1011-385 TriStar 200	British Airways *The Lakeland Rose*	
G–BGBC	L.1011-385 TriStar 200	British Airways *The Shot Silk Rose*	
G–BGBE	Jodel DR.1050	D. G. Perry	
G–BGBF	D.31A Turbulent	R. Clark	
G–BGBG	PA-28-181 Archer II	Harlow Print Ltd.	
G–BGBH	PA-23 Aztec 250F	Bruce Harrison Ltd.	
G–BGBI	Cessna F.150L	Cambridge Technical Developments (Leasing) Ltd.	
G–BGBK	PA-38-112 Tomahawk	G. F. Nash	
G–BGBM	SIPA 903	P. G. Phelps	
G–BGBN	PA-38-112 Tomahawk	R. T. Dickson & I. Conway	
G–BGBO	Cessna F.172N	Renco (Aviation) Ltd.	
G–BGBP	Cessna F.152	Brailsford Aviation Ltd.	
G–BGBR	Cessna F.172N	Steer Aviation Ltd.	
G–BGBS	PA-23 Aztec 250	T. S. Grimshaw Ltd.	
G–BGBT	Partenavia P.68B	Cosalt Public Co. Ltd.	
G–BGBU	Auster AOP.9	K. J. Garrett	
G–BGBV	Slingsby T.65A Vega	R. Richards	
G–BGBW	PA-38-112 Tomahawk	C.S.E. (Aircraft Services) Ltd.	
G–BGBX	PA-38-112 Tomahawk	C.S.E. (Aircraft Services) Ltd.	
G–BGBY	PA-38-112 Tomahawk	Cheshire Flying Services Ltd.	
G–BGBZ	R. Commander 114	R. S. Fenwick	
G–BGCA	Slingsby T.65A Vega	E. C. Neighbour	
G–BGCB	Slingsby T.65A Vega	D. J. Dawson	
G–BGCC	PA-31-325 Navajo	Noble Aviation Ltd.	
G–BGCD	Cameron A-140 balloon	Skysales Ltd. *Maxim*	
G–BGCH	PA-38-112 Tomahawk	D. & J. Fountain-Barber	
G–BGCK	AA-5A Cheetah	G. W. Plowman & Son Ltd.	
G–BGCL	AA-5A Cheetah	Goddard Aviation	
G–BGCM	AA-5A Cheetah	Publishing Innovations Leasing Ltd.	
G–BGCN	AA-5A Cheetah	Goddard Aviation	
G–BGCO	PA-44-180 Seminole	Aviation Beauport Ltd.	
G–BGCS	EMB-110P1 Bandeirante	Genair	
G–BGCV	AS.350B Ecureuil	Valley of Gleneagles Helicopters Ltd.	
G–BGCX	Taylor JT.1 Monoplane	G. M. R. Walters	
G–BGCY	Taylor JT.1 Monoplane	R. L. A. Davies	
G–BGDA	Boeing 737-236	British Airways *River Tamar*	
G–BGDB	Boeing 737-236	British Airways *River Tweed*	
G–BGDC	Boeing 737-236	British Airways *River Humber*	
G–BGDD	Boeing 737-236	British Airways *River Tees*	
G–BGDE	Boeing 737-236	British Airways *River Avon*	
G–BGDF	Boeing 737-236	British Airways *River Thames*	
G–BGDG	Boeing 737-236	British Airways *River Medway*	
G–BGDH	Boeing 737-236	British Airways *River Clyde*	
G–BGDI	Boeing 737-236	British Airways *River Ouse*	
G–BGDJ	Boeing 737-236	British Airways *River Trent*	
G–BGDK	Boeing 737-236	British Airways *River Mersey*	
G–BGDL	Boeing 737-236	British Airways *River Don*	
G–BGDN	Boeing 737-236	British Airways *River Tyne*	
G–BGDO	Boeing 737-236	British Airways *River Usk*	
G–BGDP	Boeing 737-236	British Airways *River Taff*	
G–BGDR	Boeing 737-236	British Airways *River Bann*	
G–BGDS	Boeing 737-236	British Airways *River Severn*	
G–BGDT	Boeing 737-236	British Airways *River Forth*	
G–BGDU	Boeing 737-236	British Airways *River Dee*	
G–BGDV	—	British Airways	
G–BGDW	—	British Airways	

Notes	Reg.	Type	Owner or Operator
	G–BGDX	—	British Airways
	G–BGDY	—	British Airways
	G–BGDZ	—	British Airways
	G–BGEA	Cessna F.150M	R. L. Beverley
	G–BGEC	Cameron V-77 balloon	Central Garage (Southborough) Ltd. *Crikey*
	G–BGED	Cessna U.206F	Lubair (Transport Services) Ltd.
	G–BGEE	Evans VP-1	A. Morris
	G–BGEF	Jodel D.112	F. W. J. Ellis
	G–BGEH	Monnet Sonerai II	G. K. Penson
	G–BGEI	Baby Great Lakes	D. H. Greenwood
	G–BGEK	PA-38-112 Tomahawk	Cheshire Flying Services Ltd.
	G–BGEL	PA-38-112 Tomahawk	Cheshire Flying Services Ltd.
	G–BGEM	Partenavia P.68B	Link-Hampson Ltd.
	G–BGEN	D.H.C.-6 Twin Otter 310	Loganair Ltd.
	G–BGEO	PA-31-350 Navajo Chieftain	Christian Salveson (Cold Storage) Ltd.
	G–BGEP	Cameron D-38 balloon	Cameron Balloons Ltd.
	G–BGES	Currie Wot	J. Roberts
	G–BGET	PA-38-112 Tomahawk	J. V. & J. Leasing
	G–BGEV	PA-38-112 Tomahawk	J. Apthorp Aviation & Leasing Ltd.
	G–BGEW	Nord NC.854S	R. A. Yates
	G–BGEX	Brookland Mosquito 2	D. R. C. Pugh
	G–BGFA	—	—
	G–BGFB	—	—
	G–BGFC	Evans VP-2	R. W. Eastman & J. A. Jones
	G–BGFD	PA-32-300 Cherokee Six	Fairoaks Aviation Services Ltd.
	G–BGFE	—	—
	G–BGFF	FRED Srs. 2	G. R. G. Smith
	G–BGFG	AA-5A Cheetah	Cabair Ltd.
	G–BGFH	Cessna F.182Q	TBF (Transport) Ltd.
	G–BGFI	AA-5A Cheetah	Prestwick Flying Club
	G–BGFJ	Jodel D.9 Bebe	C. M. Fitton
	G–BGFK	Evans VP-1	D. Beaumont
	G–BGFM	Rollason-Luton Beta 4	G. H. C. Jiggins
	G–BGFN	PA-25 Pawnee 235	Management Farm Services Ltd.
	G–BGFS	Westland-Bell 47G-3B1	G. S. Mason
	G–BGFT	PA-34-200T-2 Seneca	C.S.E. (Aircraft Services) Ltd.
	G–BGFV	BN-2A Islander	BN (Bembridge) Ltd.
	G–BGFX	Cessna F.152	A. W. Fay
	G–BGGA	Bellanca 7GCBC Citabria	A. G. Forshaw
	G–BGGB	Bellanca 7GCBC Citabria	R. J. W. Wood
	G–BGGC	Bellanca 7GCBC Citabria	S. Dorset Engineering Co. (Weymouth) Ltd.
	G–BGGD	Bellanca 8GCBC Scout	Bristol & Gloucestershire Gliding Club
	G–BGGE	PA-38-112 Tomahawk	C.S.E. (Aircraft Services) Ltd.
	G–BGGF	PA-38-112 Tomahawk	C.S.E. (Aircraft Services) Ltd.
	G–BGGG	PA-38-112 Tomahawk	C.S.E. (Aircraft Services) Ltd.
	G–BGGI	PA-38-112 Tomahawk	C.S.E. (Aircraft Services) Ltd.
	G–BGGJ	PA-38-112 Tomahawk	C.S.E. (Aircraft Services) Ltd.
	G–BGGK	PA-38-112 Tomahawk	C.S.E. (Aircraft Services) Ltd.
	G–BGGL	PA-38-112 Tomahawk	C.S.E. (Aircraft Services) Ltd.
	G–BGGM	PA-38-112 Tomahawk	C.S.E. (Aircraft Services) Ltd.
	G–BGGN	PA-38-112 Tomahawk	C.S.E. (Aircraft Services) Ltd.
	G–BGGO	Cessna F.152	E. Midlands School of Flying Ltd.
	G–BGGP	Cessna F.152	E. Midlands School of Flying Ltd.
	G–BGGT	Zenith CH.200	P. R. M. Nind
	G–BGGU	Wallis WA–116R-R	K. H. Wallis
	G–BGGV	Wallis WA–120 Srs. 2	K. H. Wallis
	G–BGGW	Wallis WA–122	K. H. Wallis
	G–BGGX	AB-206B JetRanger 3	Willowbrook International Ltd.
	G–BGGY	AB-206B JetRanger 3	Willowbrook International Ltd.
	G–BGHA	Cessna F.152	M. D. Ward
	G–BGHC	Saffery Hot Pants Firefly balloon	H. C. Saffery *Petunia*
	G–BGHD	Saffery Helios Blister balloon	H. C. Saffery
	G–BGHE	Convair L-13A	J. Davis
	G–BGHF	Westland WG.30	Westland Helicopters Ltd.
	G–BGHG	AS.350B Ecureuil	McAlpine Helicopters Ltd.
	G–BGHI	Cessna F.152	Taxon Ltd.
	G–BGHJ	Cessna F.172N	P.A.C.K. Enterprises (Sussex) Ltd.
	G–BGHK	Cessna F.152	Solent Flight Centre

Reg.	Type	Owner or Operator	Notes
G-BGHL	GA-7 Cougar	Peacock Salt Ltd.	
G-BGHM	Robin R1180T	Mitchell Dewey Associates	
G-BGHN	Agusta-Bell 47G-3B1	M. H. Wills	
G-BGHO	Agusta-Bell 47G-3B1	M. H. Wills	
G-BGHP	Beech 76 Duchess	Mercian Flange Co. Ltd.	
G-BGHS	Cameron N-31 balloon	Balloon Stable Ltd.	
G-BGHT	Falconair F-12	T. K. Baillie	
G-BGHU	T-6G Harvard	Gladaircraft Ltd.	
G-BGHV	Cameron V-77 balloon	E. Davies	
G-BGHW	Thunder Ax8-90 balloon	Edinburgh University Balloon Group *James Tytler*	
G-BGHX	Chasle YC-12 Tourbillon	C. Clark	
G-BGHY	Taylor JT.1 Monoplane	J. Prowse	
G-BGHZ	FRED Srs. 2	T. A. Timms	
G-BGIA	Cessna 152 II	A. G. Chrismas Ltd.	
G-BGIB	Cessna 152 II	K. D. Wickenden	
G-BGIC	Cessna 172N	T. F. Turner	
G-BGID	Westland-Bell 47G-3B1	Maltin Aviation Ltd.	
G-BGIF	AS.350B Ecureuil	Cabair Ltd.	
G-BGIG	PA-38-112 Tomahawk	Apollo Leasing Ltd.	
G-BGIH	Rand KR-2	G. & D. G. Park	
G-BGII	PA-32-300 Cherokee Six	Rosefair Electronics Ltd.	
G-BGIJ	Cameron 0-77 balloon	H. P. Carlton	
G-BGIK	Taylor JT.1 Monoplane	J. H. Medforth	
G-BGIL	AS.350B Ecureuil	McAlpine Helicopters Ltd.	
G-BGIM	AS.350B Ecureuil	Lord Glendyne	
G-BGIN	PA-31-350 Navajo Chieftain	P. G. Roberts	
G-BGIO	Benson B.8M	C. G. Johns	
G-BGIP	Colt 56A balloon	Lipton Export Ltd.	
G-BGIS	Boeing 707-321C	Scimitar Airlines Ltd.	
G-BGIU	Cessna F.172H	Metro Equipment (Chesham) Ltd.	
G-BGIV	Bell 47G-5	Helicopter Farming Ltd.	
G-BGIW	Bell 47G-2	Autair Ltd.	
G-BGIX	H.295 Super Courier	Nordic Oil Services Ltd.	
G-BGIY	Cessna F.172N	Mercian Flange Co. Ltd.	
G-BGIZ	Cessna F.152	Telepoint Ltd.	
G-BGJA	Cessna FA.152	D. G. Crabtree	
G-BGJB	PA-44-180 Seminole	Capricorn Air Ltd.	
G-BGJD	—	—	
G-BGJE	Boeing 737-236	British Airtours Ltd. *Sandpiper*	
G-BGJF	Boeing 737-236	British Airtours Ltd. *Skylark*	
G-BGJG	Boeing 737-236	British Airtours Ltd. *Kingfisher*	
G-BGJH	Boeing 737-236	British Airtours Ltd. *Wren*	
G-BGJI	Boeing 737-236	British Airtours Ltd. *Swallow*	
G-BGJJ	Boeing 737-236	British Airtours Ltd. *Kestrel*	
G-BGJK	Boeing 737-236	British Airtours Ltd. *Firecrest*	
G-BGJL	Boeing 737-236	British Airtours Ltd. *Goldfinch*	
G-BGJM	Boeing 737-236	British Airtours Ltd. *Curlew*	
G-BGJN	—	—	
G-BGJO	—	—	
G-BGJP	—	—	
G-BGJR	—	—	
G-BGJS	—	—	
G-BGJT	—	—	
G-BGJU	Cameron V-56 Balloon	D. T. Watkins *Spoils*	
G-BGJV	H.S.748 Srs. 2B	British Aerospace	
G-BGJW	GA-7 Cougar	Trent Helicopters Ltd.	
G-BGJY	—	—	
G-BGJZ	Airborne Industries AB.400 balloon	Balloon Stable Ltd. *Adva II*	
G-BGKA	P.56 Provost T.1 (XF690)	D. W. Mickleburgh	
G-BGKB	SOCATA Rallye 110ST	Air Touring Services Ltd.	
G-BGKC	SOCATA Rallye 110ST	G. Breen	
G-BGKD	SOCATA Rallye 110ST	Air Touring Services Ltd.	
G-BGKE	BAC One-Eleven 539	British Airways *County of West Midlands*	
G-BGKF	BAC One-Eleven 539	British Airways *County of Stafford*	
G-BGKG	One-Eleven 539	British Airways *County of Warwick*	
G-BGKH	Cessna A.188B AgTruck	Plains Aerial Spraying Ltd.	
G-BGKI	Cessna A.188B AgTruck	Plains Aerial Spraying Ltd.	
G-BGKJ	MBB Bo 105C	North Scottish Helicopters Ltd.	
G-BGKK	Westland-Bell 47G-3B1	GSM Helicopters Ltd.	
G-BGKM	SA.365C Dauphin	Management Aviation Ltd.	

Notes	Reg.	Type	Owner or Operator
	G–BGKO	GY-20 Minicab	R. B. Webber
	G–BGKP	MBB Bo 105C	Management Aviation Ltd.
	G–BGKR	PA-28-161 Warrior II	J. E. Cannings
	G–BGKS	PA-28-161 Warrior II	C.S.E. Aviation Ltd.
	G–BGKT	Auster AOP.9	K. H. Wallis
	G–BGKU	PA-28R-201 Arrow III	E. Farr
	G–BGKV	PA-28R-201 Arrow III	G. E. Salter Industrial Enterprises Ltd.
	G–BGKW	Evans VP-1	I. W. Black
	G–BGKX	PA-38-112 Tomahawk	Flamingo Aviation Ltd.
	G–BGKY	PA-38-112 Tomahawk	C.S.E. Aviation Ltd.
	G–BGKZ	J/5F Aiglet Trainer	R. C. H. Hibberd
	G–BGLA	PA-38-112 Tomahawk	Cambrian Flying Club
	G–BGLB	Bede BD-5B	W. Sawney
	G–BGLD	Beech 76 Duchess	A. E. C. Cohen & D. R. Brown
	G–BGLE	Saffrey S.330 Balloon	C. J. Dodd & ptnrs.
	G–BGLF	Evans VP-1	E. F. Fryer
	G–BGLG	Cessna 152	Skyviews & General Ltd.
	G–BGLH	Cessna 152	M. A. Lenihan
	G–BGLI	Cessna 152	M. A. Lenihan
	G–BGLK	Monnet Sonerai II	G. L. Kemp & J. Beck
	G–BGLN	Cessna FA.152	Shoreham Flight Simulation Ltd.
	G–BGLO	Cessna F.172N	C. H. Slaughter & Co. Ltd.
	G–BGLR	Cessna F.152	Shirlstar Container Transport Ltd.
	G–BGLS	Baby Great Lakes	D. S. Morgan
	G–BGLW	PA-34-200 Seneca	Spooner Aviation Ltd.
	G–BGLX	Cameron N-56 balloon	Sara A. G. Williams
	G–BGLZ	Stits SA-3A Playboy	Air Executive Ltd.
	G–BGMA	D.31 Turbulent	G. C. Masterson
	G–BGMB	Taylor JT.2 Titch	E. M. Bourne
	G–BGMC	D.H.C.-6 Twin Otter 310	Brymon Aviation Ltd.
	G–BGMD	D.H.C.-6 Twin Otter 310	Brymon Aviation Ltd.
	G–BGME	SIPA S.903	M. Emery (G–BCML)
	G–BGMJ	GY-201 Minicab	H. P. Burrill
	G–BGMM	PA-28-181 Archer II	Allen Technical Services Ltd.
	G–BGMP	Cessna F.172G	Norvic Racing Engines Ltd.
	G–BGMR	GY-201 Minicab	T. J. D. Hodge & A. B. Holloway
	G–BGMS	Taylor JT.2 Titch	M. A. J. Spice
	G–BGMT	MS.894E Rallye 235GT	M. E. Taylor
	G–BGMU	Westland-Bell 47G-3B1	Alec Wortley Ltd.
	G–BGMV	Scheibe SF.25B Falke	Wolds Gliding Club Ltd.
	G–BGMW	Edgeley EA-7 Optica	Edgeley Aircraft Ltd.
	G–BGMX	Enstrom F-280C Shark	Barry Sheene Racing Ltd.
	G–BGNA	Short SD3-30	Loganair Ltd.
	G–BGND	Cessna F.172N	N. F. Duke
	G–BGNL	Hiway Super Scorpion powered hang glider	G. Breen
	G–BGNN	AA-5A Cheetah	Three H Aircraft Hire Ltd.
	G–BGNO	AA-5A Cheetah	Noscar Aviation
	G–BGNP	Saffery S.200 balloon	N. H. Ponsford
	G–BGNR	Cessna F.172N	A. Hamlin
	G–BGNS	Cessna F.172N	Wickwell (UK) Ltd.
	G–BGNT	Cessna F.152	Solent Flight Centre
	G–BGNU	Beech E90 King Air	Norwich Union Fire Insurance Ltd.
	G–BGNV	GA-7 Cougar	H. Snelson (Engineers) Ltd.
	G–BGNW	Boeing 737-219ADV	Britannia Airways Ltd. *George Stephenson*
	G–BGNX	PA-28RT-201T Turbo Arrow IV	Dickson Bros. Ltd.
	G–BGNZ	Cessna FRA.150L	Cinque Ports Flying Club
	G–BGOA	Cessna FR.182Q	The Forestry Commission
	G–BGOC	Cessna F.152	Latharp Ltd.
	G–BGOD	Colt 77A balloon	Aquarious Balloon School Ltd.
	G–BGOE	Beech 76 Duchess	Ackroyd Westwood & Associates Ltd.
	G–BGOF	Cessna F.152	Kingsmetal Ltd.
	G–BGOG	PA-28-161 Warrior II	Southern Air
	G–BGOH	Cessna F.182Q	Zonex Ltd.
	G–BGOI	Cameron O-56 balloon	Balloon Stable Ltd. *Skymaster*
	G–BGOJ	Cessna F.150L	J. Edwards
	G–BGOL	PA-28R-201T Turbo Arrow III	M. G. Tyrrell & Co. Ltd.
	G–BGOM	PA-31-310 Navajo	Chesterton Development Co. Ltd.

Reg.	Type	Owner or Operator	Notes
G-BGON	GA-7 Cougar	Planet Aviation Ltd.	
G-BGOO	Colt 56 SS balloon	British Gas Corporation	
G-BGOP	Dassault Falcon 20F	Datsun (UK) Ltd.	
G-BGOR	AT-6D Harvard III	P. Mercer	
G-BGOU	AT-6C Harvard IIA	A. P. Snell & ptnrs.	
G-BGOV	AT-6C Harvard IIA	Aces High Ltd.	
G-BGOX	PA-31-350 Navajo Chieftain	Anglo Scottish Air Parcels (*Operations suspended*)	
G-BGOY	PA-31-350 Navajo Chieftain	Anglo Scottish Air Parcels (*Operations suspended*)	
G-BGOZ	Westland-Bell 47G-3B1	GSM Helicopters	
G-BGPA	Cessna 182L	Arch Motors & Manufacturing Co. Ltd.	
G-BGPB	AT-16 Harvard IV (385)	A. G. Walker & R. Lamplough	
G-BGPC	D.H.C.-6 Twin Otter 310	Loganair Ltd.	
G-BGPD	Piper L-4H Cub	P. D. Whiteman	
G-BGPE	Thunder Ax6-56 balloon	C. Wolstenholme *Sergeant Pepper*	
G-BGPF	Thunder Ax6-56Z balloon	Thunder Balloons Ltd. *Pepsi*	
G-BGPG	AA-5B Tiger	Trehaven Trust Ltd.	
G-BGPH	AA-5B Tiger	Modern Equipment Services Ltd.	
G-BGPI	Plumb BGP-1	B. G. Plumb	
G-BGPJ	PA-28-161 Warrior II	Mervit Ltd.	
G-BGPK	AA-5B Tiger	Ann Green Manufacturing Co. Ltd.	
G-BGPL	PA-28-161 Warrior II	Burnthills Aviation Ltd.	
G-BGPM	Evans VP-2	T. G. Painter	
G-BGPN	PA-18-150 Super Cub	Moore Bros. (Builders) Ltd.	
G-BGPO	PA-25 Pawnee 235	CKS Air Ltd.	
G-BGPP	PA-25 Pawnee 235	Sprayfields (Scothern) Ltd.	
G-BGPS	Aero Commander 200D	K. Davison	
G-BGPT	Parker Teenie Two	P. N. Haigh	
G-BGPU	PA-28 Cherokee 140	Air Navigation & Trading Co. Ltd.	
G-BGPV	BN-2A Islander	BN (Bembridge) Ltd.	
G-BGPW	BN-2A Islander	BN (Bembridge) Ltd.	
G-BGPX	BN-2A Islander	BN (Bembridge) Ltd.	
G-BGPY	BN-2A Islander	BN (Bembridge) Ltd.	
G-BGPZ	MS.890A Rallye Commodore	M. J. Kirk	
G-BGRA	Taylor JT.2 Titch	J. R. C. Thompson	
G-BGRB	AA-5B Tiger	Lessnow Ltd.	
G-BGRC	PA-28 Cherokee 140	Snowdon Aviation Ltd.	
G-BGRE	Beech A200 Super King Air	Dowty Group Services Ltd.	
G-BGRF	Beech 95-58P Baron	Hitchins (Hatfield) Ltd.	
G-BGRG	Beech 76 Duchess	Lambson Holdings Ltd.	
G-BGRH	Robin DR.400 2+2	A. Taylor & ptnrs.	
G-BGRI	Jodel DR.1051	R. M. McEwan	
G-BGRJ	Cessna T.310R	C.C.M. (Raven) Ltd. & Gledhill Water Storage Ltd.	
G-BGRK	PA-38-112 Tomahawk	Goodwood Terrena Ltd.	
G-BGRL	PA-38-112 Tomahawk	Goodwood Terrena Ltd.	
G-BGRM	PA-38-112 Tomahawk	Goodwood Terrena Ltd.	
G-BGRN	PA-38-112 Tomahawk	Goodwood Terrena Ltd.	
G-BGRO	Cessna F.172M	Citation Flying Services Ltd.	
G-BGRP	—	—	
G-BGRR	PA-38-112 Tomahawk	Cormack (Aircraft Services) Ltd.	
G-BGRS	Thunder Ax7-77Z balloon	P. Hassall Ltd.	
G-BGRT	Steen Skybolt	R. C. Taverson	
G-BGRU	—	—	
G-BGRV	—	—	
G-BGRW	—	—	
G-BGRX	PA-38-112 Tomahawk	Flamingo Aviation Ltd.	
G-BGRZ	BN-2A Islander	Capital Aviation Sales (UK) Ltd.	
G-BGSA	SOCATA Rallye 150GT	G. A. Schulz & ptnrs.	
G-BGSB	P.56 Provost T.1 (WV494)	J. Powell	
G-BGSC	Ayres Thrush Commander S2R-T34/500	Shoreham Flight Simulation Ltd.	
G-BGSD	Pitts S-2A Special	Rothmans International Ltd.	
G-BGSE	Pitts S-2A Special	P. H. Meeson	
G-BGSF	Pitts S-2A Special	Rothmans International Ltd.	
G-BGSG	PA-44-180 Seminole	Spooner Aviation Ltd.	
G-BGSH	PA-38-112 Tomahawk	Apollo Leasing Ltd.	
G-BGSI	PA-38-112 Tomahawk	Burnthills Plant Hire Ltd.	
G-BGSJ	Piper J-3C-65 Cub	H. A. Bridgman	
G-BGSK	AA-5A Cheetah	G. W. Plowman & Son Ltd.	
G-BGSL	AA-5A Cheetah	R. M. de Garston	

Notes	Reg.	Type	Owner or Operator
	G–BGSM	SOCATA Rallye 150GT	Tyre & Tune
	G–BGSN	Enstrom F-28C-2	Nicelynn Ltd.
	G–BGSO	PA-31-310 Navajo	Cranbury Estates Ltd.
	G–BGSS	PA-38-112 Tomahawk	RB Aviation Ltd.
	G–BGST	Thunder Ax7-65 balloon	L. H. T. Large & ptnrs. *Eclipse*
	G–BGSV	Cessna F.172N	Francis Mander Aviation
	G–BGSW	Beech F33 Debonair	Summers Transport Ltd.
	G–BGSX	Cessna F.152	Salair
	G–BGSY	GA-7 Cougar	M. Harland & Son Ltd.
	G–BGSZ	GA-7 Cougar	Cabair Ltd.
	G–BGTA	Firebird Bunce B.500 balloon	S. J. Bunce
	G–BGTB	SOCATA TB.10 Tobago	Dukeries Aviation Ltd.
	G–BGTC	Auster AOP.9	P. J. Marsham
	G–BGTD	H.S.125 Srs. 700B	Rank Xerox (UK) Ltd.
	G–BGTE	R. Commander 114A	Fletcher Rentals Ltd.
	G–BGTF	PA-44-180 Seminole	Lincoln Street Motors Birmingham Ltd.
	G–BGTG	PA-23 Aztec 250	Pearson Aviation Ltd.
	G–BGTH	PA-23 Aztec 250F	Western Flying Services Ltd.
	G–BGTI	Piper J-3C-65 Cub	W. J. Clarke & A. C. Broad
	G–BGTJ	PA-28 Cherokee 180	Serendipity Aviation
	G–BGTK	Cessna FR.182RG	Kestrel Air Services Ltd.
	G–BGTL	GY-20 Minicab	A. K. Lang
	G–BGTM	Thunder Ax6-56Z balloon	Engineering Polymers Ltd.
	G–BGTN	—	—
	G–BGTO	—	—
	G–BGTP	Robin HR.100/210	A. E. James & P. Houghton
	G–BGTR	PA-28 Cherokee 140	Keenair Services Ltd.
	G–BGTS	PA-28 Cherokee 140	Keenair Services Ltd.
	G–BGTT	Cessna 310R	Air Atlantique
	G–BGTU	BAC One-Eleven 409	Turbo Union Ltd.
	G–BGTV	Boeing 737-2T5	Orion Airways Ltd.
	G–BGTW	Boeing 737-2T5	Orion Airways Ltd.
	G–BGTX	Jodel D.117	D. A. Davidson
	G–BGTY	Boeing 737-2Q8	Orion Airways Ltd.
	G–BGTZ	Westland-Bell 47G-3B1	GSM Helicopters Ltd.
	G–BGUA	PA-38-112 Tomahawk	Tamar Aviation
	G–BGUB	PA-32-300 Cherokee Six	J. Beckers & ptnrs.
	NOTE: The G-BGUx sequence will not be issued unless specifically requested.		
	G–BGUY	Cameron V-56 balloon	G. V. Beckwith
	G–BGVA	Cessna 414A	A. Masters Ltd.
	G–BGVB	Robin DR.315	R. J. Fray
	G–BGVC	—	—
	G–BGVE	CP1310-C3 Super Emeraude	B. Arnall
	G–BGVF	Colt 77A balloon	Hot Air Balloon Co. Ltd.
	G–BGVH	Beech 76 Duchess	Laura Ashley Ltd.
	G–BGVI	Cessna F.152	Bournemouth Flying Club
	G–BGVJ	PA-28 Cherokee 180	H. Devonish
	G–BGVK	PA-28-161 Warrior II	W. E. Wordworth
	G–BGVL	PA-38-112 Tomahawk	Cambrian Flying Club
	G–BGVM	WLS Cassutt IIM	J. T. Mirley
	G–BGVN	PA-28RT-201 Arrow IV	Essex Aviation Ltd.
	G–BGVP	Thunder Ax6-56Z balloon	Hot Air Balloon Co. Ltd.
	G–BGVR	Thunder Ax6-56Z balloon	Hot Air Balloon Co. Ltd.
	G–BGVS	Cessna F.172M	P.D.A. Aviation Ltd.
	G–BGVT	Cessna R.182RG	T. Hayselden (Doncaster) Ltd.
	G–BGVU	PA-28 Cherokee 180	Solair Ltd.
	G–BGVV	AA-5A Cheetah	Peter Turnbull (York) Ltd.
	G–BGVW	AA-5A Cheetah	G. W. Plowman & Son Ltd.
	G–BGVX	Cessna P.210N	Industrial Pharmaceutical Service Ltd.
	G–BGVY	AA-5B Tiger	N. L. Fryer
	G–BGVZ	PA-28-181 Archer II	McCullough Aviation Ltd.
	G–BGWA	GA-7 Cougar	Lough Erne Aviation Ltd.
	G–BGWC	Robin DR.400/180	E. F. Braddon
	G–BGWD	Robin HR.100/285	Hordell Engineering Ltd.
	G–BGWF	PA-18-150 Super Cub	Farmair Ltd.
	G–BGWG	PA-18-150 Super Cub	Farmair Ltd.
	G–BGWH	PA-18-150 Super Cub	Farmair Ltd.
	G–BGWI	Cameron V-56 balloon	Army Balloon Club
	G–BGWJ	Sikorsky S-61N	Bristow Helicopters Ltd.
	G–BGWM	PA-28-181 Archer II	Billbon Ltd.

G–APES (Top) V.953C Merchantman of Air Bridge Carriers

G–BGYU (Centre) Embraer EMB-110PI Bandeirante of Air UK

G–BICX (Bottom) Maule M5-235C Lunar Rocket/*S. G. Richards*

Reg.	Type	Owner or Operator	Notes
G–BGWN	PA-38-112 Tomahawk	Apollo Leasing Ltd.	
G–BGWO	Jodel D.112	G. Payne & A. Case	
G–BGWP	MBB Bo 105C	Management Aviation Ltd.	
G–BGWR	Cessna U.206A	Cecil Aviation Ltd.	
G–BGWS	Enstrom F-280C Shark	G. Firbank & N. M. Grimshaw	
G–BGWT	WS-58 Wessex 60 Srs. 1	Bristow Helicopters Ltd.	
G–BGWU	PA-38-112 Tomahawk	Burnthills Aviation Ltd.	
G–BGWV	Aeronca 7AC Champion	RFC Flying Group	
G–BGWW	PA-23 Aztec 250E	Chargewell Ltd.	
G–BGWX	—	—	
G–BGWY	Thunder Ax6-56Z balloon	Dinal (Car Care) UK Ltd.	
G–BGWZ	Eclipse Super Eagle	J. S. Long	
G–BGXA	Piper J-3C-65 Cub	P. E. & J. A. Bates	
G–BGXB	PA-38-112 Tomahawk	Cambrian Flying Club	
G–BGXC	SOCATA TB.10 Tobago	A. J. Halliday	
G–BGXD	SOCATA TB.10 Tobago	Selles Dispensing Chemists Ltd.	
G–BGXE	Douglas DC-10-30	Laker Airways	
G–BGXF	Douglas DC-10-30	Laker Airways	
G–BGXG	Douglas DC-10-30	Laker Airways	
G–BGXH	Douglas DC-10-30	Laker Airways *Florida Belle*	
G–BGXI	Douglas DC-10-30	Laker Airways	
G–BGXJ	Partenavia P.68B	Wickenby Aviation Ltd.	
G–BGXK	Cessna 310R	B. H. Turner & ptnrs.	
G–BGXL	Benson B.8MV	H. A. Bancroft-Wilson	
G–BGXN	PA-38-112 Tomahawk	Air Navigation & Trading Co. Ltd.	
G–BGXO	PA-38-112 Tomahawk	Travelwell Coach & Commercial Service Ltd.	
G–BGXP	Westland-Bell 47G-3B1	Specbridge Ltd.	
G–BGXR	Robin HR.200/100	J. H. Spanton	
G–BGXS	PA-28-236 Dakota	Debian Car Hire Ltd.	
G–BGXT	SOCATA TB.10 Tobago	County Aviation Ltd.	
G–BGXU	WMB-1 balloon	C. J. Dodd & ptnrs.	
G–BGXV	Piper J-3C-65 Cub	Anvils Flying Group	
G–BGXX	Jodel DR.1051M1	B. A. Parris	
G–BGXZ	Cessna FA.152	Renco (Aviation) Ltd.	
G–BGYB	—	—	
G–BGYE	—	—	
G–BGYG	PA-28-161 Warrior II	C.S.E. (Aircraft Services) Ltd.	
G–BGYH	PA-28-161 Warrior II	C.S.E. (Aircraft Services) Ltd.	
G–BGYJ	Boeing 737-204	Britannia Airways Ltd. *Sir Barnes Wallis*	
G–BGYK	Boeing 737-204	Britannia Airways Ltd. *R. J. Mitchell*	
G–BGYL	Boeing 737-204	Britannia Airways Ltd. *Jean Batten*	
G–BGYN	PA-18-150 Super Cub	A. G. Walker	
G–BGYP	GA-7 Cougar	Bambair & Nixon Aviation	
G–BGYR	H.S.125 Srs. 600B	British Aerospace	
G–BGYS	EMB-110P2 Bandeirante	Air UK	
G–BGYT	EMB-110P1 Bandeirante	Air UK	
G–BGYU	EMB-110P2 Bandeirante	Air UK	
G–BGYV	EMB-110P1 Bandeirante	Air UK	
G–BGYW	Hughes 269C	Minster Aviation Ltd.	
G–BGYZ	Hughes 269C	G. Hewson	
G–BGZA	ICA IS-28M2	F. A. Wright	
G–BGZC	C.A.S.A. 1.131 Jungmann	J. E. Douglas	
G–BGZE	PA-38-112 Tomahawk	Manchester School of Flying Ltd.	
G–BGZF	PA-38-112 Tomahawk	Shirlstar Container Transport Ltd.	
G–BGZG	PA-38-112 Tomahawk	Kilmartin Leasing & Finance	
G–BGZH	PA-38-112 Tomahawk	C.S.E. Aviation Ltd.	
G–BGZJ	PA-38-112 Tomahawk	RB Aviation Ltd.	
G–BGZK	Westland-Bell 47G-3B1	K. McDonald	
G–BGZL	Eiri PIK-20E	J. Hulme & ptnrs.	
G–BGZN	WMB.2 Windtracker balloon	S. R. Woolfries	
G–BGZO	M.S.880B Rallye Club	W. G. R. Wunderlich	
G–BGZP	D.H.C.-6 Twin Otter 310	Brymon Aviation Ltd.	
G–BGZR	Meagher Model balloon MK.1	S. C. Meagher	
G–BGZS	Keirs Heated Air Tube	M. N. J. Kirby	
G–BGZW	PA-38-112 Tomahawk	Nottingham School of Flying Ltd.	
G–BGZX	PA-32 Cherokee Six 260	R. H. R. Rue	
G–BGZY	Jodel D.120	P. J. Sebastian	
G–BGZZ	Thunder Ax6-56 balloon	J. M. Robinson	
G–BHAA	Cessna 152	Herefordshire Aero Club Ltd.	
G–BHAB	Cessna 152	Herefordshire Aero Club Ltd.	

Notes	Reg.	Type	Owner or Operator
	G–BHAC	Cessna A.152	Herefordshire Aero Club Ltd.
	G–BHAD	Cessna A.152	Shropshire Aero Club Ltd.
	G–BHAE	Cameron V-56 balloon	K. Morris
	G–BHAF	PA-38-112 Tomahawk	C.S.E. Aviation Ltd.
	G–BHAG	Scheibe SF.25E Super Falke	British Gliding Association
	G–BHAH	Sikorsky S-61N	Management Aviation Ltd.
	G–BHAI	Cessna F.152	Channel Islands Aero Holdings Ltd.
	G–BHAJ	Robin DR.400/180	Crocker Aviation Services
	G–BHAK	PA-28RT-201 Arrow IV	Distance No Object Ltd.
	G–BHAL	Rango Saffery S.200 SS	A. M. Lindsay *Anneky Panky*
	G–BHAM	Thunder Ax6-56 balloon	D. Sampson
	G–BHAO	Beech 76 Duchess	Eagle Aircraft Services Ltd.
	G–BHAR	Westland-Bell 47G-3B1	Warban Investments Ltd.
	G–BHAT	Thunder Ax7-77 balloon	C. P. Witter Ltd. *Witter*
	G–BHAU	B.175 Britannia 253F	Redcoat Air Cargo Ltd.
	G–BHAV	Cessna F.152	A. G. Chrismas Ltd.
	G–BHAW	Cessna F.172N	J. M. Boucher
	G–BHAX	Enstrom F-28C	Flairair
	G–BHAY	PA-28RT-201 Arrow IV	Chiltern Aviation
	G–BHBA	Campbell Cricket	S. M. Irwin
	G–BHBB	Colt 77C balloon	S. D. Bellew
	G–BHBC	GA-7 Cougar	Cabair Ltd. (G-MALA/G-BGCP)
	G–BHBD	Colt 77A balloon	Colt Balloons Ltd.
	G–BHBE	Westland Bell 47G-3B1	Heliwork Ltd.
	G–BHBF	Sikorsky S-76	Bristow Helicopters
	G–BHBG	PA-32R-300 Lance	Inkermann Motor Sales Ltd.
	G–BHBH	Cessna 550 Citation II	RTZ Services Ltd.
	G–BHBI	Mooney M.20J	Express Aviation Services Ltd.
	G–BHBJ	Cameron D-96 airship	Blickframe Ltd.
	G–BHBK	Viscount V-5 ballon	B. Hargraves & B. King
	G–BHBL	L.1011-385 TriStar 200	British Airways *The Red Ensign Rose*
	G–BHBM	L.1011-385 TriStar 200	British Airways *The Piccadilly Rose*
	G–BHBN	L.1011-385 TriStar 200	British Airways *The Fragrant Star Rose*
	G–BHBO	L.1011-385 TriStar 200	British Airways *The Morning Jewel Rose*
	G–BHBP	L.1011-385 TriStar 200	British Airtours Ltd. *Osprey*
	G–BHBR	L.1011-385 TriStar 200	British Airtours Ltd. *Golden Eagle*
	G–BHBS	PA-28RT-201T Turbo Arrow IV	Zipmaster Ltd.
	G–BHBT	MA.5 Charger	R. G. & C. J. Maidment
	G–BHBU	Westland Bell 47G-3B1 (Soloy)	Heliwork Ltd.
	G–BHBV	Westland Bell 47G-3B1 (Soloy)	Helicrops Ltd.
	G–BHBW	Westland Bell 47G-3B1	Heliwork Ltd.
	G–BHBX	Agusta Bell 47G-3B1	Heliwork Ltd.
	G–BHBY	Westland Bell 47G-3B1	Heliwork Ltd.
	G–BHBZ	Partenavia P.68B	C. A. Church Ltd.
	G–BHCB	AA-5A Cheetah	Huronair
	G–BHCC	Cessna 172M	Bevan Lynch Aviation Ltd.
	G–BHCD	—	—
	G–BHCE	Jodel D.112	G. F. M. Garner
	G–BHCF	WMB.2 Windtracker balloon	C. J. Dodd & ptnrs.
	G–BHCI	Jodel D.140	Kent Gliding Club Ltd.
	G–BHCJ	H.S.748 Srs. 2A	Dan-Air Services Ltd.
	G–BHCK	AA-5A Cheetah	Crispex (Foods) Ltd.
	G–BHCM	Cessna F.172H	The English Connection Ltd.
	G–BHCO	SOCATA TB.10 Tobago	Butler Heating Associates Ltd.
	G–BHCP	Cessna F.152	Neptune Securities Ltd.
	G–BHCS	Cameron N-77 ballon	Cameron Balloons Ltd.
	G–BHCT	PA-23 Aztec 250	Colt Transport Ltd.
	G–BHCW	PA-22 Tri-Pacer 150	Tredair
	G–BHCX	Cessna F.152	Light Planes (Lancs) Ltd.
	G–BHCZ	PA-38-112 Tomahawk	Capital Aviation Sales (UK) Ltd.
	G–BHDA	Shultz balloon	G. F. Fitzjohn
	G–BHDB	Maule M5-235 Lunar Rocket	Cleanacres Ltd.
	G–BHDD	V.668 Varsity T.1 (WL626)	Loughborough & Leicester Aircraft Museum
	G–BHDE	SOCATA TB.10 Tobago	G. Brookes
	G–BHDH	Douglas DC-10-30	British Caledonian Airways *Sir Walter Scott*
	G–BHDI	Douglas DC-10-30	British Caledonian Airways
	G–BHDJ	Douglas DC-10-30	British Caledonian Airways
	G–BHDK	*Boeing B-29A-BN (461748)	Imperial War Museum
	G–BHDM	Cessna F.152 II	Tayside Aviation Ltd.

Reg.	Type	Owner or Operator	Notes
G–BHDO	Cessna F.182Q II	Air Service Training Ltd.	
G–BHDP	Cessna F.182Q II	Air Service Training Ltd.	
G–BHDR	Cessna F.152 II	Blackpool Business Flying Syndicate	
G–BHDS	Cessna F.152 II	Tayside Aviation Ltd.	
G–BHDT	SOCATA TB.10 Tobago	Capel Aviation	
G–BHDU	Cessna F.152 II	P. N. Voysey	
G–BHDV	Cameron V-77 balloon	D. E. P. Price	
G–BHDW	Cessna F.152	A. G. Chrismas Ltd.	
G–BHDX	Cessna F.172N	R. J. D. Rimmer	
G–BHDZ	Cessna F.172N	Northair Aviation Ltd.	
G–BHEC	Cessna F.152	Vectstar Ltd.	
G–BHED	Cessna FA.152	P. Skinner	
G–BHEE	Partenavia P.68B	DK Aviation Services Ltd.	
G–BHEF	Partenavia P.68B	DK Aviation Services Ltd.	
G–BHEG	Jodel D.150	P. R. Underhill	
G–BHEH	Cessna 310G	P. D. Higgs	
G–BHEI	Cessna T.182RG	Northair Aviation Ltd.	
G–BHEJ	—	—	
G–BHEK	CP.1315C-3 Super Emeraude	D. B. Winstanley	
G–BHEL	Jodel D.117	L. L. Vickers	
G–BHEM	Benson B.8M	E. Kenny	
G–BHEN	Cessna FA.152	Leicestershire Aero Club Ltd.	
G–BHEO	Cessna F.182RG	Cosworth Engineering Ltd.	
G–BHEP	Cessna 172 RG Cutlass	Memec Systems Ltd.	
G–BHER	SOCATA TB.10 Tobago	W. R. M. Dury	
G–BHET	SOCATA TB.10 Tobago	Potissue Ltd.	
G–BHEU	Thunder Ax7-65 balloon	M. H. R. Govett *Polo Moche*	
G–BHEV	PA-28R Cherokee Arrow 200	C. A. Savile	
G–BHEW	Sopwith Triplane Replica (N5430)	The Hon. Patrick Lindsay	
G–BHEX	Colt 56A balloon	A. S. Dear & ptnrs. *Super Wasp*	
G–BHEY	Pterodactyl O.R.	High School of Hang Gliding Ltd.	
G–BHEZ	Jodel D.150	E. J. Horsfall	
G–BHFA	Pterodactyl O.R.	High School of Hang Gliding Ltd.	
G–BHFB	Pterodactyl O.R.	High School of Hang Gliding Ltd.	
G–BHFC	Cessna F.152	Trehaven Trust Ltd.	
G–BHFD	D.H.C.-6 Twin Otter 310	Loganair Ltd.	
G–BHFF	Jodel D.112	J. A. Scott	
G–BHFG	SNCAN SV-4C (45)	R. J. Godfrey	
G–BHFH	PA-34-200T-2 Seneca	Heltor Ltd.	
G–BHFI	Cessna F.152	The BAE (Warton) Flying Group	
G–BHFJ	PA-28RT-201T Turbo Arrow IV	C.S.E. Aviation Ltd.	
G–BHFK	PA-28-151 Warrior	Ilkeston Car Hire Ltd.	
G–BHFL	PA-28 Cherokee 180	R. & C. Lord	
G–BHFM	Murphy S.200 balloon	M. Murphy	
G–BHFN	Eiri PIK-20E-1	B. A. Eastwell	
G–BHFP	Eiri PIK-20E	H. C. Mackinnon & ptnrs.	
G–BHFR	Eiri PIK-20E-1	G. Mackie	
G–BHFS	Robin DR.400/180	Flair (Soft Drinks) Ltd.	
G–BHFU	Saffery S.330 balloon	N. H. Ponsford *Jennie Toyhill*	
G–BHFY	Beech 95-B58 Baron	Kebbell Holdings Ltd.	
G–BHFZ	Saffery S.200 balloon	D. Morris	
G–BHGA	PA-31-310 Navajo	Marmot Ltd.	
G–BHGB	Colt 77A balloon	Colt Balloons Ltd.	
G–BHGC	PA-18-150 Super Cub	Herefordshire Gliding Club Ltd.	
G–BHGF	Cameron V-56 balloon	I. T. & H. Seddon *Biggles*	
G–BHGG	Cessna F.172N	Tosair Ltd.	
G–BHGH	—	—	
G–BHGI	—	—	
G–BHGJ	Jodel D.120	B. Brooks	
G–BHGK	Sikorsky S-76	North Scottish Helicopters Ltd.	
G–BHGM	Beech 76 Duchess	Eagle Aircraft Services Ltd.	
G–BHGN	Evans VP-1	A. R. Cameron	
G–BHGO	PA-32 Cherokee Six 260	L. J. Steward	
G–BHGP	SOCATA TB.10 Tobago	Go-Plane Ltd.	
G–BHGR	Robin DR.315	Headcorn Flying School Ltd.	
G–BHGS	PA-31-350 Navajo Chieftain	Jersey European Airways	
G–BHGT	Beech B90 King Air	Air Continental Securities Ltd.	
G–BHGU	WMB.2 Windtracker balloon	I. D. Bamber & ptnrs.	
G–BHGV	Keirs captive ballon	K. J. Faulkoner	
G–BHGW	Colt 12A balloon	Colt Balloons Ltd.	
G–BHGX	Colt 56B balloon	S. Doyle *Vomibag*	

Notes	Reg.	Type	Owner or Operator
	G-BHGY	PA-28R Cherokee Arrow 200	Luca Marketing Ltd.
	G-BHGZ	Bucker Bu.131 Jungmann	Sywell Aviation Services Ltd.
	G-BHHA	EMB-110P1 Bandeirante	Loganair Ltd.
	G-BHHB	Cameron V-77 balloon	I. G. N. Franklin
	G-BHHC	—	—
	G-BHHD	—	—
	G-BHHE	Jodel DR.1051/M1	B. E. Lowe-Lauri
	G-BHHF	—	—
	G-BHHG	Cessna F.152	Northamptonshire School of Flying Ltd.
	G-BHHH	Thunder Ax7-65 balloon	Thunder Balloons Ltd.
	G-BHHI	Cessna F.152	J. G. Fairhurst
	G-BHHJ	Cessna F.152	Leicestershire Aero Club Ltd.
	G-BHHK	Cameron N-77 balloon	S. Bridge & ptnrs.
	G-BHHN	Cameron V-77 balloon	Itchen Valley Balloon Group
	G-BHHO	PA-28 Cherokee 180	Peter Clifford Aviation Ltd.
	G-BHHP	Cameron 0-105 balloon	A. J. Machinery
	G-BHHR	Robin DR.400/180R	R. Jones
	G-BHHX	Jodel D.112	C. F. Walter
	G-BHHY	G-164 Ag Cat D	Miller Aerial Spraying Ltd.
	G-BHHZ	Rotorway Scorpion 133	P. A. Gunn & D. Willingham
	G-BHIA	Cessna F.152	Leicestershire Aero Club Ltd.
	G-BHIB	Cessna F.182Q	Northair Aviation Ltd.
	G-BHIC	Cessna F.182Q	General Building Services Ltd.
	G-BHID	SOCATA TB.10 Tobago	W. B. Pinckney & Sons Farming Co. Ltd.
	G-BHIF	Colt 160A balloon	Colt Balloons Ltd.
	G-BHIG	Colt 31A balloon	Colt Balloons Ltd.
	G-BHIH	Cessna F.172N	Watkiss Group Aviation Ltd.
	G-BHII	Cameron V-77 balloon	R. M. Davies
	G-BHIJ	Eiri PIK-20E-1	R. W. Hall & ptnrs.
	G-BHIK	Adam-RA-14 Loisirs	R. A. & J. Yates
	G-BHIL	PA-28-161 Warrior II	Simulated Flight Training Ltd.
	G-BHIM	Jodel D.112	G. P. Badham
	G-BHIN	Cessna F.152	Kestrel A Services
	G-BHIP	Thunder Ax3-1 balloon	Thunder Balloons Ltd.
	G-BHIR	PA-28R Cherokee Arrow 200	J. P. Aviation
	G-BHIS	Thunder Ax7-65 balloon	D. Clark *Yo-Yo*
	G-BHIT	SOCATA TB.9 Tampico	Prices Supreme Coaches
	G-BHIV	AS.350B Ecureuil	Marley Tile Co. Ltd.
	G-BHIW	—	—
	G-BHIX	Cameron V-56 balloon	P. S. Richardson
	G-BHIY	Cessna F.150K	W. H. Cole
	G-BHIZ	PA-31 Navajo	L. J. Steward
	G-BHJA	Cessna A.152	Capital Aviation Sales Ltd.
	G-BHJB	Cessna A.152	Capital Aviation Sales Ltd.
	G-BHJE	Beech 95-58P Baron	Warwick & Warwick (Philately) Ltd.
	G-BHJF	SOCATA TB.10 Tobago	G. S. Goodsir & ptnrs.
	G-BHJG	Cessna 172RG Cutlass	A. T. Gibbs
	G-BHJI	Mooney 201	T. R. Bamber & B. Refson
	G-BHJK	Maule M5-235C Lunar Rocket	A. J. Machinery Ltd.
	G-BHJL	—	—
	G-BHJN	Fournier RF-4D	G. G. Milton
	G-BHJO	PA-28-161 Warrior II	Nairn Flying Services Ltd.
	G-BHJP	Partenavia P-68C	Shirlstar Container Transport Ltd.
	G-BHJR	Saffery S.200 balloon	P. N. S. Bussey
	G-BHJS	Partenavia P-68B	Fairoaks Aviation Services Ltd.
	G-BHJU	Robin DR.400/2+2	Harlow Transport Services Ltd.
	G-BHJV	Cameron Rectangular SS 60 balloon	R. J. Reynolds Tobacco Co. Ltd.
	G-BHJW	Cessna F.152	Leicestershire Aero Club Ltd.
	G-BHJY	EMB-110P1 Bandeirante	Jersey European Airways
	G-BHJZ	EMB-110P1 Bandeirante	Jersey European Airways
	G-BHKA	Evans VP-1	M. L. Perry
	G-BHKB	Westland-Bell 47G-3B1 (Soloy)	Helicrops Ltd.
	G-BHKC	Westland-Bell 47G-3B1 (Soloy)	Heliwork Finance Ltd.
	G-BHKD	Westland-Bell 47G-3B1	Heliwork Ltd.
	G-BHKE	Benson B.8MV	V. C. Whitehead
	G-BHKG	Cessna F.172N	Telepoint
	G-BHKH	Cameron 0-65 balloon	D. G. Body
	G-BHKI	Cessna 402C	Evergreen Holdings Ltd.

Reg.	*Type*	*Owner or Operator*	*Notes*
G–BHKJ	Cessna 421C	Northair Aviation Ltd.	
G–BHKL	Colt Flying Bottle 1A balloon	Hot-Air Balloon Co. Ltd.	
G–BHKM	Colt 12A balloon	Hot-Air Balloon Co. Ltd.	
G–BHKN	Colt 12A balloon	Hot-Air Balloon Co. Ltd.	
G–BHKO	Colt 12A balloon	Hot-Air Balloon Co. Ltd.	
G–BHKP	Colt 12A balloon	Hot-Air Balloon Co. Ltd.	
G–BHKR	Colt 12A balloon	Hot-Air Balloon Co. Ltd.	
G–BHKS	Beech E90 King Air	The Plessey Co. Ltd.	
G–BHKT	Jodel D.112	G. V. Harfield & J. R. Nutter	
G–BHKU	AA-5A Cheetah	Crispex (Foods) Ltd.	
G–BHKV	AA-5A Cheetah	H. Snelson Engineers Ltd.	
G–BHKW	Westland-Bell 47G-3B1	Specbridge Ltd.	
G–BHKX	Beech 76 Duchess	Eagle Aircraft Services Ltd.	
G–BHKY	Cessna 310R II	Airwork Ltd.	
G–BHLA	Cessna 421C	Daleview Investments Ltd.	
G–BHLD	—	—	
G–BHLE	Robin DR.400/180	V-G Instruments Ltd.	
G–BHLF	H.S.125 Srs. 700B	The Marconi Co. Ltd.	
G–BHLG	Sinnett X3-4 Windrifter balloon	D. Sinnett	
G–BHLH	Robin DR.400/180	Trinecare Ltd.	
G–BHLI	R. Turbo Commander 690B	Flightline International Ltd.	
G–BHLJ	Saffery-Rigg S.200 balloon	I. A. Rigg	
G–BHLK	GA-7 Cougar	Cougar Flying Group	
G–BHLL	Cessna 421C	Northair Aviation Ltd.	
G–BHLM	Cessna 421C	Northair Aviation Ltd.	
G–BHLN	Cessna 441	Career Care Group Ltd.	
G–BHLO	Cessna 441	McAlpine Aviation Ltd.	
G–BHLP	Cessna 441	Northair Aviation Ltd.	
G–BHLT	D.H.82A Tiger Moth	R. L. Godwin	
G–BHLU	Fournier RF-3	G. G. Milton	
G–BHLV	CP.301A Emeraude	G. G. Milton	
G–BHLW	Cessna 120	Wichita Flying Group	
G–BHLX	AA-5B Tiger	Tiger Aviation (Jersey) Ltd.	
G–BHLY	Sikorsky S-76A	Bristow Helicopters Ltd.	
G–BHLZ	GY-30 Supercab	M. Hutton	
G–BHMA	SIPA 903	Fairwood Flying Club	
G–BHMC	M.S.880B Rallye Club	Trago Stadium Ltd.	
G–BHMD	Rand KR-2	W. D. Francis	
G–BHME	WMB.2 Windtracker balloon	I. R. Bell & ptnrs.	
G–BHMF	Cessna FA.152	Denham Flying Training School Ltd.	
G–BHMG	Cessna FA.152	Denham Flying Training School Ltd.	
G–BHMH	Cessna FA.152	Denham Flying Training School Ltd.	
G–BHMI	Cessna F.172N	Howard, Richard & Co. Ltd.	
G–BHMJ	Avenger T.200-2112 balloon	R. Light *Lord Anthony 1*	
G–BHMK	Avenger T.200-2112 balloon	P. Kinder *Lord Anthony 2*	
G–BHML	Avenger T.200-2112 balloon	L. Caulfield *Lord Anthony 3*	
G–BHMM	Avenger T.200-2112 balloon	M. Murphy *Lord Anthony 4*	
G–BHMN	SOCATA TB.10 Tobago	Cavendish Aviation Ltd.	
G–BHMO	PA-20M Cerpa Special (Pacer)	N. J. Mathias & R. J. Maxey	
G–BHMR	Stinson 108-3	J. R. Rowell	
G–BHMS	PA-34-200T-2 Seneca	BHF Petroleum Ltd.	
G–BHMT	Evans VP-1	P. E. J. Sturgeon	
G–BHMU	Colt 21A balloon	Hot Air Balloon Co. Ltd.	
G–BHMV	Bell 206A JetRanger	Bristow Helicopters Ltd.	
G–BHMW	F.27 Friendship Mk. 200	Air UK	
G–BHMX	F.27 Friendship Mk. 200	Air UK	
G–BHMY	F.27 Friendship Mk. 200	Air UK	
G–BHMZ	F.27 Friendship Mk. 200	Air UK	
G–BHNA	Cessna F.152	W. E. B. Wordsword	
G–BHNC	Cameron O-65 balloon	D. & C. Bareford	
G–BHND	Cameron N-65 balloon	Hunter & Sons (Mells) Ltd.	
G–BHNE	Boeing 727-2J4	Dan-Air Services Ltd.	
G–BHNF	Boeing 727-2J4	Dan-Air Services Ltd.	
G–BHNG	PA-23 Aztec 250	Southern Wings Ltd.	
G–BHNH	Cessna 404 Titan	Northair Aviation Ltd.	
G–BHNI	Cessna 404 Titan	Northair Aviation Ltd.	
G–BHNJ	—	—	
G–BHNK	Jodel D.120A	G. N. Smith	
G–BHNL	Jodel D.112	R. North & J. Ware	
G–BHNM	PA-44-180 Seminole	Lincoln Street Motors (Birmingham) Ltd.	
G–BHNN	PA-32R-301 Saratoga SP	Diagold Ltd.	
G–BHNO	PA-28-181 Archer II	Davison Plant Hire Co.	

Notes	Reg.	Type	Owner or Operator
	G–BHNP	Eiri PIK-20E-1	Astley's Ltd.
	G–BHNR	Cameron N-77 balloon	Bath University Hot-Air Balloon Club
	G–BHNS	PA-28R-201 Arrow III	Cluebond Ltd.
	G–BHNT	Cessna F.172N	Kestrel Air Services Ltd.
	G–BHNU	Cessna F.172N	B. Swindell (Haulage) Ltd.
	G–BHNV	Westland-Bell 47G-3B1	Specbridge Ltd.
	G–BHNW	—	—
	G–BHNX	Jodel D.117	J. A. Boyd
	G–BHNY	Cessna 425	McAlpine Aviation Ltd.
	G–BHNZ	AA-5B Tiger	Rosemount Aviation
	G–BHOA	Robin DR.400/160	M. F. Bunn
	G–BHOB	Cessna 404	Euroair Transport Ltd.
	G–BHOC	R. Commander 112A	Tuscany Ltd.
	G–BHOD	—	—
	G–BHOE	—	—
	G–BHOF	Sikorsky S-61N	Bristow Helicopters Ltd.
	G–BHOG	Sikorsky S-61N	Bristow Helicopters Ltd.
	G–BHOH	Sikorsky S-61N	Bristow Helicopters Ltd.
	G–BHOI	Westland-Bell 47G-3B1	Helicopter Hire Ltd.
	G–BHOJ	Colt 12A balloon	Hot Air Balloon Co. Ltd.
	G–BHOL	Jodel DR.1050	B. D. Deubelbeiss
	G–BHOM	PA-18 Super Cub 95	W. J. C. Scrope
	G–BHON	Rango NA.6 balloon	Rango Kite Co. *Rosamund Hine*
	G–BHOO	Thunder Ax7 balloon	D. Livesey & J. M. Purves *Scraps*
	G–BHOP	Thunder Ax3 balloon	B. W. A. Browitt
	G–BHOR	PA-28-161 Warrior II	A. Rifat
	G–BHOT	—	—
	G–BHOU	Cameron V-65 balloon	D. I. Gray-Fisk
	G–BHOV	Partenavia P.68C	Fosters Shopfitters (Southern) Ltd.
	G–BHOW	Beech 95-58P Baron	Anglo-African Machinery Ltd.
	G–BHOZ	SOCATA TB.9 Tampico	J. R. Bone
	G–BHPJ	Eagle Microlite	G. Breen
	G–BHPK	Piper J-3C-65 Cub	J. A. Verlander
	G–BHPL	C.A.S.A. 1.131 Jungmann	M. G. Jefferies
	G–BHPM	PA-18 Super Cub 95	P. Morgans
	G–BHPN	Colt 12A balloon	Colt Balloons Ltd.
	G–BHPO	Colt 12A balloon	Colt Balloons Ltd.
	G–BHPS	Jodel D.120A	C. J. & S. E. Francis
	G–BHPT	Piper J-3C-65 Cub	Sywell Aviation Services Ltd.
	G–BHPU	Sikorsky S-61N	British Caledonian Helicopters Ltd.
	G–BHPV	Cessna U.206G	Balfour Beatty Construction Ltd.
	G–BHPX	Cessna 152	A. G. Chrismas Ltd.
	G–BHPY	Cessna 152	Ashdown Publishing Ltd.
	G–BHPZ	Cessna 172N	O'Brian Properties Ltd.
	G–BHRA	R. Commander 114A	Rockwell Aviation Ltd.
	G–BHRB	Cessna F.152	Light Planes (Lancashire) Ltd.
	G–BHRC	PA-28-161 Warrior II	Sherwood Flying Club Ltd.
	G–BHRD	D.H.C.1 Chipmunk 22 (WP977)	B. C. Heywood & ptnrs.
	G–BHRE	Persephone S.200 balloon	Cupro-Sapphire Ltd.
	G–BHRF	Airborne Industries AB400 gas balloon	Balloon Stable Ltd.
	G–BHRG	—	—
	G–BHRH	Cessna FA.150K	Citation Flying Services Ltd.
	G–BHRI	Can-Can S.200 balloon	Cupro-Sapphire Ltd.
	G–BHRK	Colt Saucepan balloon	Colt Balloons Ltd.
	G–BHRL	Thunder Ax6-56Z balloon	Thunder Balloons Ltd.
	G–BHRM	Cessna F.152	Seatoller Ltd.
	G–BHRN	Cessna F.152	Channel Islands Aero Holdings (Jersey) Ltd.
	G–BHRO	R. Commander 112A	John Raymond Transport Ltd.
	G–BHRP	PA-44-180 Seminole	Rosemount Aviation
	G–BHRR	CP.301A Emeraude	W. H. Cole
	G–BHRS	ICA IS-28M2	British Aerospace Aircraft Group
	G–BHRU	Petunia S.1000 balloon	Cupro-Sapphire Ltd.
	G–BHRV	Mooney M.20J	Tecnovil Equipamentos Industriales
	G–BHRW	Jodel DR.221	J. M. Bisco & ptnrs.
	G–BHRX	Cessna 340A	Martin Butler Associates Ltd.
	G–BHRY	Colt 56A balloon	Hot Air Balloon Co. Ltd.
	G–BHSA	Cessna 152	Skyviews & General Ltd.
	G–BHSB	Cessna 172N	W. R. Craddock & Son Ltd.
	G–BHSC	—	—
	G–BHSD	Scheibe SF.25E Super Falke	Lasham Gliding Soc. Ltd.

Reg.	Type	Owner or Operator	Notes
G–BHSE	R. Commander 114	B.C.C. (Decorations) Ltd.	
G–BHSF	AA-5A Cheetah	C. A. Tour Productions Ltd. (G–BHAS)	
G–BHSG	AB-206A JetRanger	R. J. Anderson-Brown	
G–BHSI	Jodel D.9	J. H. Betton	
G–BHSJ	CP.301A Emeraude	G. Weale	
G–BHSL	C.A.S.A. 1.131 Jungmann	Cotswold Flying Group	
G–BHSM	AB-206B JetRanger 2	Dollar Air Services Ltd.	
G–BHSN	Cameron N-56 balloon	Ballooning Endeavours Ltd.	
G–BHSO	PA-23 Aztec 250	Truman Aviation Ltd.	
G–BHSP	Thunder Ax7-77Z balloon	Chicago Instruments Ltd.	
G–BHSR	—	—	
G–BHSS	Pitts S-1C Special	J. Eisdon-Davies	
G–BHST	Hughes 369D	Abbey Hill Vehicle Services	
G–BHSU	H.S.125 Srs. 700B	Shell Aircraft Ltd.	
G–BHSV	H.S.125 Srs. 700B	Shell Aircraft Ltd.	
G–BHSW	H.S.125 Srs. 700B	Shell Aircraft Ltd.	
G–BHSX	—	—	
G–BHSY	Jodel DR.1050	S. R. Orwin & T. R. Allebone	
G–BHSZ	Cessna 152	Millet Shipping Ltd.	
G–BHTA	PA-28-236 Dakota	R. Price	
G–BHTB	Monnet Sonerai II	R. H. Thring	
G–BHTC	Jodel DR.1050/M1	T. A. Carpenter	
G–BHTD	Cessna T.188C AgHusky	Dallah-ADS Ltd.	
G–BHTF	Enstrom F-28C	Southern Air	
G–BHTG	Thunder Ax6-56 balloon	F. R. & Mrs. S. H. MacDonald	
G–BHTH	T-6G Texan (2807)	Keenair Services Ltd.	
G–BHTI	SA.102.5 Cavalier	R. Cochrane	
G–BHTK	D.H.C.-6 Twin Otter 310	Loganair Ltd.	
G–BHTM	Cameron Can SS balloon	BP Oil Ltd.	
G–BHTO	—	—	
G–BHTP	PA-31T-500 Cheyenne I	C.S.E. Aviation Ltd.	
G–BHTR	Bell 206B JetRanger 3	J. S. Bloor Ltd.	
G–BHTS	—	—	
G–BHTT	Cessna 500 Citation	Lucas Industries Ltd.	
G–BHTU	—	—	
G–BHTV	Cessna 310R	Air Atlantique Ltd.	
G–BHTW	Cessna FR.172J	Apollo Manufacturing (Derby) Ltd.	
G–BHTZ	AA-5A Cheetah	W. H. Wilkins Ltd.	
G–BHUA	Douglas C-47	Aces High Ltd.	
G–BHUB	Douglas C-47	Aces High Ltd.	
G–BHUE	Jodel DR.1050	G. A. Mason	
G–BHUF	PA-18-150 Super Cub	T. A. McMullin	
G–BHUG	Cessna 172N	B. Giles	
G–BHUH	Cremer PC.14 balloon	P. A. Cremer	
G–BHUI	Cessna 152	Three Counties Aero Club Ltd.	
G–BHUJ	Cessna 172N	Three Counties Aero Club Ltd.	
G–BHUL	Beech E90 King Air	Airmore Aviation Ltd. (G–BBKM)	
G–BHUM	D.H.82A Tiger Moth	S. G. Towers	
G–BHUN	PZL-104 Wilga 35	W. Radwanski	
G–BHUO	Evans VP-2	R. A. Povall	
G–BHUP	Cessna F.152	Light Planes (Lancs) Ltd.	
G–BHUR	Thunder Ax3 balloon	B. F. G. Ribbons	
G–BHUU	PA-25 Pawnee 235	Farmwork Services (Eastern) Ltd.	
G–BHUV	PA-25 Pawnee 235	Farmwork Services (Eastern) Ltd.	
G–BHUW	Boeing N2S-5 Kaydet	Willrow Products Co. Ltd.	
G–BHUX	—	—	
G–BHUZ	—	—	
G–BHVA	—	—	
G–BHVB	PA-28-161 Warrior II	R. C. Hammond	
G–BHVC	Cessna 172RG Cutlass	Ian Willis Publicity Ltd.	
G–BHVE	Saffery S.330 balloon	P. M. Randles	
G–BHVF	Jodel D.150A	C. A. Parker	
G–BHVG	Boeing 737-2T5	Orion Airways Ltd.	
G–BHVH	Boeing 737-2T5	Orion Airways Ltd.	
G–BHVI	Boeing 737-2T5	Orion Airways Ltd.	
G–BHVM	Cessna 152	Three Counties Aero Club Ltd.	
G–BHVN	Cessna 152	Three Counties Aero Club Ltd.	
G–BHVP	Cessna 182Q	F. & S. E. Horridge	
G–BHVR	Cessna 172N	Chartered Trust Ltd.	
G–BHVS	Enstrom F-28A	Spooner Aviation Ltd.	
G–BHVT	Boeing 727-212	Dan-Air Services Ltd.	
G–BHVU	Cessna 414A	Stewart Air Ltd.	

Notes	Reg.	Type	Owner or Operator
	G–BHVV	Piper J-3C-65 Cub	I. S. Hodge
	G–BHVY	AA-5B Tiger	Amex Automobile Exports Ltd.
	G–BHVZ	Cessna 180	R. Moore
	G–BHWA	Cessna F.152	Wickenby Aviation Ltd.
	G–BHWB	Cessna F.152	Wickenby Aviation Ltd.
	G–BHWD	Hughes 369D	J. P. D. Heyward
	G–BHWE	Boeing 737-204ADV	Britannia Airways Ltd. *Sir Sidney Camm*
	G–BHWF	Boeing 737-204ADV	Britannia Airways Ltd. *Lord Brabazon of Tara*
	G–BHWG	Mahatma S.200SR balloon	H. W. Gandy *Spectrum*
	G–BHWH	Weedhopper JC-24A	G. A. Clephane
	G–BHWI	AA-5B Tiger	Elmdrake Aviation Ltd.
	G–BHWJ	AA-5A Cheetah	W. H. Wilkins Ltd.
	G–BHWK	M.S.880B Rallye Club	Tredair
	G–BHWL	—	—
	G–BHWM	M.S.880B Rallye Club	Tredair
	G–BHWN	WMB.3 Windtracker 200 balloon	C. J. Dodd & G. J. Luckett
	G–BHWO	WMB.4 Windtracker II balloon	C. J. Dodd
	G–BHWP	AA-5B Tiger	Batrade Ltd.
	G–BHWR	AA-5A Cheetah	Abraxas Aviation Ltd.
	G–BHWS	Cessna F.152	Shirlstar Container Transport Ltd.
	G–BHWW	Cessna U.206G	Sir G. P. Grant-Suttie
	G–BHWX	Cessna 404	Air & General Finance Ltd.
	G–BHWY	PA-28R-200 Cherokee Arrow	F. Johnson (Westwood Garages) [Ltd.
	G–BHWZ	PA-28-181 Archer II	Zitair Flying Club Ltd.
	G–BHXD	Jodel D.120	P. R. Powell
	G–BHXE	Thunder Ax3 balloon	C. Benning
	G–BHXG	D.H.C.-6 Twin Otter 310	Loganair Ltd.
	G–BHXI	BN-2B Islander	Euroair Transport Ltd.
	G–BHXJ	Nord 1203/2 Norecrin	R. E. Coates
	G–BHXK	PA-28 Cherokee 140	I. R. F. Hammond
	G–BHXL	Evans VP-2	J. J. Fitzgerald
	G–BHXM	Electra Flyer Eagle	C. G. Wrzesiew
	G–BHXN	Van's RV.3	P. R. Hing
	G–BHXO	Colt 12A balloon	Colt Balloons Ltd.
	G–BHXP	—	—
	G–BHXR	Thunder Ax7-65 balloon	Thunder Balloons Ltd.
	G–BHXS	Jodel D.120	K. Fox & J. Smith
	G–BHXT	Thunder Ax6-56Z balloon	Thunder Balloons Ltd.
	G–BHXU	AB-206B JetRanger 3	Castle Air Charters Ltd.
	G–BHXX	PA-23 Aztec 250	Hockstar Ltd.
	G–BHXY	Piper J-3C-65 Cub	R. K. Haldenby & T. S. Warren
	G–BHYA	Cessna R.182RG II	MLP Aviation Ltd.
	G–BHYB	Sikorsky S-76A	British Airways Helicopters Ltd.
	G–BHYC	Cessna 172RG Cutlass	Westair Flying Services Ltd.
	G–BHYD	Cessna R.172K	Sylmar Aviation Services Ltd.
	G–BHYE	PA-34-200T-2 Seneca	C.S.E. Aviation Ltd.
	G–BHYF	PA-34-200T-2 Seneca	C.S.E. Aviation Ltd.
	G–BHYG	PA-34-200T-2 Seneca	C.S.E. Aviation Ltd.
	G–BHYI	Stampe SV-4A	D. E. Starkey & A. T. Fraser
	G–BHYN	Evans VP-2	A. B. Cameron
	G–BHYO	Cameron N-77 balloon	C. Sisson
	G–BHYP	Cessna F.172M	J. Burgess & ptnrs.
	G–BHYR	Cessna F.172M	Ameys Garage Ltd.
	G–BHYS	PA-28-181 Archer II	Truman Aviation Ltd.
	G–BHYT	EMB-110P2 Bandeirante	Genair
	G–BHYU	Beech A200 Super King Air	Kenton Utilities & Developments [Ltd.
	G–BHYV	Evans VP-1	L. Chiappi
	G–BHYW	AB-206A JetRanger	Norman Bailey Helicopters Ltd.
	G–BHYX	Cessna 152	Tradecliff Ltd.
	G–BHYY	PA-28-161 Warrior II	Shirlstar Container Transport Ltd.
	G–BHZA	Piper J-3C-65 Cub	A. J. Hyatt & J. R. Wraight
	G–BHZC	R. Commander 690C	Rialto Builders Ltd.
	G–BHZD	PA-38-112 Tomahawk	Cormack (Aircraft Services) Ltd.
	G–BHZE	PA-28-181 Archer II	Cormack (Aircraft Services) Ltd.
	G–BHZF	Evans VP-2	D. Silsbury
	G–BHZG	Monnet Sonerai II	R. A. Gardiner & B. Chapman
	G–BHZH	Cessna F.152	Shoreham Flight Simulation Ltd.
	G–BHZI	Thunder Ax3 balloon	Thunder Balloons Ltd.
	G–BHZJ	Hughes Stratosphere 150 balloon	P. J. Hughes

Reg.	Type	Owner or Operator	Notes
G–BHZK	AA-5B Tiger	L. Henderson	
G–BHZL	AA-5A Cheetah	Carpfinch Ltd.	
G–BHZM	Jodel DR.1050	P. B. Kingsford	
G–BHZN	AA-5B Tiger	Peacock Salt Ltd.	
G–BHZO	AA-5A Cheetah	J. Brett	
G–BHZU	Piper J-3C-65 Cub	S. R. Clarke	
G–BHZV	Jodel D.120A	W. C. Forster	
G–BHZX	Thunder Ax7-65A balloon	Thermark (Plastic Processing) Ltd.	
G–BHZY	Monnet Sonerai II	C. A. Keech	
G–BHZZ	Air Britain Manchester Eagle 8 balloon	Air Britain Manchester Branch	
G–BIAA	SOCATA TB.9 Tampico	Air Touring Services Ltd.	
G–BIAB	SOCATA TB.9 Tampico	Meridian Aviation	
G–BIAC	M.S.894E Rallye Minerva	Trans Europe Air Charter Ltd.	
G–BIAH	Jodel D.112	N. M. Bloom	
G–BIAI	WMB.2 Windtracker balloon	I. Chadwick	
G–BIAJ	BAe 146-100	British Aerospace	
G–BIAK	SOCATA TB.10 Tobago	Trent Combustion Components Ltd.	
G–BIAL	Rango NA.8 balloon	A. M. Lindsay	
G–BIAM	SOCATA TB.10 Tobago	Jabrest Ltd.	
G–BIAO	Evans VP-2	J. Stephenson	
G–BIAP	PA-16 Clipper	I. M. Callier & P. J. Bish	
G–BIAR	Rigg Skyliner II balloon	I. A. Rigg	
G–BIAT	—	—	
G–BIAU	—	—	
G–BIAV	Sikorsky S–76A	British Airways Helicopters Ltd.	
G–BIAW	Sikorsky S–76A	British Airways Helicopters Ltd.	
G–BIAX	Taylor JT.2 Titch	G. F. Rowley	
G–BIAY	AA-5 Traveler	Brackley Motors Ltd.	
G–BIBA	SOCATA TB.9 Tampico	Channel Aviation Ltd.	
G–BIBB	Mooney M.20C	S. F. & G. G. Dray	
G–BIBC	Cessna 310R	Airwork Ltd.	
G–BIBD	Rotec Rally 2B	A. Clarke	
G–BIBE	EMB-110P1 Bandeirante	Loganair Ltd.	
G–BIBF	A12 Sport balloon	T. J. Smith	
G–BIBG	Sikorsky S-76A	British Caledonian Helicopters Ltd.	
G–BIBH	—	—	
G–BIBJ	Enstrom F-280C Shark	W. W. Kedrick & Sons Ltd.	
G–BIBK	Taylor JT.2 Titch	T. Corbett	
G–BIBL	Taylor JT.2 Titch	J. Sharp	
G–BIBM	AA-5A Cheetah	Ace Star Ltd.	
G–BIBN	Cessna FA.150K	P. H. Lewis	
G–BIBO	Cameron V-65 balloon	Southern Balloon Group	
G–BIBP	AA-5A Cheetah	Peacock Salt Ltd.	
G–BIBR	—	—	
G–BIBS	Cameron P-20 balloon	Cameron Balloons Ltd.	
G–BIBT	AA-5B Tiger	Grumman Travel (Surrey) Ltd.	
G–BIBU	Morris Ax7-77 balloon	K. Morris	
G–BIBV	WMB.3 Windtracker balloon	P. B. Street	
G–BIBW	Cessna F.172N	Northair Aviation Ltd.	
G–BIBX	WMB.2 Windtracker balloon	I. A. Rigg	
G–BIBY	Beech F33A Bonanza	Carl Peterson Ltd.	
G–BIBZ	Thunder Ax3 balloon	Thunder Balloons Ltd.	
G–BICA	—	—	
G–BICB	Rotec Rally 2B	J. D. Lye & A. P. Jones	
G–BICC	Vulture Tx3 balloon	C. P. Clitheroe	
G–BICD	Auster 5	J. A. S. Baldry & ptnrs.	
G–BICE	AT-6C Harvard IIA	C. M. L. Edwards	
G–BICF	GA-7 Cougar	Apollo Leasing Ltd.	
G–BICG	Cessna F.152	R. M. Clarke	
G–BICI	Cameron R-833 balloon	Ballooning Endeavours Ltd.	
G–BICJ	Monnet Sonerai II	J. R. S. Heaton	
G–BICL	Cessna 425	Northair Aviation Ltd.	
G–BICM	Colt 56A balloon	T. A. R. & S. Turner	
G–BICN	F.8L Falco	R. J. Barber	
G–BICO	Neale Mitefly Model balloon	T. J. Neale	
G–BICP	Robin DR.360	I. H. Thompson	
G–BICR	Jodel D.120A	S. W. C. Hall & ptnrs.	
G–BICS	Robin R.2100A	Tredair	
G–BICT	Evans VP-1	A. C. Coombe & D. L. Tribe	
G–BICU	Cameron V-56 balloon	I. S. Clark	
G–BICV	Boeing 737-2L9	Dan-Air Services Ltd.	

Notes	Reg.	Type	Owner or Operator
	G–BICW	PA-28-161 Warrior II	Palestar Aviation
	G–BICX	Maule M5-235C Lunar Rocket	Capital Aviation Sales (UK) Ltd.
	G–BICY	PA-23 Apache 160	Allen Technical Services Ltd.
	G–BICZ	Hill Hummer	J. Simpson
	G–BIDA	M.S.880B Rallye Club 100ST	Oakley Motor Units Ltd.
	G–BIDB	BAe 167 Strikemaster	British Aerospace
	G–BIDC	Bell 212	Helicopter Hire Ltd.
	G–BIDD	Evans VP-1	A. R. Bender
	G–BIDE	CP.301A Emeraude	D. Elliott
	G–BIDF	Cessna F.172P	J. Baumhardt Associates
	G–BIDG	Jodel D.150A	D. R. Gray
	G–BIDH	Cessna 152	Bevan Lynch Aviation Ltd.
	G–BIDI	PA-28R-201 Arrow III	Starline Engineering Ltd.
	G–BIDJ	PA-18-150 Super Cub	Marchington Gliding Club
	G–BIDK	PA-18-150 Super Cub	Light Aircraft Engineering Ltd.
	G–BIDL	Osprey 1G balloon	A. P. Chown
	G–BIDM	Cessna F.172H	D. H. MacDonald
	G–BIDN	P.57 Sea Prince T.1	Atlantic & Caribbean Aviation Ltd.
	G–BIDO	CP.301A Emeraude	T. J. N. H. Palmer & G. W. Oliver
	G–BIDP	PA-28-181 Archer II	MAP Aviation Ltd.
	G–BIDR	—	—
	G–BIDS	—	—
	G–BIDT	Cameron A375 balloon	Ballooning Endeavours Ltd.
	G–BIDU	Cameron V-77 balloon	E. Eleazor
	G–BIDV	Colt Ax2 balloon	International Distillers & Vintners (House Trade) Ltd.
	G–BIDW	Sopwith 1½ strutter replica (A8226)	RAF Museum
	G–BIDX	Jodel D.112	H. N. Nuttall & R. P. Walley
	G–BIDY	WMB.2 Windtracker balloon	D. M. Campion
	G–BIDZ	Colt 21A balloon	Hot Air Balloon Co. Ltd.
	G–BIEB	Agusta-Bell 47G-3B1	P. Colby
	G–BIEC	AB-206A JetRanger 2	Autair Helicopters Ltd.
	G–BIED	Beech F90 King Air	United Biscuits (Foods) Ltd.
	G–BIEE	Beech C90 King Air	Eagle Aircraft Services Ltd.
	G–BIEF	Cameron V-77 balloon	D. S. Bush
	G–BIEG	PA-32-301 Saratoga	Spooner Aviation Ltd.
	G–BIEH	Sikorsky S-76A	Management Aviation Ltd.
	G–BIEJ	Sikorsky S-76A	Bristow Helicopters Ltd.
	G–BIEK	WMB.4 Windtracker balloon	P. B. Street
	G–BIEL	WMB.4 Windtracker balloon	A. T. Walden
	G–BIEM	D.H.C.-6 Twin Otter 310	Loganair Ltd.
	G–BIEN	Jodel D.120A	M. Hanley & ptnrs.
	G–BIEO	—	—
	G–BIER	Rutan Long-Eze	V. Mossor
	G–BIES	Maule M5-235C Lunar Rocket	William Procter Farms
	G–BIET	Cameron O-77 balloon	G. M. Westley
	G–BIEU	AA-5A Cheetah	Gatelee Printing (Wokingham) Ltd.
	G–BIEV	AA-5A Cheetah	Abraxas Aviation Ltd.
	G–BIEW	Cessna U.206G	G. D. Atkinson
	G–BIEX	Andreasson BA-4B	H. P. Burrill
	G–BIEY	PA-28-151 Warrior	A. F. Aviation Ltd.
	G–BIEZ	Beech F90 King Air	Eagle Aircraft Services Ltd.
	G–BIFA	Cessna 310R-II	Michian Aviation Ltd.
	G–BIFB	PA-28 Cherokee 150	C. J. Reed
	G–BIFC	Colt 14A balloon	Colt Balloons Ltd.
	G–BIFD	R. Commander 114	D. H. MacDonald
	G–BIFE	Cessna A.185F	Conguess Aviation Ltd.
	G–BIFF	AA-5A Cheetah	Mission Control Music Ltd.
	G–BIFM	—	—
	G–BIFN	Benson B.8M	K. Willows
	G–BIFO	Evans VP-1	P. Raggett
	G–BIFP	Colt 56C balloon	J. Philp
	G–BIFR	—	—
	G–BIFS	Beech C90 King Air	Fromeat (International) Co. Ltd.
	G–BIFT	Cessna F.150L	Practavia Ltd.
	G–BIFU	Short Skyhawk balloon	D. K. Short
	G–BIFV	Jodel D.150	J. H. Kirkham
	G–BIFW	Scruggs BL.2 Wunda balloon	D. Morris
	G–BIFY	Cessna F.150L	Practavia Ltd.
	G–BIFZ	Partenavia P.68C	Abbey Hill Vehicle Services
	G–BIGB	Bell 212	Bristow Helicopters Ltd.

Reg.	Type	Owner or Operator	Notes
G–BIGC	Cameron O-42 balloon	C. M. Moroney	
G–BIGE	Champion Cloudseeker balloon	A. Foster	
G–BIGD		Cameron Balloons Ltd.	
G–BIGF	Thunder Ax7-77 balloon	M. D. Stever & C. A. Allen	
G–BIGG	Saffery S.200 balloon	A. F. Langhelt	
G–BIGH	Piper L-4H Cub	Cormack (Aircraft Services) Ltd.	
G–BIGI	Mooney M.20J	Kingswinford Engineering Co. Ltd.	
G–BIGJ	Cessna F.172M	Essex Aero Services Ltd.	
G–BIGK	Taylorcraft BC-12	J. Rowell	
G–BIGL	Cameron O-65 balloon	A. H. L. Alpin	
G–BIGM	Avenger T.200-2112 balloon	M. Murphy	
G–BIGN	Attic Srs. 1 balloon	G. Nettleship	
G–BIGP	Benson B.8.	R. H. S. Cooper	
G–BIGR	Avenger T.200-2112 balloon	R. Light	
G–BIGU	Benson B.8M	J. R. Martin	
G–BIGX	Benson B.8M	J. R. Martin	
G–BIGY	Cameron V-65 balloon	Dante Balloon Group	
G–BIGZ	Scheibe SF.25B Falke	K. Ballington	
G–BIHA	—	—	
G–BIHB	Scruggs BL.2 Wunda balloon	D. Morris	
G–BIHC	Scruggs BL.2 Wunda balloon	P. D. Kiddell	
G–BIHD	Robin DR.400/160	G. R. Pope & ptnrs.	
G–BIHE	Cessna FA.152	Inverness Flying Services Ltd.	
G–BIHF	SE-5A Replica	K. J. Garrett	
G–BIHG	PA-28 Cherokee 140	T. Parmenter	
G–BIHH	Sikorsky S-61N	British Caledonian Helicopters Ltd.	
G–BIHI	Cessna 172M	J. H. A. Rogers	
G–BIHN	Skyship 500 airship	Airship Industries Ltd.	
G–BIHO	D.H.C.-6 Twin Otter 310	Brymon Aviation Ltd.	
G–BIHP	Van Den Bemden gas balloon	J. J. Harris	
G–BIHR	WMB.2 Windtracker balloon	A. F. Langhelt	
G–BIHS	T.6G Harvard	Aces High Ltd.	
G–BIHT	PA-17 Vagabond	G. D. Thomson	
G–BIHU	Saffery S.200 balloon	A. F. Langhelt	
G–BIHV	WMB.2 Windtracker balloon	T. H. W. Bradley	
G–BIHW	Aeronca A65TAC	W. F. Crozier & ptnrs.	
G–BIHX	Benson B.8M	C. C. Irvine	
G–BIHY	Isaacs Fury	D. E. Olivant	
G–BIIA	Fournier RF-3	M. K. Field	
G–BIIB	Cessna F.172M	Citation Flying Services Ltd.	
G–BIIC	Scruggs BL.2 Wunda balloon	S. J. Hodder & D. Cockerill	
G–BIID	PA-18 Super Cub 95	Nolson Aviation Ltd.	
G–BIIE	Cessna F.172P	Shoreham Flight Simulation Ltd.	
G–BIIF	Fournier RF-4D	L. P. Holmes (G–BVET)	
G–BIIG	Thunder Ax-6-56Z balloon	Thunder Balloons Ltd.	
G–BIIH	Scruggs BL.2T Turbo balloon	B. M. Scott	
G–BIIJ	Cessna F.152	Timecost Consultants Ltd.	
G–BIIK	M.S.883 Rallye 115	G. A. Bentley	
G–BIIL	Thunder Ax6-56 balloon	G. W. Reader	
G–BIIM	Scruggs BL.2A Wunda balloon	K. D. Head	
G–BIIN	BN-2B Islander	Pilatus BN Ltd.	
G–BIIO	BN-2B Islander	Pilatus BN Ltd.	
G–BIIT	PA-28-161 Warrior II	Tayside Aviation Ltd.	
G–BIIV	PA-28-181 Archer II	R. Watkins	
G–BIIW	Rango NA.10 balloon	Rango Kite Co.	
G–BIIX	Rango NA.12 balloon	Rango Kite Co.	
G–BIIZ	Great Lakes 2T-1A Sport Trainer	P. Lindsay	
G–BIJA	Scruggs BL.2A Wunda balloon	P. L. E. Bennett	
G–BIJB	PA-18-150 Super Cub	Essex Gliding Club	
G–BIJC	AB-206B JetRanger 3	Peter Colby Commercial Ltd.	
G–BIJD	Bo 208C Junior	D. J. Dulborough	
G–BIJE	—	—	
G–BIJS	Luton L.A.4A Minor	I. J. Smith	
G–BIJT	AA-5A Cheetah	G. W. Plowman & Son Ltd.	
G–BIJU	CP.301 Emeraude	C. J. Norman & M. Howard (G–BHTX)	
G–BIJV	Cessna F.152	Civil Service Flying Club Ltd.	
G–BIJW	Cessna F.152	Civil Service Flying Club Ltd.	
G–BIJX	Cessna F.152	Civil Service Flying Club Ltd.	
G–BIJY	—	—	
G–BIJZ	Skyventurer Mk. I balloon	R. Sweeting	
G–BIKA	Boeing 757-236	British Airways	
G–BIKB	Boeing 757-236	British Airways	

Notes	Reg.	Type	Owner or Operator
	G–BIKC	Boeing 757-236	British Airways
	G–BIKE	PA-28R Cherokee Arrow 200	R. V. Webb Ltd.
	G–BIKD	Boeing 757-236	British Airways
	G–BIKF	Boeing 757-236	British Airways
	G–BIKG	Boeing 757-236	British Airways
	G–BILA	Daletol DM.165L Viking	R. Lamplough
	G–BILB	WMB.2 Windtracker balloon	A. F. Langhelt
	G–BILD	Piper J-3C-65 Cub	W. R. Pickett (G–KERK)
	G–BILE	Scruggs BL.2B balloon	P. D. Ridout
	G–BILF	Practavia Sprite 125	G. Harfield
	G–BILG	Scruggs BL.2B balloon	P. D. Ridout
	G–BILI	Piper J-3C-65 Cub	A. S. Nixon
	G–BILJ	Cessna FA.152	Shoreham Flight Simulation Ltd.
	G–BILK	Cessna FA.152	Spectrum Computor Systems Ltd.
	G–BILL	PA-25 Pawnee 235	Bowker Air Services Ltd.
	G–BILM	—	—
	G–BILN	—	—
	G–BILO	—	—
	G–BILP	Cessna 152	Skyviews & General Ltd.
	G–BILR	Cessna 152	Skyviews & General Ltd.
	G–BILS	Cessna 152	Skyviews & General Ltd.
	G–BILT	Cessna F.172P	Denham Flying Training School Ltd.
	G–BILU	Cessna 172RG	Denham Flying Training School Ltd.
	G–BILV	Cessna FA.152	Denham Flying Training School Ltd.
	G–BILW	Colt 60 balloon	Hot Air Balloon Co. Ltd.
	G–BILX	Colt 31A balloon	Hot Air Balloon Co. Ltd.
	G–BILZ	Taylor JT.1 Monoplane	G. Beaumont
	G–BIMA	A.300B4-203 Airbus	Laker Airways *Metro*
	G–BIMB	A.300B4-203 Airbus	Laker Airways *Orient Express*
	G–BIMC	A.300B4-203 Airbus	Laker Airways *Inter City Express*
	G–BIMD	A.300B4-203 Airbus	Laker Airways
	G–BIME	A.300B4-203 Airbus	Laker Airways
	G–BIMF	A.300B4-203 Airbus	Laker Airways
	G–BIMG	A.300B4-203 Airbus	Laker Airways
	G–BIMH	A.300B4-203 Airbus	Laker Airways
	G–BIMI	—	Laker Airways
	G–BIMJ	—	Laker Airways
	G–BIMK	Tiger T.200 Srs. 1 balloon	M. K. Baron
	G–BIML	Turner Super T.40A	R. T. Callow
	G–BIMM	PA-18 Super Cub 135	D. S. & I. M. Morgan
	G–BIMN	Steen Skybolt	C. R. Williamson
	G–BIMO	Stampe SV-4C	R. K. G. Hannington
	G–BIMR	—	—
	G–BIMS	—	—
	G–BIMT	Cessna FA.152	B. Ashley
	G–BIMU	Sikorsky S-61N	British Caledonian Helicopters Ltd.
	G–BIMV	Sikorsky S-61N	Management Aviation Ltd.
	G–BIMW	D.H.C.-6 Twin Otter 310	Aurigny Air Services Ltd.
	G–BIMX	Rutan Vari-Eze	A. S. Knowles
	G–BIMZ	Beech 76 Duchess	Barrein Engineers Ltd.
	G–BINA	Saffery S.9 balloon	A. P. Bashford
	G–BINB	WMB.2A Windtracker balloon	S. R. Woolfries
	G–BINC	Tour de Calais balloon	Cupro Sapphire Ltd.
	G–BIND	M.S.894E Rallye 235	G. Archer
	G–BINE	Scruggs BL.2A Wunda balloon	M. Gilbey
	G–BINF	Saffery S.200 balloon	T. Lewis
	G–BING	Cessna F.172P	J. E. M. Patrick
	G–BINI	Scruggs BL.2C balloon	S. R. Woolfries
	G–BINJ	Rango NA.12 balloon	M. R. Haslam
	G–BINK	—	—
	G–BINL	Scruggs BL.2B balloon	P. D. Ridout
	G–BINM	Scruggs BL.2B balloon	P. D. Ridout
	G–BINN	Unicorn UE.1A balloon	Unicorn Group
	G–BINO	Evans VP-1	J. I. Visser
	G–BINP	Rango NA.9 balloon	N. H. Ponsford
	G–BINR	Unicorn UE.1A balloon	Unicorn Group
	G–BINS	Unicorn UE.2A balloon	Unicorn Group
	G–BINT	Unicorn UE.1A balloon	Unicorn Group
	G–BINU	Saffery S.200 balloon	T. Lewis
	G–BINV	Saffery S.200 balloon	R. S. Harris
	G–BINW	Scruggs BL.2B balloon	P. G. Mackin
	G–BINX	Scruggs BL.2B balloon	P. D. Ridout
	G–BINY	Oriental balloon	J. L. Morton

Reg.	Type	Owner or Operator	Notes
G–BINZ	Rango NA.8 balloon	T. J. Sweeting & M. O. Davies	
G–BIOA	Hughes 369D	Weetabix Ltd.	
G–BIOB	Cessna F.172P	Hunting Surveys & Consultants Ltd.	
G–BIOC	Cessna F.150L	T. E. Abell	
G–BIOI	Jodel DR.1051-M	R. Cochrane	
G–BIOJ	—	—	
G–BIOK	Cessna F.152	Rogers Aviation Sales Ltd.	
G–BIOL	Colt 77A balloon	Colt Balloons Ltd.	
G–BIOM	Cessna F.152	Shetland Flying Club	
G–BION	Cameron V-77 balloon	Elliott's Pharmacy Ltd.	
G–BIOO	Unicorn UE.2B balloon	Unicorn Group	
G–BIOP	Scruggs BL.2D balloon	J. P. S. Donnellan	
G–BIOR	M.S.880B Rallye Club	Actwood Ltd.	
G–BIOS	Scruggs BL.2B balloon	D. Eaves	
G–BIOT	Benson B.8M	A. J. Bundark	
G–BIOU	Jodel D.117A	M. S. Printing & Graphics Machinery Ltd.	
G–BIOV	—	—	
G–BIOW	Slingsby T.67A	Slingsby Engineering Ltd.	
G–BIOX	Potter Crompton PRO.1 balloon	G. M. Potter	
G–BIOY	PAC-14 Special Shape balloon	P. A. Cremer	
G–BIOZ	Rotorway 133 Executive	G. Meech	
G–BIPA	AA-5B Tiger	Light Aircraft Engineering Ltd.	
G–BIPB	Weedhopper JC-24B	E. H. Moroney	
G–BIPC	PAC-14 Hefferlump balloon	P. A. Cremer	
G–BIPF	Scruggs BL.2C balloon	D. Morris	
G–BIPG	Global Mini balloon	P. Globe	
G–BIPH	Scruggs BL.2B balloon	C. M. Dewsnap	
G–BIPI	Everett Blackbird Mk. 1	R. J. Everett (Engineering) Ltd.	
G–BIPJ	PA-36-375 Brave	G. B. Pearce	
G–BIPK	Saffery S.200 balloon	P. J. Kelsey	
G–BIPL	AA-5A Cheetah	Parspex Ltd.	
G–BIPM	Flamboyant Ax7-65 balloon	Pepsi Cola International Ltd.	
G–BIPN	Fournier RF-3	Syerston Soaring Group	
G–BIPO	Mudry/CAARP CAP.20LS-200	Personal Plane Services Ltd.	
G–BIPP	Beech 200 Super King Air	Eagle Aircraft Services Ltd.	
G–BIPR	Sikorsky S-76A	Bristow Helicopters Ltd.	
G–BIPS	M.S.880B Rallye 100ST	P. A. Shenton & D. P. Griffiths	
G–BIPT	Jodel D.112	C. K. Farley	
G–BIPU	AA-5B Tiger	Aero Group 78	
G–BIPV	AA-5B Tiger	I. D. Longfellow	
G–BIPW	Avenger T.200-2112 balloon	A. F. Langhelt	
G–BIPX	Saffery S.9 balloon	J. R. Havers	
G–BIPY	Bensen B.8	A. J. Wood	
G–BIPZ	McCandless Mk. 4 gyroplane	B. McIntyre	
G–BIRA	SOCATA TB.9 Tampico	Air Touring Services Ltd.	
G–BIRB	M.S.880B Rallye 100ST	The Rallye Flying Club	
G–BIRC	PA-28 Cherokee 140	Inkerman Aircraft Services	
G–BIRD	Pitts S-1C Special	R. N. York	
G–BIRE	Colt 56 Bottle balloon	Hot Air Balloon Co. Ltd.	
G–BIRF	H.S.748 Srs. 2B	British Aerospace Public Ltd. Co.	
G–BIRG	M.S.880B Rallye Club	Air Touring Services Ltd.	
G–BIRH	PA-18 Super Cub 135	I. R. F. Hammond	
G–BIRI	C.A.S.A. 1.131E Jungmann	L. B. Jefferies	
G–BIRJ	Cessna F.172P	W. H. & J. Rogers Group Ltd.	
G–BIRK	Avenger T.200-2112 balloon	D. Harland	
G–BIRL	Avenger T.200-2112 balloon	R. Light	
G–BIRM	Avenger T.200-2112 balloon	P. Higgins	
G–BIRN	Short SD3-30	Loganair Ltd.	
G–BIRO	Cessna 172P	M. C. Grant	
G–BIRP	Arena Mk. 17 Skyship balloon	A. S. Viel	
G–BIRR	—	—	
G–BIRS	Cessna 182P	D. P. Cranston (G–BBBS)	
G–BIRT	Robin R.1180T	W. D'A. Hall	
G–BIRU	H.S.125 Srs. 700B	MAM Aviation Ltd.	
G–BIRV	Bensen B.8MV	R. Hart	
G–BIRW	M.S.505 Criquet (F+IS)	G. A. Warner	
G–BIRX	Scruggs RS.5000 balloon	J. H. Searle	
G–BIRY	Cameron V-77 balloon	J. J. Winter	
G–BIRZ	Zenair CH.250	S. M. Kawalski	
G–BISA	Hase IIIT balloon	M. A. Hase	
G–BISB	Cessna F.152	Brailsford Aviation Ltd.	

Notes	Reg.	Type	Owner or Operator
	G–BISC	Robinson R-22	The Howell Helicopter Co.
	G–BISD	Enstrom F-280C-UK-2	W. H. & J. J. Gadsby
	G–BISE	Enstrom F-280C-UK-2	Richard Knight Aviation
	G–BISF	Robinson R-22	Trent Helicopters Ltd.
	G–BISG	FRED Srs. 3	R. A. Coombe
	G–BISH	Cameron 0-42 balloon	Zebedee Balloon Service
	G–BISI	Robinson R-22	Trent Helicopters Ltd.
	G–BISJ	Cessna 340A	Castle Aviation
	G–BISK	R. Commander 112B	Rogers Auto's Ltd.
	G–BISL	Scruggs BL.2B balloon	P. D. Ridout
	G–BISM	Scruggs BL.2B balloon	P. D. Ridout
	G–BISN	Boeing Vertol 234LR Chinook	British Airways Helicopters Ltd.
	G–BISO	Boeing Vertol 234LR Chinook	British Airways Helicopters Ltd.
	G–BISP	Boeing Vertol 234LR Chinook	British Airways Helicopters Ltd.
	G–BISR	—	—
	G–BISS	Scruggs BL.2C balloon	P. D. Ridout
	G–BIST	Scruggs BL.2C balloon	P. D. Ridout
	G–BISU	B.170 Freighter 31M	Instone Air Line Ltd.
	G–BISV	Cameron 0-65 balloon	Hylyne Rabbits Ltd.
	G–BISW	Cameron 0-65 balloon	Hylyne Rabbits Ltd.
	G–BISX	Colt 56A balloon	Long John International Ltd.
	G–BISY	Scruggs BL.2C balloon	P. T. Witty
	G–BISZ	Sikorsky S-76A	Bristow Helicopters Ltd.
	G–BITA	PA-18-150 Super Cub	N. T. & M. Smith
	G–BITB	—	—
	G–BITE	SOCATA TB.10 Tobago	I. M. White
	G–BITF	Cessna F.152	Bristol & Wessex Aeroplane Club
	G–BITG	Cessna F.152	Bristol & Wessex Aeroplane Club
	G–BITH	Cessna F.152	Bristol & Wessex Aeroplane Club
	G–BITI	Scruggs RS.5000 balloon	A. E. Smith
	G–BITJ	—	—
	G–BITK	FRED Srs. 2	B. J. Miles
	G–BITL	Horncastle LL-901 balloon	M. J. Worsdell
	G–BITM	Cessna F.172P	Pleusque Ltd.
	G–BITN	Short Albatross balloon	D. K. Short
	G–BITO	Jodel D.112	A. Dunbar
	G–BITR	Sikorsky S-76A	Bristow Helicopters Ltd.
	G–BITS	Drayton B-56 balloon	M. J. Betts
	G–BITT	Bo 208C Junior	J. E. Alexander
	G–BITV	Short SD3-30	Inter City/Guernsey Airlines
	G–BITX	Short SD3-30	Inter City/Guernsey Airlines
	G–BITY	FD.31T balloon	A. J. Bell
	G–BITZ	Cremer Sandoe PACDS. 14 balloon	P. A. Cremer & C. D. Sandoe
	G–BIUA	BN-2A Islander	Pilatus BN Ltd.
	G–BIUB	BN-2A Islander	Pilatus BN Ltd.
	G–BIUD	BN-2A Islander	Pilatus BN Ltd.
	G–BIUE	BN-2A Islander	Pilatus BN Ltd.
	G–BIUG	BN-2A Islander	Pilatus BN Ltd.
	G–BIUH	BN-2A Islander	Pilatus BN Ltd.
	G–BIUI	Cessna F.152	Northair Aviation Ltd.
	G–BIUK	—	—
	G–BIUL	Cameron 60 SS balloon	Engineering Appliances Ltd.
	G–BIUM	Cessna F.152	Brailsford Aviation Ltd.
	G–BIUN	Cessna F.152	Brailsford Aviation Ltd.
	G–BIUO	R. Commander 112A	E. T. Green
	G–BIUP	SNCAN NC.854S	J. M. Kirk
	G–BIUR	Boeing 727-155C	Dan-Air Services Ltd.
	G–BIUS	Cessna 310R	Smiths (Harlow) Ltd.
	G–BIUT	Scruggs BL.2C balloon	N. J. Ball
	G–BIUU	PA-23 Aztec 250	Kingsmetal Ltd.
	G–BIUV	H.S.748 Srs. 2A	Dan-Air Services Ltd. (G–AYYH)
	G–BIUW	PA-28-161 Warrior II	Staeng Ltd.
	G–BIUX	PA-28-161 Warrior II	C.S.E. Aviation Ltd.
	G–BIUY	PA-28-181 Archer II	Aviation C.N.C. Ltd.
	G–BIUZ	Vickers Slingsby T.67A	Slingsby Engineering Ltd.
	G–BIVA	Robin R.2112	Cotswold Aero Club Ltd.
	G–BIVB	Jodel D.112	C. & T. Wright
	G–BIVC	Jodel D.112	The Harrier Flying Group
	G–BIVD	—	—
	G–BIVE	—	—
	G–BIVF	CP.301C-3 Emeraude	J. Casker
	G–BIVG	Thunder Ax3 balloon	Thunder Balloons Ltd.

G–BIIZ (Top) Great Lakes 2T-1A Sport Trainer

G–BIRN (Centre) Short SD3-30 of Loganair

G–BISU (Bottom) Bristol 170 Freighter 31M of Instone Air Line/*A. S. Wright*

Reg.	Type	Owner or Operator	Notes
G–BIVH	Thunder Ax7-77 Bottle balloon	Thunder Balloons Ltd.	
G–BIVI	Cremer PAC.500 airship	P. A. Cremer	
G–BIVJ	Cessna F.152	Northair Aviation Ltd.	
G–BIVK	Bensen B.8	J. G. Toy	
G–BIVL	Bensen B.8	T. E. Davies	
G–BIVM	—	—	
G–BIVN	—	—	
G–BIVO	G.164D Ag-Cat	Miller Aerial Spraying Ltd.	
G–BIVP	AS.350B Ecureuil	Slea Aviation Ltd.	
G–BIVR	Featherlight Mk. 1 balloon	A. P. Newman & N. P. Kemp	
G–BIVS	Featherlight Mk. 2 balloon	J. M. J. Roberts & S. R. Rushton	
G–BIVT	Saffery S.80 balloon	L. F. Guyot	
G–BIVU	AA-5A Cheetah	MLP Aviation Ltd.	
G–BIVV	AA-5A Cheetah	MLP Aviation Ltd.	
G–BIVW	–	–	
G–BIVX	Saffery S.80 balloon	P. T. Witty	
G–BIVY	Cessna 172N	J. & D. Gordon	
G–BIVZ	D.31A Turbulent	N. Jones	
G–BIWA	Stevendon Skyreacher balloon	S. D. Barnes	
G–BIWB	Scruggs RS.5000 balloon	P. D. Ridout	
G–BIWC	Scruggs RS.5000 balloon	P. D. Ridout	
G–BIWD	Scruggs RS.5000 balloon	D. Eaves	
G–BIWE	Scruggs BL.2D balloon	M. D. Saunders	
G–BIWF	Warren balloon	P. D. Ridout	
G–BIWG	Zelenski Mk. 2 balloon	P. D. Ridout	
G–BIWH	Cremer Super Fliteliner balloon	G. Lowther	
G–BIWI	Cremer WS.1 balloon	P. A. Cremer	
G–BIWJ	Unicorn UE.1A balloon	A. F. Langhelt	
G–BIWK	Cameron V-65 balloon	I. R. Williams & R. G. Bickerdale	
G–BIWL	PA-32-301 Saratoga	R. Arnold & ptnrs.	
G–BIWM	Cameron A-530 balloon	Cameron Balloons Ltd.	
G–BIWN	Jodel D.112	A. N. Houghton	
G–BIWO	Scruggs RS.5000 balloon	D. Morris	
G–BIWP	Mooney M.20J	Tropair Cooling Ltd.	
G–BIWR	Mooney M.20F	C. W. Yarnton	
G–BIWS	Cessna 182R	Rogers Aviation Sales Ltd.	
G–BIWU	Cameron V-65 balloon	J. T. Whicker & J. W. Unwin	
G–BIWV	Cremer PAC-550T balloon	P. A. Rutherford	
G–BIWW	AA-5 Traveler	R. Davis	
G–BIWX	AT-16 Harvard IV	R. Lamplough	
G–BIWY	Westland WG.30	Westland Helicopters Ltd.	
G–BIWZ	—	—	
G–BIXA	SOCATA TB.9 Tampico	Air Touring Services Ltd.	
G–BIXB	SOCATA TB.9 Tampico	Air Touring Services Ltd.	
G–BIXD	BN-2B Islander	Pilatus BN Ltd.	
G–BIXE	BN-2B Islander	Pilatus BN Ltd.	
G–BIXF	BN-2B Islander	Pilatus BN Ltd.	
G–BIXG	BN-2B Islander	Pilatus BN Ltd.	
G–BIXH	Cessna F.152	W. H. & J. Rogers Group Ltd.	
G–BIXI	Cessna 172RG Cutlass	J. F. P. Lewis	
G–BIXJ	Saffery S.40 balloon	T. M. Pates	
G–BIXK	Rand KR-2	R. G. Cousins	
G–BIXL	P-51D Mustang	R. Lamplough	
G–BIXM	Beech C90 King Air	Eagle Aircraft Services Ltd.	
G–BIXN	Boeing A.75N1 Stearman	Evans Estates Ltd.	
G–BIXP	V.S. Spitfire IX	R. Lamplough	
G–BIXR	Cameron A-140 balloon	Skysales Ltd.	
G–BIXS	Avenger T.200-2112 balloon	M. Stuart	
G–BIXT	Cessna 182R	W. Lipka	
G–BIXU	AA-5B Tiger	Cabair Ltd.	
G–BIXV	Bell 212	Bristow Helicopters Ltd.	
G–BIXW	Colt 56 balloon	Lighter-Than-Air Ltd.	
G–BIXX	Pearson Srs. 2 balloon	D. Pearson	
G–BIXY	Piper J-3C-90 Cub	M. Johnson	
G–BIXZ	Grob G-109	K. E. White	
G–BIYA	Short SD3-30	Short Bros. Ltd.	
G–BIYD	Short SD3-30	Short Bros. Ltd.	
G–BIYE	Short SD3-30	Short Bros. Ltd.	
G–BIYF	Short SD3-30	Short Bros. Ltd.	
G–BIYG	Short SD3-30	Short Bros. Ltd.	

Notes	Reg.	Type	Owner or Operator
	G-BIYH	Short SD3-30	Short Bros. Ltd.
	G-BIYI	Cameron Viva 65 balloon	Sarnia Balloon Group
	G-BIYJ	PA-19 Super Cub 95	S. R. Wilkeston
	G-BIYK	Isaacs Fury	R. S. Martin
	G-BIYM	PA-32R-301 Saratoga SP	Marlow Chemical Co. Ltd.
	G-BIYN	Pitts S-1S Special	R. P. Lewis
	G-BIYO	PA-31-310 Turbo Navajo	Airmore Aviation Ltd.
	G-BIYP	—	—
	G-BIYR	PA-18 Super Cub 135	Nalson Aviation Ltd.
	G-BIYT	Colt Cloudhopper balloon	E. T. Houten
	G-BIYU	Fokker S.11-1 Instructor	H. R. Smallwood
	G-BIYV	Cremer 14.700-15 balloon	G. Lowther & ptnrs.
	G-BIYW	Jodel D.112	C. J. Watson
	G-BIYX	PA-28 Cherokee 140	S. R. Kilby
	G-BIYY	PA-19 Super Cub 95	A. E. & W. J. Taylor
	G-BIYZ	Cessna 182R	Northair Aviation Ltd.
	G-BIZA	AB-206 JetRanger 3	Alan Mann Helicopters Ltd.
	G-BIZB	AB-206 JetRanger 3	Alan Mann Helicopters Ltd.
	G-BIZC	AB-206 JetRanger 3	Alan Mann Helicopters Ltd.
	G-BIZD	AB-206 JetRanger 3	Alan Mann Helicopters Ltd.
	G-BIZE	SOCATA TB.9 Tampico	Air Touring Services Ltd.
	G-BIZF	Cessna F.172P	Lectime Ltd.
	G-BIZG	Cessna F.152	Lectime Ltd.
	G-BIZH	—	—
	G-BIZI	Robin DR.400/120	Headcorn Flying School Ltd.
	G-BIZJ	Nord 3202	Keenair Services Ltd.
	G-BIZK	Nord 3202	Keenair Services Ltd.
	G-BIZL	Nord 3202	Keenair Services Ltd.
	G-BIZM	Nord 3202	Keenair Services Ltd.
	G-BIZN	Slingsby T.67A	Slingsby Engineering Ltd.
	G-BIZO	PA-28R Cherokee Arrow 200	Aviation Advisory Services Ltd.
	G-BIZP	Pilatus PC.6-B2/H2 Porter	Peterborough Parachute Centre Ltd.
	G-BIZR	SOCATA TB.9 Tampico	Air Touring Services Ltd.
	G-BIZT	Bensen B.80D	J. Ferguson
	G-BIZU	Thunder Ax6-56Z balloon	Thunder Balloons Ltd.
	G-BIZV	PA-19 Super Cub 95	W. R. Pickett
	G-BIZW	Champion 7GCBC Citabria	R. Windley
	G-BIZX	Beech B200 Super King Air	Eagle Aircraft Services Ltd.
	G-BIZY	Jodel D.112	G. J. Allen & ptnrs.
	G-BIZZ	—	—
	G-BJAA	Unicorn UE.1A balloon	K. H. Turner
	G-BJAB	Ayres S2R Thrush Commander	Ag-Air
	G-BJAC	Boeing Vertol 234LR Chinook	British Airways Helicopters Ltd.
	G-BJAD	FRED Srs. 2	C. Allison
	G-BJAE	Starck AS.80 Lavadoux	D. J. & S. A. E. Phillips
	G-BJAF	Piper J-3C-65 Cub	K. Badger
	G-BJAG	PA-28-181 Archer II	C.S.E. Aviation Ltd.
	G-BJAH	Unicorn UE.1A balloon	A. D. Hutchings
	G-BJAJ	AA-5B Tiger	Batrade Ltd.
	G-BJAK	Mooney M.20C	J. B. Wilson
	G-BJAL	C.A.S.A. 1.131E Jungmann	Buccaneer Aviation Ltd.
	G-BJAM	Westland-Bell 47G-3B1	Trent Helicopters Ltd.
	G-BJAN	SA.102-5 Cavelier	J. Powlesland
	G-BJAO	Bensen B.8	G. L. Stockdale
	G-BJAP	D.H.82A Tiger Moth	J. Pothecary
	G-BJAR	Unicorn UE.3A balloon	Unicorn Group
	G-BJAS	Rango NA.9 balloon	A. Lindsay
	G-BJAT	Pilatus P2-05	Hon. P. Lindsay
	G-BJAU	PZL-104 Wilga 35	Anglo Polish Sailplanes Ltd.
	G-BJAV	GY-80 Horizon 160	R. Pickett
	G-BJAW	Cameron V-65 balloon	G. W. McCarthy
	G-BJAX	Pilatus P2-05	R. G. Hanna
	G-BJAY	Piper J-3C-65 Cub	J. W. Wrisdale
	G-BJAZ	Thunder Ax7-77 balloon	W. F. Brain
	G-BJBA	Cessna 152	Namemay Ltd.
	G-BJBB	Cessna 152	Namemay Ltd.
	G-BJBD	BN-2B Islander	Pilatus BN Ltd.
	G-BJBE	BN-2B Islander	Pilatus BN Ltd.
	G-BJBF	BN-2B Islander	Pilatus BN Ltd.
	G-BJBG	BN-2B Islander	Pilatus BN Ltd.
	G-BJBH	BN-2B Islander	Pilatus BN Ltd.
	G-BJBI	Cessna 414A	Northair Aviation Ltd.

Reg.	Type	Owner or Operator	Notes
G-BJBJ	Boeing 737-2T5	Orion Airways Ltd.	
G-BJBK	PA-19 Super Cub 95	J. D. Campbell	
G-BJBL	Unicorn UE.1A balloon	Unicorn Group	
G-BJBM	Monnet Sonerai I	J. Pickrell & ptnrs.	
G-BJBN	Ball JB.980 balloon	J. D. Ball	
G-BJBO	Jodel DR.250/160	W. F. Telfer	
G-BJBP	Beech A200 Super King Air	Air Ecosse Ltd. (G-HLUB)	
G-BJBR	Robinson R-22	Sloane Helicopters Ltd.	
G-BJBS	Robinson R-22	Sloane Helicopters Ltd.	
G-BJBT	Robinson R-22	Sloane Helicopters Ltd.	
G-BJBU	PA-23 Aztec 250	Harvest Air Ltd.	
G-BJBV	PA-28-161 Warrior II	C.S.E. Aviation Ltd.	
G-BJBW	PA-28-161 Warrior II	C.S.E. Aviation Ltd.	
G-BJBX	PA-28-161 Warrior II	C.S.E. Aviation Ltd.	
G-BJBY	PA-28-161 Warrior II	C.S.E. Aviation Ltd.	
G-BJBZ	Rotorway 133 Executive	Rotorway (UK) Ltd.	
G-BJCA	PA-28-161 Warrior II	Link Services Ltd.	
G-BJCC	Unicorn UE.1A balloon	R. J. Pooley	
G-BJCD	Bede BD-5BH	Brockmoor-Bede Aircraft (UK) Ltd.	
G-BJCE	Cessna F.172P	A & G Aviation Ltd.	
G-BJCF	CP.1310-C3 Super Emeraude	M. W. Wooldridge & A. T. Fines	
G-BJCG	Fournier RF-5	Executive Air Sport	
G-BJCH	Ocset I balloon	B.H.M.E.D. Balloon Group	
G-BJCI	PA-18-150 Super Cub	M. T. A. Sands	
G-BJCJ	PA-28-181 Archer II	Frost & Frost	
G-BJCL	Morane Saulnier M.S.230	B. J. S. Grey	
G-BJCM	FRED Srs. 2	J. C. Miller	
G-BJCN	—	—	
G-BJCO	—	—	
G-BJCP	Unicorn UE.2B balloon	Unicorn Group	
G-BJCR	Partenavia P.68C	Alvair Aviation Ltd.	
G-BJCS	Meagher Mk. 2 balloon	S. A. Fowler	
G-BJCT	Boeing 737-204ADV	Britannia Airways Ltd.	
G-BJCU	Boeing 737-204ADV	Britannia Airways Ltd.	
G-BJCV	Boeing 737-204ADV	Britannia Airways Ltd.	
G-BJCW	PA-31R-301 Saratoga SP	C.S.E. Aviation Ltd.	
G-BJCX	—	—	
G-BJCY	Slingsby T.67A	Slingsby Engineering Ltd.	
G-BJCZ	—	—	
G-BJDB	SC.7 Skyvan Srs. 3	Short Bros. Ltd.	
G-BJDC	SC.7 Skyvan Srs. 3	Short Bros. Ltd.	
G-BJDD	SC.7 Skyvan Srs. 3	Short Bros. Ltd.	
G-BJDE	Cessna F.172M	Sibson Aero Sales Ltd.	
G-BJDF	M.S.880B Rallye 100T	D. M. Leonard	
G-BJDG	SOCATA TB.10 Tobago	Sutton Windows Ltd.	
G-BJDI	Cessna FR.182	Northair Aviation Ltd.	
G-BJDJ	H.S.125 Srs. 700B	Consolidated Contractors (UK) Services Ltd.	
G-BJDK	European E.14 balloon	Aeroprint Tours	
G-BJDL	Rango NA.9 balloon	D. Lawrance	
G-BJDM	SA.102-5 Cavelier	J. D. McCracken	
G-BJDN	AA-5A Cheetah	Cabair Ltd.	
G-BJDO	AA-5A Cheetah	Cabair Ltd.	
G-BJDP	Cremer Cloudcruiser balloon	P. J. Petitt & M. J. Harper	
G-BJDR	Fokker S.11-1 Instructor	J. D. Read	
G-BJDS	British Bulldog balloon	A. J. Cremer	
G-BJDT	SOCATA TB.9 Tampico	Air Touring Services Ltd.	
G-BJDU	Scruggs BL.2B-2 balloon	C. D. Ibell	
G-BJDV	Kingram balloon	T. J. King & S. Ingram	
G-BJDW	Cessna F.172M	E. P. Collier	
G-BJDX	Scruggs BL.2D-2 balloon	A. R. Maple	
G-BJDY	Unicorn UE.4A balloon	Unicorn Group	
G-BJDZ	Unicorn UE.1A balloon	A. P. & K. E. Chown	
G-BJEA	BN-2B Islander	Pilatus BN Ltd.	
G-BJEB	BN-2B Islander	Pilatus BN Ltd.	
G-BJEC	BN-2B Islander	Pilatus BN Ltd.	
G-BJED	BN-2B Islander	Pilatus BN Ltd.	
G-BJEE	BN-2B Islander	Pilatus BN Ltd.	
G-BJEF	BN-2B Islander	Pilatus BN Ltd.	
G-BJEG	BN-2B Islander	Pilatus BN Ltd.	
G-BJEH	BN-2B Islander	Pilatus BN Ltd.	
G-BJEI	PA-19 Super Cub 95	H. J. Cox & D. Platt	
G-BJEJ	BN-2B Islander	Pilatus BN Ltd.	

Notes	Reg.	Type	Owner or Operator
	G–BJEK	BN-2B Islander	Pilatus BN Ltd.
	G–BJEL	Nord NC.854	H. C. Cox
	G–BJEM	Cube balloon	A. J. Cremer
	G–BJEN	Scruggs RS.5000 balloon	N. J. Richardson
	G–BJEO	PA-34-200T-3 Seneca	C.S.E. Aviation Ltd.
	G–BJEP	Scruggs RS.5000 balloon	A. P. & K. E. Chown
	G–BJER	Scruggs RS.5000 balloon	A. P. & K. E. Chown
	G–BJES	Scruggs RS.5000 balloon	J. E. Christopher
	G–BJET	Cessna 425	Gatwick Air Taxis Ltd.
	G–BJEU	Scruggs BL.2D-2 balloon	G. G. Kneller
	G–BJEV	Aeronca 11AC Chief	M. A. Musselwhite
	G–BJEW	Cremer balloon	C. D. Sandoe
	G–BJEX	Bo 208C Junior	G. D. H. Crawford
	G–BJEY	BHMED Srs. 1 balloon	D. R. Meades & J. S. Edwards
	G–BJEZ	Cameron O-105 balloon	Cameron Balloons Ltd.
	G–BJFA	AA-5A Cheetah	Cabair Ltd.
	G–BJFB	Mk. 1A balloon	Aeroprint Tours
	G–BJFC	European E.8 balloon	P. D. Ridout
	G–BJFD	BHMED Srs. 1 balloon	D. G. Dance & I. R. Bell
	G–BJFE	PA-19 Super Cub 95 (L-18C)	C. C. Lovell
	G–BJFF	Enstrom F-28C-UK-2	Southern Air
	G–BJFG	Enstrom F-280C-UK-2	Southern Air
	G–BJFH	Boeing 737-2S3	Air Europe Ltd. *Roma*
	G–BJFI	Bell 47G-2A1	Harvest Air Ltd.
	G–BJFJ	Mk. 1A balloon	A. P. & K. E. Chown
	G–BJFK	Short SD3-30	Inter City/Guernsey Airlines
	G–BJFL	Sikorsky S-76A	Bristow Helicopters Ltd.
	G–BJFM	Jodel D.120	I. R. Ferrars
	G–BJFN	Mk. IV balloon	Windsor Balloon Group
	G–BJFO	Mk. II balloon	Windsor Balloon Group
	G–BJFP	Mk. III balloon	Windsor Balloon Group
	G–BJFR	Mk. IV balloon	Windsor Balloon Group
	G–BJFS	Mk. IV balloon	Windsor Balloon Group
	G–BJFT	Mk. IV balloon	Windsor Balloon Group
	G–BJFU	Mk. IV balloon	Windsor Balloon Group
	G–BJFV	Mk. V balloon	Windsor Balloon Group
	G–BJFW	Mk. V balloon	Windsor Balloon Group
	G–BJFX	Mk. V balloon	Windsor Balloon Group
	G–BJFY	Mk. I balloon	Windsor Balloon Group
	G–BJFZ	Mk. II balloon	Windsor Balloon Group
	G–BJGA	Mk. IV balloon	Windsor Balloon Group
	G–BJGB	Mk. I balloon	Windsor Balloon Group
	G–BJGC	Mk. IV balloon	Windsor Balloon Group
	G–BJGD	Mk. IV balloon	Windsor Balloon Group
	G–BJGE	Thunder Ax3 balloon	Thunder Balloons Ltd.
	G–BJGF	Mk. I balloon	D. & D. Eaves
	G–BJGG	Mk. 2 balloon	D. & D. Eaves
	G–BJGH	Slingsby T.67A	Slingsby Engineering Ltd.
	G–BJGI	H.S.748 Srs. 2B	British Aerospace Public Ltd. Co.
	G–BJGJ	Cessna 441	Northair Aviation Ltd.
	G–BJGK	Cameron V-77 balloon	A. T. Willmer
	G–BJGL	Cremer balloon	G. Lowther
	G–BJGM	Unicorn UE.1A balloon	D. Eaves & P. D. Ridout
	G–BJGN	Scruggs RS.5000 balloon	K. H. Turner
	G–BJGO	Cessna 172N	Powers Aviation Ltd.
	G–BJGP	Mk. 1C balloon	A. P. & K. E. Chown
	G–BJGR	Mk. 2A balloon	A. P. & K. E. Chown
	G–BJGS	Cremer balloon	C. A. Larkins
	G–BJGT	—	—
	G–BJGW	M.H.1521M Broussard	G. A. Warner
	G–BJGX	Sikorsky S-76A	Bristow Helicopters Ltd.
	G–BJGY	Cessna F.172P	Lectime Ltd.
	G–BJGZ	—	—
	G–BJHA	Cremer balloon	G. Cope
	G–BJHB	—	—
	G–BJHC	Swan 1 balloon	C. A. Swan
	G–BJHD	Mk. 3B balloon	S. Meagher
	G–BJHE	Osprey 1B balloon	R. B. Symonds & J. M. Hopkins
	G–BJHF	Osprey 1B balloon	K. R. Bundy
	G–BJHG	Cremer balloon	P. A. Cremer & H. J. A. Green
	G–BJHI	Osprey 1B balloon	A. P. & K. E. Chown
	G–BJHJ	Osprey 1C balloon	D. Eaves
	G–BJHK	EAA Acro Sport	J. H. Kimber

Reg.	Type	Owner or Operator	Notes
G-BJHL	Osprey 1C balloon	E. Bartlett	
G-BJHM	Osprey 1B balloon	W. P. Fulford	
G-BJHN	Osprey 1B balloon	J. E. Christopher	
G-BJHO	Osprey 1C balloon	G. G. Kneller	
G-BJHP	Osprey 1C balloon	N. J. Richardson	
G-BJHR	Osprey 1B balloon	J. E. Christopher	
G-BJHS	S.25 Sunderland V	Sunderland Ltd.	
G-BJHT	Thunder Ax7-65 balloon	Thunder Balloons Ltd.	
G-BJHU	Osprey 1C balloon	G. G. Kneller	
G-BJHV	Voisin Replica	M. P. Sayer	
G-BJHW	Osprey 1C balloon	N. J. Richardson	
G-BJHX	Osprey 1C balloon	A. B. Gulliford	
G-BJHY	Osprey 1B balloon	T. J. King & S. Ingram	
G-BJHZ	Osprey 1B balloon	M. Christopher	
G-BJIA	Allport balloon	D. J. Allport	
G-BJIB	D.31 Turbulent	N. H. Lemon	
G-BJIC	Dodo 1A balloon	P. D. Ridout	
G-BJID	Osprey 1B balloon	P. D. Ridout	
G-BJIE	Sphinx balloon	P. T. Witty	
G-BJIF	Bensen B.8M	H. Redwin	
G-BJIG	Slingsby T.67A	Slingsby Engineering Ltd.	
G-BJIH	Solent balloon	K. Andrews & S. C. Jerrim	
G-BJII	Sphinx balloon	I. French	
G-BJIJ	Osprey 1B balloon	R. Hownsell	
G-BJIK	Osprey 1B balloon	A. P. & K. E. Chown	
G-BJIL	Cessna 550 Citation II	RTZ Services Ltd.	
G-BJIM	—	—	
G-BJIN	Osprey 1B balloon	A. P. & K. E. Chown	
G-BJIO	Osprey 1B balloon	A. P. & K. E. Chown	
G-BJIP	Osprey 1B balloon	A. P. & K. E. Chown	
G-BJIR	Cessna 550 Citation II	Royco Homes Ltd.	
G-BJIS	Mk. 1 balloon	P. Paine	
G-BJIT	Bell 212	Bristow Helicopters Ltd.	
G-BJIU	Bell 212	Bristow Helicopters Ltd.	
G-BJIV	PA-18-150 Super Cub	M. T. A. Sands	
G-BJIW	T-1 balloon	S. Holland & G. Watmore	
G-BJIX	T-1 balloon	S. Holland & G. Watmore	
G-BJIY	—	—	
G-BJIZ	PA-42 Cheyenne III	C.S.E. Aviation Ltd.	
G-BJJA	Kingram 01 balloon	A. P. & K. E. Chown	
G-BJJB	Kingram 01 balloon	A. P. & K. E. Chown	
G-BJJC	Dodo Mk. 1 balloon	A. P. & K. E. Chown	
G-BJJD	Dodo Mk. 1 balloon	A. P. & K. E. Chown	
G-BJJE	Dodo Mk. 3 balloon	D. Eaves	
G-BJJF	Dodo Mk. 4 balloon	D. Eaves	
G-BJJG	Dodo Mk. 5 balloon	D. Eaves	
G-BJJH	Dodo Mk. 4 balloon	A. P. & K. E. Chown	
G-BJJI	SAS balloon	R. Hounsell & M. R. Rooke	
G-BJJJ	Bitterne balloon	A. P. & K. E. Chown	
G-BJJK	Bitterne balloon	R. Hounsell & M. R. Rooke	
G-BJJL	SAS balloon	M. R. Rooke	
G-BJJM	Bitterne balloon	A. P. & K. E. Chown	
G-BJJN	Cessna F.172M	C. P. Hawker	
G-BJJO	Bell 212	Bristow Helicopters Ltd.	
G-BJJP	Bell 212	Bristow Helicopters Ltd.	
G-BJJR	Bell 212	Bristow Helicopters Ltd.	
G-BJJS	Sphinx balloon	C. N. Childs	
G-BJJT	Mabey balloon	M. W. Mabey	
G-BJJU	Sphinx balloon	T. M. Bates	
G-BJJV	Beech B200 Super King Air	Eagle Aircraft Services Ltd.	
G-BJJW	Mk. B balloon	S. Meagher	
G-BJJX	Mk. B balloon	S. Meagher	
G-BJJY	Mk. B balloon	S. Meagher	
G-BJJZ	Unicorn UE.1A balloon	R. Woodley	
G-BJKA	SA.365C Dauphin 2	Management Aviation Ltd.	
G-BJKB	SA.365C Dauphin 2	Management Aviation Ltd.	
G-BJKC	Mk. B balloon	S. Meagher	
G-BJKD	Mk. B balloon	S. Meagher	
G-BJKE	Mk. A balloon	D. Addison	
G-BJKF	SOCATA TB.9 Tampico	Air Touring Services Ltd.	
G-BJKG	Mk. A balloon	D. Addison	
G-BJKH	Mk. A balloon	D. Addison	
G-BJKI	Mk. A balloon	D. Addison	

Notes	Reg.	Type	Owner or Operator
	G–BJKJ	Mk. A balloon	D. Addison
	G–BJKK	Mk. A balloon	D. Addison
	G–BJKL	Mk. A balloon	D. Addison
	G–BJKM	Mk. II balloon	S. Meagher
	G–BJKN	Mk. I balloon	D. Addison
	G–BJKO	Mk. I balloon	D. Addison
	G–BJKP	Mk. 7 balloon	D. Addison
	G–BJKR	Mk. I balloon	D. Addison
	G–BJKS	Mk. I balloon	D. Addison
	G–BJKT	Mk. B balloon	S. Meagher
	G–BJKU	Osprey IB balloon	S. A. Dalmas & P. G. Tarr
	G–BJKV	Osprey IF balloon	B. Diggle
	G–BJKW	Willis Aera II	J. K. S. Willis
	G–BJKX	Cessna F.152	Westair Flying Services Ltd.
	G–BJKY	Cessna F.152	Westair Flying Services Ltd.
	G–BJKZ	Osprey IF balloon	M. J. N. Kirby
	G–BJLA	Osprey IB balloon	D. Lawrence
	G–BJLB	Nord NC.854S	M. J. Barmby
	G–BJLC	Monnet Sonerai IIL	J. P. Whitham
	G–BJLD	Eagle 8 Mk. 2 balloon	R. M. Richards
	G–BJLE	Osprey IB balloon	I. Chadwick
	G–BJLF	Unicorn UE.IC balloon	I. Chadwick
	G–BJLG	Unicorn UE.IB balloon	I. Chadwick
	G–BJLH	PA-19 Super Cub 95	Keenair Services Ltd.
	G–BJLI	D.H.82A Tiger Moth	R. G. Grocott
	G–BJLJ	Cameron D-50 balloon	Cameron Balloons Ltd.
	G–BJLL	Short SD3-30	Short Bros. Ltd.
	G–BJLM	Short SD3-30	Short Bros. Ltd.
	G–BJLN	Featherlight Mk. 3 balloon	A. P. Newman & T. J. Sweeting
	G–BJLO	PA-31-310 Navajo	Airmore Aviation Ltd.
	G–BJLP	Featherlight Mk. 3 balloon	N. P. Kemp & M. O. Davies
	G–BJLR	Featherlight Mk. 3 balloon	M. O. Davies & S. R. Roberts
	G–BJLS	Cessna 340A	J. L. Shaw
	G–BJLT	Featherlight Mk. 3 balloon	J. M. J. Roberts & C. C. Marshall
	G–BJLU	Featherlight Mk. 3 balloon	T. J. Sweeting & N. P. Kemp
	G–BJLV	Sphinx balloon	L. F. Guyot
	G–BJLW	Gleave CJ-I balloon	C. J. Gleave
	G–BJLX	Cremer balloon	P. W. May
	G–BJLY	Cremer balloon	P. Cannon
	G–BJLZ	Cremer balloon	S. K. McLean
	G–BJMA	Colt 21A balloon	Colt Balloons Ltd.
	G–BJMB	Osprey IB balloon	S. Meagher
	G–BJMC	Osprey Mk. 4A balloon	A. P. Chown
	G–BJMD	Osprey Mk. 5 balloon	A. P. Chown & K. R. Bundy
	G–BJME	Osprey Mk. 5 balloon	A. P. Chown
	G–BJMF	Osprey Mk. 4C balloon	A. P. Chown
	G–BJMG	European E.26C balloon	D. Eaves & A. P. Chown
	G–BJMH	Osprey Mk. 3A balloon	D. Eaves
	G–BJMI	European E.84 balloon	D. Eaves
	G–BJMJ	Bensen B.8M	P. R. Snowden
	G–BJMK	Cremer balloon	B. J. Larkins
	G–BJML	—	—
	G–BJMM	Cremer balloon	M. J. Larkins
	G–BJMN	Beech C90 King Air	Airmore Aviation Ltd.
	G–BJMO	Taylor JT.I Monoplane	R. C. Mark
	G–BJMP	Brugger Colibri M.B.2	F. Skinner
	G–BJMR	Cessna 310R	A-One Transport (Leeds) Ltd.
	G–BJMS	AT-16 Harvard IV	N. Grey
	G–BJMT	Osprey Mk. IE balloon	M. J. Sheather
	G–BJMU	European E.157 balloon	A. C. Mitchell
	G–BJMV	BAC One-Eleven 531FS	Dan-Air Services Ltd.
	G–BJMW	Thunder Ax8-105 balloon	Thunder Balloons Ltd.
	G–BJMX	Jarre JR.3 balloon	P. D. Ridout
	G–BJMY	AS.350B Ecureuil	McAlpine Helicopters Ltd.
	G–BJMZ	European EA.8A balloon	P. D. Ridout
	G–BJNA	Arena Mk. 117P balloon	P. D. Ridout
	G–BJNB	WAR F4U Corsair	A. V. Francis
	G–BJNC	Osprey Mk. IE balloon	G. Whitehead
	G–BJND	Osprey Mk. IE balloon	A. Billington & D. Whitmore
	G–BJNE	Osprey Mk. IE balloon	D. R. Sheldon
	G–BJNF	Cessna F.152	Lectime Ltd.
	G–BJNG	Slingsby T.67A	Slingsby Engineering Ltd.
	G–BJNH	Osprey Mk. IE balloon	D. A. Kirk

Reg.	Type	Owner or Operator	Notes
G–BJNI	Osprey Mk. 1C balloon	M. J. Sheather	
G–BJNJ	—	—	
G–BJNK	Sikorsky S-76A	Management Aviation Ltd.	
G–BJNL	Evans VP-2	K. Morris	
G–BJNM	Westland Sea King Mk. 4	Westland Helicopters Ltd.	
G–BJNN	PA-38-112 Tomahawk	C.S.E. Aviation Ltd.	
G–BJNO	AA-5B Tiger	Cabair Ltd.	
G–BJNP	Rango NA.32 balloon	N. H. Ponsford	
G–BJNR	SA. Bulldog 120 Srs. 123	British Aerospace Public Ltd. Co.	
G–BJNS	SA. Bulldog 120 Srs. 123	British Aerospace Public Ltd. Co.	
G–BJNT	SA. Bulldog 120 Srs. 123	British Aerospace Public Ltd. Co.	
G–BJNU	SA. Bulldog 120 Srs. 123	British Aerospace Public Ltd. Co.	
G–BJNV	SA. Bulldog 120 Srs. 123	British Aerospace Public Ltd. Co.	
G–BJNW	EAA Sport Biplane P.2	A. R. Thompson & M. J. Barton	
G–BJNX	Cameron 0–65 balloon	B. J. Petteford	
G–BJNY	Aeronca 11CC Chief	R. A. C. Hoppenbrouwers	
G–BJNZ	PA-23 Aztec 250	Park Management Ltd. (G–FANZ)	
G–BJOA	PA-28-181 Archer II	C.S.E. Aviation Ltd.	
G–BJOB	Jodel D.140C	B. E. Cotton	
G–BJOC	Colt 240A balloon	Colt Balloons Ltd.	
G–BJOD	Sportster HA-2M	H. J. Goddard	
G–BJOE	Jodel D.120A	J. R. Ramshaw	
G–BJOF	Partenavia P.68B	Kingsmetal Ltd.	
G–BJOG	BN-2T Islander	Pilatus BN Ltd.	
G–BJOH	BN-2T Islander	Pilatus BN Ltd.	
G–BJOI	Isaacs Special	J. O. Isaacs	
G–BJOJ	BN-2B Islander	Pilatus BN Ltd.	
G–BJOK	BN-2B Islander	Pilatus BN Ltd.	
G–BJOL	BN-2B Islander	Pilatus BN Ltd.	
G–BJOM	BN-2B Islander	Pilatus BN Ltd.	
G–BJON	BN-2B Islander	Pilatus BN Ltd.	
G–BJOO	BN-2B Islander	Pilatus BN Ltd.	
G–BJOP	BN-2B Islander	Pilatus BN Ltd.	
G–BJOR	BN-2B Islander	Pilatus BN Ltd.	
G–BJOS	BN-2B Islander	Pilatus BN Ltd.	
G–BJOT	Jodel D.117	G. J. Anderson & ptnrs.	
G–BJOU	BN-2B Islander	Pilatus BN Ltd.	
G–BJOV	—	—	
G–BJOW	—	—	
G–BJOX	—	—	
G–BJOY	H.S.125 Srs. 600B	H. Goodman (G–BBEP)	
G–BJOZ	Scheibe SF.25B Falke	P. W. Hextall	
G–BJPA	Osprey Mk. 3A balloon	N. D. Brabham	
G–BJPB	Osprey Mk. 4A balloon	C. B. Rundle	
G–BJPC	Cremer 1 gyroplane	P. A. Cremer	
G–BJPD	Osprey Mk. 4D balloon	E. L. Fuller	
G–BJPE	Osprey Mk. 1E balloon	M. A. Hase	
G–BJPF	Setco Mk. 5LA balloon	S. A. Hassell	
G–BJPG	Osprey Mk. 4D balloon	A. P. Chown	
G–BJPH	Osprey Mk. 3G balloon	K. R. Bundy	
G–BJPI	BD.5-G	M. D. McQueen	
G–BJPJ	Osprey Mk. 3A	K. R. Bundy	
G–BJPK	Osprey Mk. 1B balloon	G. M. Hocquard	
G–BJPL	Osprey Mk. 4A balloon	M. Vincent	
G–BJPM	Bursell PW.1 balloon	I. M. Holdsworth	
G–BJPN	JK Mk. 1 balloon	A. Kaye & J. Corcoran	
G–BJPO	BC balloon	S. Browne & J. Cheetham	
G–BJPP	Rango NA.8 balloon	A. P. Chown	
G–BJPR	Avenger T.200-2112 balloon	A. P. Chown & S. A. Hassell	
G–BJPS	Osprey Mk. 4B balloon	S. A. Hassell	
G–BJPT	Osprey Mk. 3G balloon	A. P. Chown	
G–BJPU	Osprey Mk. 4B balloon	P. Globe	
G–BJPV	Haigh balloon	M. J. Haigh	
G–BJPW	Osprey Mk. 1C balloon	P. J. Cooper & M. Draper	
G–BJPX	Phoenix balloon	Cupro Sapphire Ltd.	
G–BJPY	Cremer balloon	P. A. Cremer & P. V. M. Green	
G–BJPZ	Osprey Mk. 1C balloon	C. E. Newman	
G–BJRA	Osprey Mk. 4B balloon	E. Osborn	
G–BJRB	European E.254 balloon	D. Eaves	
G–BJRC	European E.84R balloon	D. Eaves	
G–BJRD	European E.84R balloon	D. Eaves	
G–BJRE	European E.3 balloon	A. P. Chown	

Notes	Reg.	Type	Owner or Operator
	G–BJRF	Saffery S.80 balloon	C. F. Chipping
	G–BJRG	Osprey Mk. 4B balloon	A. de Gruchy
	G–BJRH	—	—
	G–BJRI	Osprey Mk. 4D balloon	G. G. Kneller
	G–BJRJ	Osprey Mk. 4D balloon	G. G. Kneller
	G–BJRK	Osprey Mk. 1E balloon	G. G. Kneller
	G–BJRL	Osprey Mk. 4B balloon	G. G. Kneller
	G–BJRM	Cremer balloon	A. P. Chown & K. R. Bundy
	G–BJRN	Graham balloon	D. G. Goose
	G–BJRO	Osprey Mk. 4D balloon	M. Christopher
	G–BJRP	Cremer balloon	M. Williams
	G–BJRR	Cremer balloon	M. Wallbank
	G–BJRS	Cremer balloon	M. Wallbank
	G–BJRT	BAC One-Eleven 528	British Caledonian Airways Ltd.
	G–BJRU	BAC One-Eleven 528	British Caledonian Airways Ltd.
	G–BJRV	Cremer balloon	G. Clarke
	G–BJRW	Cessna U.206G	A. I. Walgate & Son Ltd.
	G–BJRX	RMB Mk. 1 balloon	R. J. MacNeil
	G–BJRY	PA-28-151 Warrior	Gordon-Air Ltd.
	G–BJRZ	Partenavia P.68C	Alvair Aircraft Sales Ltd.
	G–BJSA	—	—
	G–BJSB	Cessna A.185F	Harvest Air Ltd.
	G–BJSC	Osprey Mk. 4D balloon	N. J. Richardson
	G–BJSD	Osprey Mk. 4D balloon	N. J. Richardson
	G–BJSE	Osprey Mk. 1E balloon	J. E. Christopher
	G–BJSF	Osprey Mk. 4B balloon	N. J. Richardson
	G–BJSG	V.S. Spitfire LF.IXE	B. J. S. Grey
	G–BJSH	Sindlinger Hurricane $\frac{5}{8}$ scale replica	A. F. Winstanley
	G–BJSI	Osprey Mk. 1E balloon	N. J. Richardson
	G–BJSJ	Osprey Mk. 1E balloon	M. Christopher
	G–BJSK	Osprey Mk. 4B balloon	J. E. Christopher
	G–BJSL	Flamboyant Ax7-65 balloon	Pepsi Cola International Ltd.
	G–BJSM	Bursell Mk. 1 balloon	M. C. Bursell
	G–BJSN	Beech B200 Super King Air	Eagle Aircraft Services Ltd.
	G–BJSO	Boeing 737-266	Monarch Airlines Ltd.
	G–BJSP	—	—
	G–BJSR	Osprey Mk. 4B balloon	C. F. Chipping
	G–BJSS	Allport balloon	D. J. Allport
	G–BJST	—	—
	G–BJSU	Bensen B.8M	J. D. Newlyn
	G–BJSV	—	—
	G–BJSW	Thunder Ax7-65 balloon	Thunder Balloons Ltd.
	G–BJSX	Unicorn UE-1C balloon	N. J. Richardson
	G–BJSY	—	—
	G–BJSZ	—	—
	G–BJTA	Osprey Mk. 4B balloon	C. F. Chipping
	G–BJTB	—	—
	G–BJTC	—	—
	G–BJTD	Colt AS-90 airship	Colt Balloons Ltd.
	G–BJTE	PA-38-112 Tomahawk	C.S.E. Aviation Ltd.
	G–BJTF	Skyrider Mk. 1 balloon	D. A. Kirk
	G–BJTG	Osprey Mk. 4B balloon	M. Millen
	G–BJTH	Kestrel AC Mk. 1 balloon	G. Whitehead
	G–BJTI	Woodie K2400J Mk. 2 balloon	M. J. Woodward
	G–BJTJ	Osprey Mk. 4B balloon	G. Hocquard
	G–BJTK	Taylor JT.1 Monoplane	P. J. Hart
	G–BJTL	H.S.748 Srs. 2B	British Aerospace
	G–BJTM	H.S.748 Srs. 2B	British Aerospace
	G–BJTN	Osprey Mk. 4B balloon	M. Vincent
	G–BJTO	Piper L-4H Cub	K. R. Nunn
	G–BJTP	PA-19 Super Cub 95	J. T. Parkins
	G–BJTS	Osprey Mk. 4B balloon	G. Hocquard
	G–BJTT	Sphinx SP.2 balloon	N. J. Godfrey
	G–BJTU	Cremer Cracker balloon	D. R. Green
	G–BJTV	M.S.880B Rallye Club	J. M. Kirk
	G–BJTW	European E.107 balloon	C. J. Brealey
	G–BJTX	—	—
	G–BJTY	Osprey Mk. 4B balloon	A. E. de Gruchy
	G–BJTZ	Osprey Mk. 4A balloon	M. J. Sheather
	G–BJUA	Sphinx SP.12 balloon	T. M. Pates
	G–BJUB	BVS Special 01 balloon	P. G. Wild

Reg.	Type	Owner or Operator	Notes
G-BJUC	—	—	
G-BJUD	Robin DR.400/180R	Southern Sailplanes Ltd.	
G-BJUE	Osprey Mk. 4B balloon	M. Vincent	
G-BJUF	SH Aerostatics Srs. 2 balloon	S. Haldenby	
G-BJUG	SOCATA TB.9 Tampico	Air Touring Services Ltd.	
G-BJUH	Unicorn UE.1C balloon	A. P. Chown	
G-BJUI	Osprey Mk. 4B balloon	B. A. de Gruchy	
G-BJUJ	Short SD3–30	Short Bros. Ltd.	
G-BJUK	Short SD3-30	Short Bros. Ltd.	
G-BJUL	Short SD3-30	Short Bros. Ltd.	
G-BJUM	Unicron UE.1C balloon	S. A. Hassell	
G-BJUN	Unicorn UE.1C balloon	K. R. Bundy	
G-BJUO	Unicorn UE.4B balloon	Unicorn Group	
G-BJUP	Osprey Mk. 4B balloon	W. J. Pill	
G-BJUR	—	—	
G-BJUS	PA-38-112 Tomahawk	C.S.E. Aviation Ltd.	
G-BJUT	—	—	
G-BJUU	Osprey Mk. 4B balloon	M. Vincent	
G-BJUV	Cameron V-20 balloon	Cameron Balloons Ltd.	
G-BJUW	Osprey Mk. 4B balloon	C. F. Chipping	
G-BJUX	Bursell balloon	I. M. Holdsworth	
G-BJUY	Colt Ax-77 balloon	Colt Balloons Ltd.	
G-BJUZ	BAT Mk. II balloon	A. R. Thompson	
G-BJVA	BAT Mk. I balloon	B. L. Thompson	
G-BJVB	Cremcorn Ax1.4 balloon	P. A. Cremer & I. Chadwick	
G-BJVC	—	—	
G-BJVE	—	—	
G-BJVF	Thunder Ax3 balloon	Thunder Balloons Ltd.	
G-BJVG	Thunder Ax8-105 balloon	Thunder Balloons Ltd.	
G-BJVH	Cessna F.182Q	Phillips Plasters (Leeds) Ltd.	
G-BJVI	Osprey Mk. 4D balloon	S. M. Colville	
G-BJVJ	Cessna F.152	Northair Aviation Ltd.	
G-BJVK	—	—	
G-BJVL	Hermes balloon	Cupro Sapphire Ltd.	
G-BJVM	Cessna 172N	M. A. Cullen	
G-BJVN	—	—	
G-BJVO	Cameron D-50 balloon	Cameron Balloons Ltd.	
G-BJVP	Cessna 550 Citation II	Jetsave Ltd.	
G-BJVR	PA-38-112 Tomahawk	C.S.E. Aviation Ltd.	
G-BJVS	CP.1315C-3 Super Emeraude	Victor Sierra Aero Club	
G-BJVT	Cessna F.152	Rogers Aviation Ltd.	
G-BJVU	Thunder Ax6-56 balloon	Thunder Balloons Ltd.	
G-BJVV	Robin R.1180	Headcorn Flying School Ltd.	
G-BJVW	—	—	
G-BJVX	—	—	
G-BJVY	—	—	
G-BJVZ	—	—	
G-BJWA	Short SD3-30	Short Bros. Ltd.	
G-BJWB	H.S.125 Srs. 700B	Opencity Ltd.	
G-BJWC	—	—	
G-BJWD	Zenith CH.300	D. Winton	
G-BJWE	—	—	
G-BJWF	—	—	
G-BJWG	Beech B200 Super King Air	Eagle Aircraft Services Ltd.	
G-BJWH	—	—	
G-BJWI	—	—	
G-BJWJ	—	—	
G-BJWK	—	—	
G-BJWL	—	—	
G-BJWM	—	—	
G-BJWT	Wittman W.10 Tailwind	J. F. Bakewell & R. A. Shelley	
G-BKAA	H.S.125 Srs. 700B	Aravco Ltd.	
G-BKAT	Pitts S-1C Special	I. M. G. Senior & J. G. Harper	
G-BKAY	R. Commander 114	B.K. Aviation Ltd.	
G-BKCL	PA-30 Twin Comanche 160C	Jubilee Airways Ltd. (G–AXSP)	
G-BKCM	Bell 206B JetRanger 3	Patgrove Ltd.	
G-BKDE	Kendrick I	J. K. Rushton	
G-BKDF	Kendrick II	J. K. Rushton	
G-BKGF	Saxon II	Saxon Aircraft Co. Ltd.	
G-BKJW	PA-23 Aztec 250	Alan Williams Enterprises Ltd.	
G-BKOB	Z.326 Trener Master	W. G. V. Hall	
G-BKPC	Cessna A.185F	Black Knights Parachute Centre	

Out-of-sequence registrations

Notes	Reg.	Type	Owner or Operator
	G–BLAC	Cessna FA.152	Westair Flying Services Ltd.
	G–BLAL	Cessna F.150K	Smith School of Flying Ltd.
	G–BLCE	Cessna 402C	Cecil Aviation Ltd.
	G–BLCG	SOCATA TB.10 Tobago	A. M. & Mrs. H. T. Vaughan-Read (G–BHES)
	G–BLEW	Cessna F.182Q	Interair Aviation Ltd.
	G–BLGS	SOCATA Rallye 180GT	Lasham Gliding Soc. Ltd.
	G–BLGW	F.27 Friendship Mk. 200	Air UK
	G–BLHN	Robin HR.100/285	H. M. Bonquiere
	G–BLJM	Beech 95-B55 Baron	Playtime Ltd.
	G–BLPP	Cameron V-77 balloon	L. P. Purfield *Merlin*
	G–BLPT	R. Turbo Commander 690B	Tozer, Kemsley & Millbourne Holdings Ltd. (G–LACY)
	G–BLRJ	Jodel DR.1051	M. P. Hallam
	G–BLSL	Cessna 310R	Rassler Aero Services
	G–BLST	Cessna 421C	Cecil Aviation Ltd.
	G–BLUE	Colting Ax-77 balloon	Lighter-Than-Air Ltd. *Bluebird*
	G–BMAA	Douglas DC-9-15	British Midland Airways Ltd. *Dovedale* (G–BFIH)
	G–BMAB	Douglas DC-9-15	British Midland Airways Ltd. *Ulster*
	G–BMAC	Douglas DC-9-15	British Midland Airways Ltd.
	G–BMAF	Cessna 180F	Missionary Aviation Fellowship (G–BDVR)
	G–BMAL	Sikorsky S-76A	Management Aviation Ltd.
	G–BMAS	F.27 Friendship Mk. 200	British Midland Airways Ltd.
	G–BMAT	V.813 Viscount	British Midland Airways Ltd. (G–AZLT)
	G–BMAV	AS.350B Ecureuil	Masselaz Helicopters Ltd.
	G–BMAX	FRED Srs. 2	P. Cawkwell
	G–BMCA	Beech A200 Super King Air	Marchwiel Plant & Engineering Co. Ltd.
	G–BMCL	Cessna 550 Citation II	Micro Consultants Ltd.
	G–BMCP	Maule M5-235C Lunar Rocket	M. C. Peirce
	G–BMEC	Boeing 737-2S3	Air Europe Ltd. *Joy*
	G–BMEL	PA-23 Aztec 250	Eagle Surface Coatings Ltd.
	G–BMFD	PA-23 Aztec 250	Bomford & Evershed Ltd. (G–BGYY)
	G–BMHC	Cessna U.206F	Mechanical Handling Consultants (Aviation) Ltd.
	G–BMHG	Boeing 737-2S3	Air Europe Ltd. *Adam*
	G–BMID	Jodel D.120	A. W. Cooke
	G–BMIP	Jodel D.112	M. T. Kinch
	G–BMJH	Hughes 369D	Hughes of Beaconsfield Ltd.
	G–BMLM	Beech 95-58 Baron	Mowlem Construction (Plant Hire) Ltd. (G–BBJF)
	G–BMON	Boeing 737-2K9	Monarch Airlines Ltd.
	G–BMOR	Boeing 737-2S3	Air Europe Ltd. *Eve*
	G–BMSB	V.S. Spitfire IX	M. S. Bayliss
	G–BMSF	PA-38-112 Tomahawk	Manchester School of Flying Ltd.
	G–BMSM	Boeing 737-253	Air Europe Ltd. *Roma*
	G–BMUD	Cessna 182P	Drill-Aid (Method & Materials) Ltd.
	G–BMVV	Rutan Vari-Viggen	G. B. Morris
	G–BMYU	Jodel D.120	G. H. Wilde
	G–BNHP	Saffery S.330 balloon	N. H. Ponsford *Alpha II*
	G–BNJF	PA-32RT-300 Turbo Lance II	N. J. Froment & Co. Ltd.
	G-BNOC	EMB-110P1 Bandeirante	Fairflight Ltd./Air Ecosse
	G–BNPD	PA-23 Aztec 250	Northern Pig Development Co. Ltd.
	G–BNSH	Sikorsky-S-76A	Management Aviation Ltd.
	G–BOAA	Concorde 102	British Airways (G–N94AA)
	G–BOAB	Concorde 102	British Airways (G–N94AB)
	G–BOAC	Concorde 102	British Airways (G–N81AC)
	G–BOAD	Concorde 102	British Airways (G–N94AD)
	G–BOAE	Concorde 102	British Airways (G–N94AE)
	G–BOAF	Concorde 102	British Airways (G–N94AF/G–BFKX)
	G–BOAG	Concorde 102	British Airways (G–BFKW)
	G–BOAT	Cessna 310R	Birchwood Boat International Ltd.
	G–BOBI	Cessna 152	R. Macbeth (G–BHJD)
	G–BOBY	Monnet Sonerai II	R. G. Hallam
	G–BOLT	R. Commander 114	Hooper & Jones Ltd.

G–BJCR (Top) Partenavia P.68C

G–BMAS (Centre) Fokker F.27 Friendship Mk. 200 of British Midland Airways/*S. G. Richards*

G–JJSG (Bottom) Gates Learjet 35A

Notes	Reg.	Type	Owner or Operator
	G–BOMB	Cassutt Racer	R. W. L. Breckell
	G–BOND	Sikorsky S-76	North Scottish Helicopters Ltd.
	G–BONE	Pilatus P2-06	Aeromech Ltd.
	G–BOOB	Cameron N-65 balloon	E. M. Ten Houten
	G–BOOK	Pitts S-1S Special	B. K. Lecomber
	G–BOOM	Hunter T.7	B. R. Kay
	G–BOSL	Boeing 737-2U4ADV	Britannia Airways Ltd *Sir Frank Whittle*
	G–BOSS	PA-34-200T Seneca	Elmsleigh Autos (Contract Hire) Ltd.
	G–BPAJ	D.H.82A Tiger Moth	P. A. Jackson (G–AOIX)
	G–BPAM	Jodel D.150A	A. J. Symes-Bullen
	G–BPAR	PA-31-350 Navajo Chieftain	Burnthills Plant Hire Ltd.
	G–BPAV	FRED Srs. 2	P. A. Valentine
	G–BPAZ	Pazmany PL.2	T. Davies & P. H. Chamberlain
	G–BPBN	BN-2T Islander	Pilatus BN Ltd. (G–BCMY)
	G–BPBP	Brugger Colibri Mk. II	B. Perkins
	G–BPDW	Eezybuild Envoy Mk. 2	P. D. Wheatland
	G–BPEG	Currie Wot	D. M. Harrington
	G–BPFA	Swallow GK-2	G. Knight & D. G. Pridham
	G–BPMB	Maule M5-235C Lunar Rocket	P. M. Breton
	G–BPMN	Super Coot Model A	P. Napp
	G–BPPN	Cessna F.182Q	Hunt Norris Ltd.
	G–BPUF	Thunder Ax6-56Z balloon	Buf-Puf Balloon Group *Buf-Puf*
	G–BPYN	Piper J-3C-65 Cub	D. W. Stubbs & ptnrs.
	G–BPZD	Nord NC.858S	Toon Ghose Aviation Ltd.
	G–BRAD	Beech 95-B55 Baron	Finrad Ltd.
	G–BRAF	V.S. Spitfire 18 (SM969)	D. W. Arnold
	G–BRAG	Taylor JT.2 Titch	A. R. Greenfield
	G–BRAL	G.159 Gulfstream I	Ford Motor Co. Ltd.
	G–BRAT	Jodel D.120A	C. T. Carlisle & ptnrs.
	G–BRED	Canadair CL-44D-4	Redcoat Air Cargo *James* (G–AZKJ)
	G–BREF	Cessna 421C	Refair Ltd.
	G–BREL	Cameron 0-77 balloon	BICC Research & Engineering Ltd.
	G–BREN	Cameron Jeans Special Shape balloon	Leaveglen Ltd.
	G–BREW	PA-31-350 Navajo Chieftain	Whitbread & Co. Ltd.
	G–BRFC	P.57 Sea Prince T.1 (WP321)	Rural Flying Corps
	G–BRGH	FRED Srs. 2	F. G. Hallam
	G–BRGV	PA-31-350 Navajo Chieftain	Centreline Air Services Ltd.
	G–BRGW	GY-201 Minicab	R. G. White
	G–BRHD	PA-23 Aztec 250F	Omega Consultants Ltd.
	G–BRIC	Cameron V-65 balloon	Astra Pharmaceuticals Ltd.
	G–BRIK	Tipsy Nipper 3	C. W. R. Piper
	G–BRIT	Cessna 421C	Britannia Airways Ltd.
	G–BRIX	PA-32-301 Saratoga SP	Taylor Maxwell & Co. Ltd.
	G–BRJP	Boeing 737-2S3	Air Europe Ltd.
	G–BRMA	*WS-51 Dragonfly Mk. 5 (WG719)	British Rotocraft Museum
	G–BRMB	*B.192 Belvedere Mk. I (XG452)	British Rotorcraft Museum
	G–BRMH	Bell 206B JetRanger 2	R.M.H. Stainless Ltd. (G–BBUX)
	G–BROM	ICA IS-28M2	British Aerospace Public Ltd. Co.
	G–BRON	Beech A200 Super King Air	Owledge Ltd. (G–BFEA)
	G–BRSL	Cameron N-56 balloon	Balloon Stable Ltd. *Boris*
	G–BRUX	PA-44-180 Seminole	Hambrair Ltd.
	G–BRWG	Maule M5-235C Lunar Rocket	R. W. Gaskell
	G–BRYA	D.H.C.-7 Dash Seven	Brymon Aviation Ltd.
	G–BRYB	D.H.C.-7 Dash Seven	Brymon Aviation Ltd.
	G–BRYC	D.H.C.-7 Dash Seven	Brymon Aviation Ltd.
	G–BSAA	H.S.125 Srs. 3B	Aravco Ltd.
	G–BSAL	G.1159 Gulfstream 2	Shell Aviation Ltd.
	G–BSBH	Short SD3-30	Short Bros. Ltd.
	G–BSDL	SOCATA TB.10 Tobago	Systems Designers Aviation Ltd.
	G–BSEL	Slingsby T-61G	Slingsby Engineering Ltd.
	G–BSFC	PA-38-112 Tomahawk	Sherwood Flying Club Ltd.
	G–BSFL	PA-23 Axtec 250	T. Kilroe & Sons Ltd.
	G–BSFT	PA-31-300 Navajo	Simulated Flight Training Ltd. (G–AXYC)
	G–BSFZ	PA-25 Pawnee 235	Skegness Air Taxi Services Ltd. (G–ASFZ)
	G–BSHL	H.S.125 Srs. 600B	S.H. Services Ltd. (G–BBMD)
	G–BSIS	Pitts S-1S Special	R. A. Mills

Reg.	Type	Owner or Operator	Notes
G–BSIX	Cessna 401	Culinair Ltd. (G–CAFE/G–AWXM)	
G–BSKY	Douglas DC-8-55F	—	
G–BSMC	PA-32-301T Turbo Saratoga	Lake Green Garage Ltd.	
G–BSPC	Jodel D.140C	B. E. Cotton	
G–BSPE	Cessna F.172P	Sceptre Precision Engineers	
G–BSSL	Beech B80 Queen Air	Parker & Heard Ltd. (G–BFEP)	
G–BSST	*Concorde 002	Fleet Air Arm Museum	
G–BSUS	Taylor JT.1 Monoplane	R. Parker	
G–BSVP	PA-23 Aztec 250F	Sun Valley Poultry Ltd.	
G–BSVT	EMB-110P2 Bandeirante	Fairflight Ltd./Air Ecosse (G–BWTV)	
G–BTAL	Cessna F.152	R. & R. Aviation	
G–BTBM	*Grumman TBM-3W2 Avenger	War Birds of GB Ltd., Blackbushe	
G–BTDK	Cessna 421B	Texalon International Ltd.	
G–BTEA	Cameron A-105 balloon	Southern Balloon Group	
G–BTFC	Cessna F.152 II	Tayside Aviation Ltd.	
G–BTFH	Cessna 414A	TBF (Transport) Ltd.	
G–BTGS	Smyth Sidewinder	T. G. Soloman	
G–BTHL	PA-31-350 Navajo Chieftain	Burnthills Plant Hire Ltd.	
G–BTHS	PA-23 Aztec 250F	Byloads Ltd.	
G–BTIE	SOCATA TB.10 Tobago	TIE Project Development Ltd.	
G–BTJM	Taylor JT.2 Titch	T. J. Miller	
G–BTLE	PA-31-350 Navajo Chieftain	Merlix Air Ltd.	
G–BTOM	PA-38-112 Tomahawk	Channel Aviation Ltd.	
G–BTSC	Evans VP-2	D. W. Burrell	
G–BTSF	Evans VP-2	D. W. Burrell	
G–BTSH	Evans VP-2	B. P. Irish	
G–BTUG	SOCATA Rallye 180T	Lasham Gliding Soc. Ltd.	
G–BTWA	Bell 206B JetRanger 2	C. Hughesdon	
G–BTWT	D.H.C.-6 Twin Otter 310	Loganair Ltd.	
G–BUCC	C.A.S.A. 1.131R Jungmann	Buccaneer Aviation Ltd.	
G–BUCK	C.A.S.A. 1.131E Jungmann	Buccaneer Aviation Ltd.	
G–BUFF	Jodel D.112	D. J. Buffham	
G–BUMP	PA-28-181 Archer II	Autofarm Ltd.	
G–BUNA	SNCAN SV-4C	J. Pearce	
G–BURD	Cessna F.172N	Volstatic Aviation Ltd.	
G–BURT	PA-28-161 Warrior II	A. T. Howarth	
G–BUSA	AS.355F Twin Squirrel	Barratt Developments Ltd.	
G–BUSY	Thunder Ax6-56A balloon	B. R. & Mrs. M. Boyle *Busy Bodies*	
G–BUZZ	AB-206B JetRanger 2	Western Air (Scotland) Ltd.	
G–BVMM	Robin HR.200/100	Stately & Co. Ltd.	
G–BVPI	Evans VP-1	N. L. E. & R. A. Dupee	
G–BVPM	Evans VP-2	P. Marigold	
G–BWEC	Cassutt-Colson Variant	W. E. Colson	
G–BWFC	Boeing Vertol 234LR	British Airways Helicopters Ltd.	
G–BWFJ	Evans VP-1	W. F. Jones	
G–BWIG	G.17S replica	K. Wigglesworth	
G–BWJB	Thunder Ax8-105 balloon	Justerini & Brooks Ltd. *Whiskey J. & B.*	
G–BWKK	Auster AOP.9 (XP279)	G. F. Kilsby	
G–BWKS	Lake LA.4-200 Buccaneer	Newell Aircraft & Tool Co. Ltd. (G–BDDI)	
G–BWRB	D.H.C.-6 Twin Otter 310	Brymon Aviation Ltd.	
G–BXNW	SNCAN SV-4C	S. W. Atkins	
G–BXYZ	R. Turbo Commander 690C	British Airports Authority	
G–BYRD	Mooney M.20K	Express Aviation Services Ltd.	
G–BYSE	AB-206B JetRanger 2	Bewise Ltd. (G–BFND)	
G–BZAC	Sikorsky S-76	British Airways Helicopters Ltd.	
G–BZBH	Thunder Ax6-65 balloon	R. S. Whittaker & P. E. Sadler	
G–BZBY	Colt 56 Buzby balloon	Post Office Communications	
G–BZFL	Cessna 401	Suton Productions Ltd. (G–AWSF)	
G–BZKK	Cameron V-56 balloon	P. J. Green & C. Bosley *Gemini II*	
G–CALL	PA-23 Axtec 250F	Aircraft Mart	
G–CBIA	BAC One-Eleven 416	Air UK (G–AWXJ) *Island Ensign*	
G–CBIL	Cessna 182K	S.B. Aviation International Ltd.	
G–CCAR	Cameron N-77 balloon	Colt Car Co. Ltd. *Colt*	
G–CCCC	Cessna 172H	P. D. Higgs	
G–CCOZ	Monnet Sonerai II	P. R. Cozens	
G–CCUB	Piper J-3C-65 Cub	Cormack (Aircraft Services) Ltd.	
G–CDAH	Taylor Super Coot A	D. A. Hood	
G–CDBI	PA-23 Aztec 250	Spacegrand Ltd.	
G–CDGA	Taylor JT.1 Monoplane	D. G. Anderson	
G–CDGL	Saffery S.330 balloon	C. J. Dodd & G. J. Luckett *Penny*	
G–CEGA	PA-34-200T-2 Seneca	Cega Aviation Ltd.	

Notes	Reg.	Type	Owner or Operator
	G–CELT	EMB-110P2 Bandeirante	Fairflight Ltd./Air Ecosse
	G–CENT	Cessna T.210L	CGH Managements Ltd.
	G–CETA	Cessna E.310Q	CETA Video Ltd. (G–BBIM)
	G–CFLY	Cessna 172F	M. J. Stephens
	G–CGFC	PA-38-112 Tomahawk	Cormack (Aircraft Services) Ltd.
	G–CGHM	PA-28 Cherokee 140	CGH Managements Ltd.
	G–CHEV	EMB-110P2 Bandeirante	Fairflight Ltd./Air Ecosse
	G–CHIK	Cessna F.152	Wickwell (UK) Ltd. (G–BHAZ)
	G–CHOP	Westland Bell 47G-3B1	Alec Wortley Ltd.
	G–CITY	PA-31-350 Navajo Chieftain	City Air Links Ltd.
	G–CJAN	PA-28-181 Archer II	Cormack (Aircraft Services) Ltd.
	G–CJBC	PA-28 Cherokee 180	J. B. Cave
	G–CJDH	Pitts S.1S Special	C. J. D. Hackett
	G–CJHI	Bell 206B JetRanger	Tudorbury Air Services Ltd. (G–BBFB)
	G–CJIM	Taylor JT.1 Monoplane	J. Crawford
	G–CLAN	PA-31-350 Navajo Chieftain	British Caledonian Airways
	G–CLEA	PA-28-161 Warrior II	Cleaver Marine Ltd.
	G–CLEF	Beech 95-58P Baron	Springfield Motel
	G–CLEM	Bo 208A2 Junior	G. Clements (G–ASWE)
	G–CLUT	Clutton Special	E. Clutton & A. Tabenor
	G–CLUX	Cessna F.172N	N. J. Hebditch
	G–CNSI	Beech 200 Super King Air	Conoco (UK) Ltd. (G-OSKA)
	G–COAL	Bell 206B JetRanger 3	NSM Aviation Ltd.
	G–COCO	Cessna F.172M	Consumer Concern Ltd.
	G–COLL	Enstrom F-280C	Capel & Co. (Printers) Ltd.
	G–COMM	PA-23 Aztec 250	Commair Aviation Ltd. (G–AZMG)
	G–COOL	Cameron O-31 balloon	Swire Bros. *Sprite*
	G–COOP	Cameron N-31 balloon	Balloon Stable Ltd. *Co-op*
	G–COPE	Enstrom F-280C	A. Cope
	G–COPS	Piper J-3C-65 Cub	P. B. Hanson
	G–COTT	Cameron 60 SS balloon	Nottingham Building Soc.
	G–CPAC	PA-28R Cherokee Arrow 200	Countryside Properties Ltd.
	G–CPFC	Cessna F.152	Central Air Services
	G–CPTS	AB-206B JetRanger 2	A. R. B. Aspinall
	G–CRAN	Robin R.1180T	Headcorn Flying School Ltd.
	G–CRIL	R. Commander 112B	Castle Reeves Investments Ltd.
	G–CRIS	Taylor JT.1 Monoplane	C. J. Bragg
	G–CSBM	Cessna F.150M	Coventry (Civil) Aviation Ltd.
	G–CSFC	Cessna 150L	D. G. Jones & A. T. Jay
	G–CSNA	Cessna 421C	Armstrong Aviation Ltd.
	G–CSSC	Cessna F.152	P. G. Kellow
	G–CSZA	V.807 Viscount	—
	G–CSZB	V.804 Viscount	— (G–AOXU)
	G–CTLN	EMB-110P1 Bandeirante	Centreline Air Services Ltd.
	G–CTRN	Enstrom F-28C-UK	W. E. Taylor & Son Ltd.
	G–CUBB	PA-18-150 Super Cub	W. B. Hosie
	G–CUBI	PA-18-135 Super Cub	G. McLean & ptnrs.
	G–CULT	Cameron D-38 airship	Colt Car Co. Ltd.
	G–CWOT	Currie Wot	D. A. Lord
	G–CXMF	G.1159 Gulfstream 2	Fay Air (Jersey) Ltd.
	G–DAAH	PA-28R-201T Turbo Arrow IV	A. A. Hunter
	G–DACA	P.57 Sea Prince T.1	Atlantic & Caribbean Aviation Ltd.
	G–DAJW	K & S Jungster I	A. J. Walters
	G–DAKS	Dakota 3 (TS423)	Aviation Enterprises
	G–DAND	SOCATA TB.10 Tobago	Dandy Caravans
	G–DANN	Stampe SV-4B	D. R. Scott-Songhurst
	G–DART	Rollason Beta B2	M. G. Ollis
	G–DATA	EMB-110P2 Bandeirante	Fairflight Ltd./Air Ecosse (G–BGNK)
	G–DAVE	Jodel D.112	D. A. Porter
	G–DAVI	Cessna TU.206G	G. A. Dommett
	G–DAVY	Evans VP-2	D. Morris
	G–DAWN	Cessna T.210M	The Earl Jermyn
	G–DBBI	H.S.125 Srs. 700B	Eagle Leasing Ltd.
	G–DCAN	PA-38-112 Tomahawk	Airways Aero Associations Ltd.
	G–DCAT	G.164D Ag-Cat	Miller Aerial Spraying Ltd.
	G–DCIO	Douglas DC-10-30	British Caledonian Airways *Flora McDonald – The Scottish Heroine*
	G–DCKK	Cessna F.172N	Manders Demolition Ltd.
	G–DCOL	PA-30 Twin Commanche 160	W. C. Smeaton (G–BGSU)
	G–DDDV	Boeing 737-2S3	Air Europe Ltd.
	G–DELI	Thunder Ax7-77 balloon	C. Delius
	G–DEVA	PA-23 Aztec 250	Eyelure Ltd.

Reg.	Type	Owner or Operator	Notes
G-DFLY	PA-38-112 Tomahawk	Airways Aero Associations Ltd.	
G-DFTS	Cessna FA.152	Denham Flying Training School Ltd.	
G-DFUB	Boeing 737-2K9	Monarch Airlines Ltd.	
G-DHLD	Beech B60 Duke	T. S. Grimshaw Ltd.	
G-DICK	Thunder Ax6-56Z balloon	Thunder Balloons Ltd.	
G-DINA	AA-5B Tiger	Simon Derereli Print Ltd.	
G-DIPS	Taylor JT.1 Monoplane	B. J. Halls	
G-DIRT	Thunder Ax7-77Z balloon	Thunder Balloons Ltd.	
G-DIVE	BN-2A-26 Islander	Secretary of State (G-BEXA)	
G-DJBE	Cessna 550 Citation II	DJB Engineering Ltd.	
G-DJHH	Cessna 550 Citation II	Armstrong Aviation Ltd.	
G-DJIM	DHCA-1	J. Crawford	
G-DJMS	PA-28-181 Archer II	D. J. McSorley	
G-DMAN	H.S.125 Srs. 600B	McAlpine Aviation Ltd.	
G-DMCH	Hiller UH-12E	D. McK. Carnegie & ptnrs.	
G-DOGS	Cessna R.182RG	Newbranch Ltd.	
G-DOLL	Thunder Ax6-56 S.I. balloon	E. Ten Houten	
G-DOVE	Cessna 182Q	Dove Air Ltd.	
G-DRAY	Taylor JT.1 Monoplane	L. J. Dray	
G-DTOO	PA-38-112 Tomahawk	Airways Aero Associations Ltd.	
G-DUET	Wood Duet	C. Wood	
G-DUKE	Beech B60 Duke	New Equipment Ltd.	
G-DUNN	Zenair CH.250	A. Dunn	
G-DUVL	Cessna F.172N	Duval Studios Ltd.	
G-DWMI	Bell 206L-1 LongRanger	Glenwood Helicopters Ltd.	
G-DYOU	PA-38-112 Tomahawk	Airways Aero Associations Ltd.	
G-EAGL	Cessna 421C	Systime Ltd.	
G-EASI	Short SD3-30	Eastern Airways Ltd. (G-BITW)	
G-EBJI	Hawker Cygnet Replica	A. V. Francis	
G-ECCO	GA-7 Cougar	J. E. Clark	
G-ECGC	Cessna F.172N-II	Leicestershire Aero Club Ltd.	
G-ECMA	PA-31-325 Navajo	Elliott Bros. (London) Ltd.	
G-ECOX	Pietenpol Aircamper GN.1	H. C. Cox	
G-EDDY	PA-28RT-201 Arrow IV	Supaglide Ltd.	
G-EDEN	SOCATA TB.10 Tobago	Tobago Air Services	
G-EDHE	PA-24 Comanche 180	E. D. Hughes (G-ASFH)	
G-EDIF	Evans VP-2	R. Simpson	
G-EEEE	Slingsby T.31 Motor Glider	R. F. Selby	
G-EENY	GA-7 Cougar	Martin Butler Associates Ltd.	
G-EEUP	SNCAN SV-4C	Gladaircraft Co. Ltd.	
G-EEZE	Rutan Vari-Eze	A. J. Nurse	
G-EGGS	Robin DR.400/180	R. Foot	
G-EGLE	Christen Eagle II	Airmore Aviation Ltd.	
G-EHAP	Sportavia-Pützer RF.7	M. J. Revill	
G-EIIR	Cameron N-77 balloon	Major C. J. T. Davey *Silver Jubilee*	
G-ELLY	AT-6D Harvard III (133867)	E. D. Sallingboe	
G-EMKM	Jodel D.120A	C. R. Davey	
G-EMMA	Cessna F.182Q	Watkiss Group Aviation	
G-EMMS	PA-38-112 Tomahawk	Surrey & Kent Flying Club Ltd.	
G-EMMY	Rutan Vari-Eze	M. J. Tooze	
G-ENIE	Nipper T.66 Srs. 3	A. J. Waller	
G-ENII	Cessna F.172M	M. S. Knight	
G-ENOA	Cessna F.172F	Genoa Precision Engineers Ltd. (G-ASZW)	
G-ENSI	Beech F33A Bonanza	F. B. Gibbons & Sons Ltd.	
G-EOFF	Taylor JT.2 Titch	G. Wylde	
G-EORG	PA-38-112 Tomahawk	Cormack (Aircraft Services) Ltd.	
G-EORR	AS.350B Ecureuil	Colt Car Co. Ltd. (G-FERG/ G-BGCW)	
G-EPDI	Cameron N-77 balloon	E.P.D. Containers & Supply Co. Ltd. & Pegasus Aviation Ltd. *Pegasus*	
G-ERDB	Hawker Cygnet Replica	R. D. Bertram	
G-ERIC	R. Commander 112TC	DJF (Jersey) Ltd.	
G-ERMS	Thunder AS33 Airship	Thunder Balloons Ltd.	
G-ERTY	D.H.82A Tiger Moth	E. R. Thomas (G-ANDC)	
G-ESAL	AB-206B JetRanger 3	Esal Commodities Ltd. (G-BHXW)	
G-ETUP	Cessna F.150L	Crepeshaw Ltd.	
G-EURO	Cessna 310R	Carex Exhaust Centres Ltd.	
G-EVAN	Taylor JT.2 Titch	E. Evans	
G-EWBJ	SOCATA TB.10 Tobago	Crocker Air Services	
G-EXEC	PA-34-200 Seneca	Red Dragon Travel Ltd.	
G-EXEX	Cessna 404	Owledge Ltd.	
G-EXIT	SOCATA Rallye 235GT	G-Exit Ltd.	

Notes	Reg.	Type	Owner or Operator
	G–EZLT	Rutan Vari-Eze	L. J. Turner
	G–EZOS	Rutan Vari-Eze	O. Smith
	G–FADS	PA-23 Aztec 250	A. C. Stanley Ltd.
	G–FAIR	SOCATA TB.10 Tobago	Fairfield Aviation Ltd.
	G–FALC	Aviamilano F.8L Falco	P. W. Hunter (G–AROT)
	G–FALL	Cessna 182L	K. J. Toyer
	G–FANL	Cessna FR.172K XP-II	Francis & Lewis Ltd.
	G–FANS	BN-2A-3 Islander	Dowty Rotol Ltd.
	G–FARM	SOCATA Rallye 235GT	M. J. Jardine-Paterson
	G–FARR	Jodel D.150	G. H. Farr
	G–FAST	Cessna F.337G	Shirlstar Container Transport Ltd.
	G–FAYE	Cessna F.150M	Cheshire Air Training School (Merseyside) Ltd.
	G–FBDC	Cessna 340A	Food Brokers Ltd. (G–BFJS)
	G–FBWH	PA-28R Cherokee Arrow 180	B. W. Hopkins
	G–FCAS	PA-23 Aztec 250	Wareprod Engineering Co. Ltd.
	G–FERY	Cessna 550 Citation II	European Ferries Ltd. (G–DJBI)
	G–FFEN	Cessna F.150M	E. P. Collier
	G–FHAS	Scheibe SF.25E Super Falke	Fourth Harrow Aviation
	G–FILM	SE.313B Alouette II	Alan Mann Helicopters Ltd. (G–BANR)
	G–FIRE	V. S. Spitfire XIVc	S. Flack
	G–FISH	Cessna 310R-II	Boston Deep Sea Fisheries Ltd.
	G–FIVE	H.S.125 Srs. 1	British Air Ferries Ltd. (G–ASEC)
	G–FIZZ	PA-28-161 Warrior II	Air Navigation & Trading Co. Ltd.
	G–FLEA	SOCATA TB.10 Tobago	Francis Barker & Son Ltd.
	G–FLIK	Pitts S.1. Special	R. P. Millinship
	G–FLIP	Cessna FA.152	Citation Flying Services Ltd.
	G–FLIX	Cessna E.310P	Peter Long International Services Ltd. (G–AZFL)
	G–FLPI	R. Commander 112A	Tuscany Ltd.
	G–FLYI	PA-34-200 Seneca	G.R.T. Catering Ltd. (G–BHVO)
	G–FMFC	EMB-110P2 Bandeirante	Fairflight Ltd./Air Ecosse
	G–FOCK	Focke-Wulf Fw.190-A	P. R. Underhill
	G–FOIL	PA-31-310 Navajo	Air Foyle Ltd.
	G–FORD	SNCAN SV-4C	P. Meeson
	G–FOTO	PA-23 Aztec 250	Davis Gibson Advertising Ltd. (G–BJDH/G–BDXV)
	G–FOUR	H.S.125 Srs. 3B	Group 4 Aviation Ltd. (G–AVRE)
	G–FOUX	AA-5A Cheetah	Baryn Finance Ltd.
	G–FOXY	Cessna F.172M	Trehaven Trust Ltd.
	G–FOYL	PA-23 Aztec 250	Air Foyle Ltd. (G–AVNK)
	G–FRAG	PA-32-300 Cherokee Six	R. Goodwin & Co. Ltd.
	G–FRED	FRED Srs. 2	R. Cox
	G–FRJB	Britten Sheriff SA-1	Air Bembridge (I.O.W.) Ltd.
	G–FROG	Hughes 369HS	K. Sutcliff & Sons Ltd.
	G–FSPL	PA-32R-300 Lance	R. E. Husband & A. D. Widdows
	G–FTTA	PA-31-350 Navajo Chieftain	Tyne-Tees Airways
	G–FUEL	Robin DR.400/180	R. Darch
	G–FUJI	Fuji FA.200-180	S. T. N. Rollesby
	G–FUND	Thunder Ax7-65Z balloon	Thunder Balloons Ltd.
	G–FURY	Sea Fury T.XI	S. Flack
	G–FUZZ	PA-19 Super Cub 95	G. W. Cline
	G–FVEE	Monnet Sonerai I	D. R. Sparke
	G–FXIV	V.S.379 Spidfire FR.XIV	M. Wickenden & ptnrs.
	G–GACA	P.57 Sea Prince T.1	Atlantic & Caribbean Aviation Ltd.
	G–GALE	PA-34-200-2 Seneca	Gale Construction Co. Ltd.
	G–GBAO	Robin R.1180TD	J. Kay-Movat
	G–GBSC	Beech E90 King Air	General Relays (Crewe) Ltd.
	G–GBSL	Beech 76 Duchess	George Barlow & Sons Ltd. (G–BGVG)
	G–GCAT	PA-28 Cherokee 140B	Cheshire Air Training School Ltd. (G–BFRH)
	G–GCKI	Mooney M.20K	Imperial Group Ltd.
	G–GDAM	PA-18 Super Cub 135	Citation Flying Services Ltd.
	G–GEAR	Cessna FR.182Q	Ranelagh Garage Ltd.
	G–GEEP	Robin R.1180T	Organic Concentrates Ltd.
	G–GEES	Cameron N-77 balloon	Mark Jarvis Ltd. *Mark Jarvis*
	G–GEOF	Pereira Osprey 2	G. Crossley
	G–GFAL	Douglas DC-10-10	Laker Airways Ltd. *Northern Belle*
	G–GFLY	Cessna F.150K	Citation Flying Services Ltd.
	G–GGAE	H.S.125 Srs. 3B/RA	Expotec (UK) Ltd.

Reg.	Type	Owner or Operator	Notes
G–GGGG	Thunder Ax77-77A balloon	Test Valley Balloon Group	
G–GHNC	AA-5A Cheetah	G. Humphrey & N. Chamberlain	
G–GIGI	M.S.893A Rallye Commodore	J. J. Knight (G–AYVX)	
G–GILL	Cessna 402C	Gill Aviation Ltd.	
G–GINA	AS.350B Ecureuil	Endeavour Aviation Ltd.	
G–GIRL	Cessna 421C	Lewis Shops Group	
G–GKNB	Beech 200 Super King Air	GKN Group Services Ltd.	
G–GLUE	Cameron N-65 balloon	M. F. Glue	
G–GMSI	SOCATA TB.9 Tampico	Dukeries Aviation Ltd.	
G–GOGO	Hughes 369D	A. W. Alloys Ltd.	
G–GOLD	Thunder Ax6-56A balloon	John Terry & Sons Ltd.	
G–GOMM	PA-32R-300 Lance	Vicmead Ltd.	
G–GOOS	Cessna F.182Q	R. Clark (Airtransport) Ltd.	
G–GOSH	Cessna 404	Euroair Transport Ltd.	
G–GOSS	Jodel DR.221	M. I. Goss	
G–GRAY	Cessna 172N	Buddale Ltd.	
G–GROW	Cameron N-77 balloon	Derbyshire Building Soc.	
G–GSKY	Douglas DC-10-10	Laker Airways Ltd. *California Belle*	
G–GSMP	Eiri PIK-20E-1	GSMP Ltd.	
G–GRIF	R. Commander 112TCA	P. Griffiths (G–BHXC)	
G–GTPL	Mooney M.20K	Valencienne Ltd. (G-BHOS)	
G–GWYN	Cessna F.172M	G. P. Owen	
G–HADI	G.1159 Gulfstream 2	Arab Express Ltd.	
G–HALL	PA-22 Tri-Pacer 160	F. P. Hall (G–ARAH)	
G–HALP	SOCATA TB.10 Tobago	D. Halpern (G–BITD)	
G–HANS	Robin DR.400 2+2.	Stahl Engineering Co. Ltd.	
G–HAPR	*B.171 Sycamore HC.14 (XG547)	British Rotorcraft Museum	
G–HARV	PA-23 Aztec 250	T. S. Grimshaw Ltd.	
G–HAWK	H.S. 1182 Hawk	British Aerospace	
G–HBUS	Bell 206B LongRanger	Willowbrook International Ltd.	
G–HEAD	Colt 56 balloon	Colt Balloons Ltd.	
G–HEAT	Bell 206B JetRanger 3	Heliwork Ltd.	
G–HELI	*Saro Skeeter Mk.12 (XM556)	British Rotorcraft Museum	
G–HELY	Agusta 109A	Barratt Developments Ltd.	
G–HENS	Cameron N-65 balloon	Harrells Dairies Ltd.	
G–HEWS	Hughes 369D	John Carroll Construction (Southern) Ltd.	
G–HFCI	Cessna F.150L	Horizon Flying Club	
G–HFCT	Cessna F.152	Martin J. Storey Ltd.	
G–HGGS	EMB-110P1 Bandeirante	Euroair Transport Ltd.	
G–HHOI	H.S.125 Srs. 700B	Trust House Forte Airport Services Ltd. (G–BHTJ)	
G–HIFI	PA-28R-201 Arrow III	First Class Furniture Ltd. (G–BFTB)	
G–HIGH	Cessna FT.337GP	P. L. Builder	
G–HILL	Cessna U.206F	HGH Air Charter (London) Ltd.	
G–HILR	Hiller UH-12E	G. & S. G. Neal (Helicopters) Ltd.	
G–HLFT	SC.5 Belfast	HeavyLift Cargo Airlines Ltd.	
G–HMCG	BN-2A-26 Islander	Renco (Aviation) Ltd. (G–BESH)	
G–HOLS	Warner Special	J. O. C. Warner	
G–HOLT	Taylor JT.1 Monoplane	K. D. Holt	
G–HOME	Colt 77A balloon	Anglia Balloon School *Tardis*	
G. HOOK	Hughes 369D	Auto Alloys (Helicopters) Ltd.	
G–HOOV	Cameron N-56 balloon	Balloon Stable Ltd. *Hoover*	
G–HOPE	Beech F33A Bonanza	Hope Jones (Electrical Spares) Ltd.	
G–HORN	Cameron V-77 balloon	G. J. E. Horn	
G–HOSE	Cessna 152 II	Toon Ghose Aviation Ltd.	
G–HOST	Cameron N-77 balloon	S. Williams & ptnrs. *Blythe Spirit*	
G–HOTS	Thunder Colt AS-80 airship	Thunder Balloons Ltd.	
G–HOUL	FRED Srs. 2	D. M. M. Richardson	
G–HOUS	Double Sky Chariot balloon	Anglia Balloons Ltd.	
G–HRLM	Brugger MB.2 Colibri	R. A. Harris	
G–HUFF	Cessna 182P	Robert Herbert (Holdings) Ltd.	
G–HUGH	PA-32RT-300T Turbo Lance II	Mann Aviation Sales Ltd. (G–IFLY)	
G–HULL	Cessna F.150M	Oldment Ltd.	
G–HUNT	Hunter F.51	M. R. Carlton	
G–HUSH	Hughes 269C	Auto Alloys (Helicopters) Ltd.	
G–HWBK	Agusta A.109A	Willowbrook International Ltd.	
G–HYDE	AB-206B JetRanger 3	Hyde Helicopters Ltd.	
G–IAIN	Cessna P.210N	Lorne Stewart Ltd.	
G–IANS	R. Turbo Commander 690B	Parahold Ltd.	
G–IANT	Cessna 404	Euroair Transport Ltd.	

Notes	Reg.	Type	Owner or Operator
	G–IASI	Beech 95-58 Baron	Interafrica Aircraft Ltd.
	G–IBFW	PA-28R-201 Arrow III	B. Walker & Co. (Dursley) Ltd. & J. & C. Ward (Holdings) Ltd.
	G–IBIS	H.S.125 Srs. 3BRA	Ibis Ltd. (G–AXPU)
	G–ICES	Thunder Ax6-56 balloon	Lighter-Than-Air Ltd.
	G–ICRU	Bell 206A JetRanger	Warmco (Manchester) Ltd.
	G–ICUB	Piper J-3C-65 Cub	G. Cormack
	G–IDDY	D.H.C.-1 Super Chipmunk	N. A. Brendish (G–BBMS)
	G–IDWR	Hughes 369HS	Ryburn Air Ltd. (G–AXEJ)
	G–IIIA	Swearingen Merlin IIIB	Peregrine Air Services Ltd.
	G–IIIB	Swearingen Merlin IIIB	Willowbrook International Ltd.
	G–IKIS	Cessna 210M	Lorne Stewart Ltd.
	G–ILLI	Cessna RF. 182RG	Escadale Properties Ltd.
	G–ILLY	PA-28-181 Archer II	A. G. & K. M. Spiers
	G–IMBE	PA-31-310 Navajo	T. Beattie Edwards & Co. Ltd. (G–BXYB/G–AXYB)
	G–INMO	PA-31-310 Navajo	Sabaru (UK) Ltd.
	G–INNY	SE-5A Replica	R. M. Ordish
	G–IOSI	Jodel DR.1051	R. G. E. Simpson & A. M. Alexander
	G–IPPM	SA.102-5 Cavalier	I. D. Perry & P. S. Murfitt
	G–IPRA	Beech A200 Super King Air	J. H. Ritblat (G–BGRD)
	G–IPSY	Rutan Vari-Eze	R. A. Fairclough
	G–IRLS	Cessna FR.172J	Starvillas Ltd.
	G–IVAN	Rutan Vari-Eze	I. Shaw
	G–IWPL	Cessna F172M	Reedy Supplies Ltd.
	G–JADE	Beech 95-58 Baron	Liaison & Consultant Services Ltd.
	G–JAKE	D.H.C.1 Chipmunk 22	Ewart & Co. (Studio) Ltd. (G–BBMY)
	G–JAKO	Cessna TU.206G	New Hatherley Garage
	G–JAKY	PA-31-325 Navajo	Malcolm Air Ltd.
	G–JAMC	Bell 222	Heart of England Helicopters Ltd.
	G–JAMI	Bell 206L LongRanger	Highland Personnel Ltd.
	G–JANE	Cessna 340A	Standard Aviation Ltd.
	G–JANS	Cessna FR.172J	Jans of London
	G–JANY	AS.350B Ecureuil	P. E. Cadbury (G–BHCN)
	G–JASP	PA-23 Aztec 250	Landsurcon (Air Survey) Ltd.
	G–JAWS	Enstrom F-280C Shark	Earl Developments Ltd.
	G–JBUS	FRED Srs. 2	R. V. Joyce
	G–JCTI	Partenavia P.68B	Rockville Motors Ltd. (G–OJOE)
	G–JDST	PA-31-350 Navajo Chieftain	Jack Tighe Ltd.
	G–JEAN	Cessna 500 Citation	Castlewood Air Services Ltd.
	G–JEFF	PA-38-112 Tomahawk	Channel Aviation Ltd.
	G–JENS	SOCATA Rallye 100ST	B. H. Burnet (G–BDEG)
	G–JENY	Baby Great Lakes	J. M. C. Pothecary
	G–JETA	Cessna 550 Citation II	IDS Aircraft Ltd.
	G–JETB	Cessna 550 Citation II	IDS Aircraft Ltd.
	G–JETC	Cessna 550 Citation II	IDS Aircraft Ltd.
	G–JFWI	Cessna F.172N	J. F. Willis
	G–JGCL	Cessna 414A	Johnson Group Management Services Ltd.
	G–JGFF	AB-206B JetRanger 3	Mailam Ltd.
	G–JILL	R. Commander 112TCA	Kennair (General Aviation) Ltd.
	G–JIMS	Cessna 340A-II	Granpack Ltd. (G–PETE)
	G–JIMY	PA-28 Cherokee 140	J. C. Kumar (G–AYUG)
	G–JJSG	Learjet 35A	Smurfit Ltd.
	G–JMCA	PA-31-350 Navajo Chieftain	Fluorocarbon Ltd.
	G–JMCC	Beech 95-58 baron	Mowlem Construction (Plant hire) Ltd.
	G–JMFW	Taylor JT.1 Monoplane	G. J. M. F. Winder
	G–JMWT	SOCATA TB.10 Tobago	McNulty's World Travel
	G–JOAN	AA-5B Tiger	Oldment Ltd. (G–BFML)
	G–JOEY	BN-2A Mk. III-2 Trislander	Aurigny Air Services (G–BDGG)
	G–JOHN	PA-28R-201T Turbo Arrow III	Fairoaks Flight Centre
	G–JOLY	Cessna 120	J. A. & G. E. Baker
	G–JONE	Cessna 172M	A. G. Jones
	G–JONS	PA-31-350 Navajo Chieftain	Berdan Enterprises Ltd.
	G–JOON	Cessna 182D	Allen Technical Services
	G–JOSE	Cessna U.206G	Safari Skylink Enterprises Ltd.
	G–JOSI	CP.1310-C3 Super Emeraude	J. G. O'Donnell (G–BCHP)
	G–JRCM	Hawker Fury Mk. I Replica	J. R. C. Morgan
	G–JRCT	Cessna 550 Citation II	IDS Aircraft Ltd.
	G–JRMM	R. Turbo Commander 690B	R. B. Tyler (Plant) Ltd.
	G–JSGM	Cessna P.210N	The Strathmarine Flying Group

Reg.	Type	Owner or Operator	Notes
G–JSSD	SA. Jetstream 3001	British Aerospace (G–AXJZ)	
G–JTCA	PA-23 Aztec 250	J. D. Tighe & Co. Ltd. (G–BBCU)	
G–JUDI	AT-6 Harvard III (FX301)	A. Haig-Thomas	
G–JUDY	AA-5A Cheetah	Pioneer Welders and Fabrications	
G–JUMP	AB-206B JetRanger 3	Hickstead Ltd.	
G–JUNE	PA-28-161 Warrior II	Allen Technical Services Ltd.	
G–JURG	R. Commander 114	Maronco Ltd.	
G–JWIV	Jodel DR.1051	J. W West	
G–KACT	Cessna 421B	Northair Aviation Ltd.	
G–KAFC	Cessna 152	Gordon King (Aviation) Ltd.	
G–KAIR	PA-28-161 Warrior II	Academy Lithoplates Ltd.	
G–KATH	Cessna P.210N	K. & F. L. Services Ltd.	
G–KBPI	PA-28-161 Warrior II	Four By Four Aviation Ltd. (G–BFSZ)	
G–KCIG	Sportavia RF5B	Executive Air Sport Ltd.	
G–KDIX	Jodel D.9 Bebe	K. Barlow	
G–KEEN	Stolp Starduster Too	A. A. Hodgson	
G–KEMZ	BN-2T Islander	Pilatus BN Ltd. (G–BIUJ)	
G–KENS	PA-32-300 Cherokee Six	Ken Heanes Ltd.	
G–KERC	Nord NC.854S	Kirk Aviation	
G–KERK	Piper J-3C-65 Cub	Kirk Aviation	
G–KERR	Cessna FR.172K-XP	A. G. Chrismas Ltd.	
G–KEYS	PA-23 Aztec 250	Ferguson Aviation	
G–KFIT	Beech F90 King Air	Carex Exhaust Centres Ltd. (G–BHUS)	
G–KIDS	PA-34-220T-3 Seneca	Holding & Barnes Ltd.	
G–KILO	Boeing 747-236F	British Airways Cargo *British Trader*	
G–KING	PA-38-112 Tomahawk	Gordon King (Aviation) Ltd.	
G–KIRK	Piper J-3C-65 Cub	M. Kirk	
G–KLAY	Enstrom F-280C Shark	Klay Electronics Ltd. (G–BGZD)	
G–KRIS	Maule M5-235C Lunar Rocket	Lord Howard de Walden	
G–KUKU	Pfalzkuku (BS676)	A. D. Lawrence	
G–KWAX	Cessna 182E Skylane	E. W. Duck	
G–KWIK	Partenavia P.68B	Process Machinery Ltd.	
G–KYAK	Yakolev C-11	R. Lamplough	
G–LADA	PA-32 Cherokee Six 300D	Bolton Car Centre (G–AYWK)	
G–LADE	PA-32 Cherokee Six 300E	Appleby Glade Ltd.	
G–LAKI	Jodel DR.1050	V. Panteli	
G–LANA	SOCATA TB.10 Tobago	Jerrard, Saunders, Donn (Solicitors)	
G–LANE	Cessna F.172N	Mark Laing Aviation Ltd.	
G–LASH	Monnet Sonerai II	A. Lawson	
G–LASS	Rutan Vari-Eze	Marconi Space & Defence Systems Ltd.	
G–LATC	EMB-110P1 Bandeirante	Euroair Transport Ltd.	
G–LAZE	Jodel DR.1050	Culmor Windows Ltd.	
G–LDYS	Colt 56A balloon	A. Green	
G–LEAM	PA-28-236 Dakota	Clutchstar Ltd. (G–BHLS)	
G–LEAR	Learjet 35A	David Pratt & ptnrs. Ltd. (G–ZEST)	
G–LEAU	Cameron N-31 balloon	Balloon Stable Ltd.	
G–LENS	Thunder Ax7-77Z balloon	Island Airship Co. Ltd.	
G–LEON	PA-31-350 Navajo Chieftain	Air Charter (Scotland) Ltd.	
G–LEXI	Cameron N-77 balloon	Lex Mead Bristol Ltd. *Lex*	
G–LFCA	Cessna F.152	Aerolease	
G–LFIX	V.S. Spitfire LF.IX	Island Trading Ltd.	
G–LIDE	PA-31-350 Navajo Chieftain	Mont Arthur Finance Ltd.	
G–LIFE	Thunder Ax6-56Z balloon	Schroder Life Assurance Ltd.	
G–LIFT	Bell 47G-3	Dolphin Property (Management) Ltd.	
G–LIMA	R. Commander 114	W. M. P. Miller	
G–LINK	Sikorsky S-61N	British Caledonian Airways	
G–LINT	Pitts S.1S Special	P. L. Moss	
G–LION	PA-18-135 Super Cub	Citation Flying Services Ltd.	
G–LITE	R. Commander 112A	Rhoburt Ltd.	
G–LOAG	Cameron N-77 balloon	Matthew Gloag & Son Ltd.	
G–LONG	Bell 206L LongRanger	Air Hanson Ltd.	
G–LONS	Enstrom F-280C-UK-2	G. Miller (G–BDIB)	
G–LOOK	Cessna F.172M	Laarbruch Flying Club	
G–LOOP	Pitts S-1C Special	P. Meeson	
G–LOTI	Bleriot XI (replica)	M. L. Beach	
G–LOVO	Cessna 414A	Lovaux Ltd. (G–KENT)	
G–LOWE	Monnet Sonerai II	J. A. Lowe	
G–LRII	Bell 206L LongRanger	Castle Air Charters Ltd.	
G–LSMI	Cessna F.152	Lincoln Street Motors (Birmingham) Ltd.	

Notes	Reg.	Type	Owner or Operator
	G–LUCK	Cessna F.150M	M. Carrington
	G–LUCY	PA-30 Twin Comanche 160	C.R.D. Tool & Engineering Ltd. (G–AVCP)
	G–LUNA	PA-32RT-300T Turbo Lance II	Everest Aviation Ltd.
	G–LYDE	Eiri PIK-20E	G. S. Wilson
	G–LYNN	PA-32RT-300 Lance II	Headon Developments Ltd. (G–BGNY)
	G–LYNX	Westland WG.13 Lynx	Westland Helicopters Ltd.
	G–MACH	SIAI-Marchetti SF.260	R. A. Sareen
	G–MACK	PA-28R Cherokee Arrow 200	J. Pottier
	G–MADI	Cessna 310R II	Talley Aviation Ltd.
	G–MAGI	AS.350B Ecureuil	Micro Consultants Ltd. (G–BHLR)
	G–MAGS	Cessna 340A	Goldstar Publications Ltd.
	G–MAGY	AS.350B Ecureuil	Micro Consultants Ltd. (G–BIYC)
	G–MAIL	D.H.C.-6 Twin Otter 310	Fairflight Ltd./Air Ecosse
	G–MALA	PA-28-181 Archer II	M. A. Lenihan & Associates (G–BIIU)
	G–MALC	AA-5 Traveler	Air Coventry Ltd. (G–BCPM)
	G–MALK	Cessna F.172N	M. H. Malik
	G–MANX	FRED Srs. 2	P. Williamson
	G–MARG	PA-31-350 Navajo Chieftain	M. Ferguson (Newtownards) Ltd.
	G–MARK	Cessna F.337H	Denis Ferranti Hoverknights Ltd.
	G–MARY	Cassutt Special I	J. Chadwick
	G–MAUL	Maule M5-235C Lunar Rocket	Capital Aviation Sales (UK) Ltd.
	G–MAWL	Maule M4-210C Rocket	E. P. Beck
	G–MAXI	PA-34-200T Seneca	Leisure Aviation
	G–MAXY	Cessna 210L	Ripmax Models Ltd.
	G–MAYO	PA-28-161 Warrior II	Ashfronts Ltd. (G–BFBG)
	G–MBAA	Hiway Skytrike Mk. 2	Hiway Hang Gliders Ltd.
	G–MBAB	Hovey Whing-Ding II	R. F. Morton
	G–MBAC	Pterodactyl Srs. 2	D. L. Giles
	G–MBAD	Weedhopper	P. W. Bailey
	G–MBAE	Lazair	H. A. Leek
	G–MBAF	R.J. Swift 3	C. G. Wrzesien
	G–MBAG	Skycraft Scout	B. D. Jones
	G–MBAH	D.H. Microlight	D. Harker
	G–MBAI	Typhoon Tripacer 250	W. G. Lamyman
	G–MBAJ	Chargus T.250	G. C. Jones
	G–MBAK	Eurowing Spirit	J. S. Potts
	G–MBAL	Hiway Demon	M. Blewitt
	G–MBAM	Skycraft Scout 2	Air Vitesse Ltd.
	G–MBAN	Eagle Microlight	R. W. Millward
	G–MBAO	Rotec Rally 2B	R. Mayo
	G–MBAP	Rotec Rally 2B	P. D. Lucas
	G–MBAR	Scout Microlight	L. C. Gianicolo
	G–MBAS	Typhoon Tripacer 250	T. J. Birkbeck
	G–MBAT	Hiway Skytrike	M. R. Gardiner
	G–MBAU	Hiway Skytrike	M. R. Gardiner
	G–MBAV	Weedhopper	L. F. Smith
	G–MBAW	Pterodactyl	M. L. Smith
	G–MBAX	Hiway Skytrike	D. Clarke
	G–MBAY	Skycraft Scout	J. Colloff & G. T. Wilkinson
	G–MBAZ	Rotec Rally 2B	Western Skysports Ltd.
	G–MBBA	Lazair	P. Roberts
	G–MBBB	Skycraft Scout 2	A. J. & B. Chalkley
	G–MBBC	Chargus T.250	R. R. G. Close-Smith
	G–MBBD	Pterodactyl	R. Penfold
	G–MBBE	Skyranger Microlight	A. S. Coombes
	G–MBBF	Chargus Titan 38	Chargus Gliding Co.
	G–MBBG	Weedhopper JC-24B	F. S. Beckett
	G–MBBH	Sealander 160	J. A. Evans & J. B. Wincott
	G–MBBI	Mirage Microlight	B. H. Trunkfield & A. A. Howard
	G–MBBJ	—	—
	G–MBBK	Mirage Microlight	M. G. Selley
	G–MBBL	Lightning Microlight	I. M. Grayland
	G–MBBM	Eipper Quicksilver MX	A. M. Reid
	G–MBBN	Eagle Microlight	S. Taylor & D. Williams
	G–MBBO	Rotec Rally 2B	A. J. Doggett
	G–MBBP	Chotia Weedhopper	G. L. Moon
	G–MBBR	Weedhopper JC-24B	J. G. Wallens
	G–MBBS	Chargus T.250	P. R. De Fraine
	G–MBBT	Tripacer 330	The Post Office
	G–MBBU	Savage Microlight	R. Venton-Walters
	G–MBBV	Rotec Rally 2B	Blois Aviation Ltd.

Reg.	Type	Owner or Operator	Notes
G-MBBW	Flexiform Hilander	B. R. Underwood	
G-MBBX	Chargus Skytrike	S. A. Geary	
G-MBBY	Flexiform Sealander	J. E. Heath	
G-MBBZ	Volmer Jensen VJ-24W	D. G. Cook	
G-MBCA	Chargus Cyclone T.250	E. M. Jelonek	
G-MBCB	Lightning Microlight	P. G. Huxham	
G-MBCC	Flexiform Sealander Trike	P. W. Robinson	
G-MBCD	La Mouette Atlas	M. G. Dean	
G-MBCE	Eagle Rainbow	A. E. Sheadown	
G-MBCF	Fledgeling Microlight	P. J. Davies	
G-MBCG	Tripacer T.250	A. G. Parkinson	
G-MBCH	Hiway Skytrike	R. Carr	
G-MBCI	Hiway Skytrike	J. R. Bridge	
G-MBCJ	Mainair Triflyer	J. R. North	
G-MBCK	Eipper Quicksilver MX	G. W. Rowbotham	
G-MBCL	Hiway Demon Triflyer	P. G. Kavanagh	
G-MBCM	Hiway Demon 175	D. M. Mudie	
G-MBCN	Hiway Super Scorpion	M. J. Hadland	
G-MBCO	Flexiform Sealander Buggy	P. Bennett	
G-MBCP	Flexiform Sealander Triflyer	P. Bennett	
G-MBCR	Mirage Microlight	Pleasurecraft Ltd.	
G-MBCS	Eagle Microlight	Pleasurecraft Ltd.	
G-MBCT	Eagle Microlight	Pleasurecraft Ltd.	
G-MBCU	Eagle Microlight	Pleasurecraft Ltd.	
G-MBCV	Lightning Microlight	R. J. Pattinson	
G-MBCW	Hiway Demon 175	C. Foster & S. B. Elwis	
G-MBCX	Hornet Microlight	L. J. Houghton	
G-MBCY	Eagle Microlight	M. B. Fewkes	
G-MBCZ	Chargus Skytrike 160	R. M. Sheppard	
G-MBDA	Rotec Rally 2B	Blois Aviation Ltd.	
G-MBDB	Typhoon Microlight	D. J. Smith	
G-MBDC	Hornet Microlight	R. R. Wolfenden & G. Priestley	
G-MBDD	Skyhook Skytrike	J. H. Clarke & R. M. Larrimore	
G-MBDE	Flexiform Skytrike	R. S. Peaks	
G-MBDF	Rotec Rally 2B	J. R. & B. T. Jordan	
G-MBDG	Goldwing Microlight	N. W. Beadle & ptnrs.	
G-MBDH	Hiway Demon Triflyer	A. T. Delaney	
G-MBDI	Flexiform Sealander	K. Bryan	
G-MBDJ	Flexiform Sealander Triflyer	L. H. Phillips	
G-MBDK	Typhoon Triflyer	D. J. Atkinson	
G-MBDL	Lone Ranger Microlight	Aero & Engineering Services Ltd.	
G-MBDM	Sigma Trike	A. R. Prentice	
G-MBDN	—	—	
G-MBDO	Flexiform Sealander Trike	G. M. Fletcher	
G-MBDP	—	—	
G-MBDR	U.A.S. Stormbuggy	J. S. Long	
G-MBDS	Mitchell Wing B.10	G. W. Holland	
G-MBDT	Eagle Microlight	I. D. Stokes	
G-MBDU	Chargus Titan 38	Property Associates Ltd.	
G-MBDV	Pterodactyl Microlight	D. J. Thomas	
G-MBDW	Tripacer Skytrike A	M. Unsworth	
G-MBDX	—	—	
G-MBDY	—	—	
G-MBDZ	Eipper Quicksilver MX	M. Ridsdale	
G-MBEA	Hornet Nimrod	B. Berry	
G-MBEB	Hiway Skytrike 250	R. MacDonald	
G-MBEC	Hiway Scorpion	C. S. Wates	
G-MBED	Chargus Titan 38	Humming Bird Aviation Ltd.	
G-MBEE	Hiway Super Scorpion Skytrike 160	P. H. Risdale & ptners.	
G-MBEF	Eipper Quicksilver MX	Pegasus Ltd.	
G-MBEG	Eipper Quicksilver MX	Pegasus Ltd.	
G-MBEH	Eagle Microlight	Pegasus Ltd.	
G-MBEI	Eagle Microlight	Pegasus Ltd.	
G-MBEJ	Eagle Microlight	Pegasus Ltd.	
G-MBEK	Eagle Microlight	Pegasus Ltd.	
G-MBEL	Eagle Microlight	Pegasus Ltd.	
G-MBEM	Eagle Microlight	Pegasus Ltd.	
G-MBEN	Eipper Quicksilver MX	Pegasus Ltd.	
G-MBEO	—	—	
G-MBEP	Eagle Microlight	R. W. Lavender	
G-MBER	Skyhook Sailwings TR-1	Skyhook Sailwings Ltd.	
G-MBES	Skyhook Sailwings TR-2	Skyhook Sailwings Ltd.	

Notes	Reg.	Type	Owner or Operator
	G–MBET	Mistral Trainer	J. W. V. Edmunds
	G–MBEU	Hiway Demon T.250	D. R. Gazey
	G–MBEV	Chargus Titan 38	London Ultra-light Flight Centre
	G–MBEW	Solar Buggy	J. R. Aspinall
	G–MBEX	Hiway Demon Skytrike	P. Johnson
	G–MBEY	Hiway Super Scorpion I	A Edees
	G–MBEZ	Pteradactyl Ptraveller II	P. A. Smith
	G–MBFA	Hiway Skytrike 250	I. G. S. Cunningham
	G–MBFB	Skyhook Powertrike	T. D. L. Williams & ptnrs.
	G–MBFC	Hiway Skytrike	F. C. Coulson
	G–MBFD	Hummingbird Microlight	Micro Aviation Ltd.
	G–MBFE	—	—
	G–MBFF	Hiway Scorpion	H. Redwin
	G–MBFG	Skyhook Sabre	A. H. Trapp
	G–MBFH	Hiway Skytrike	R. M. Buck
	G–MBFI	Hiway Skytrike II	J. R. Brabbs
	G–MBFJ	Chargus Typhoon T. 250	A. W. Knowles
	G–MBFK	Hiway Demon	K. T. Vinning
	G–MBFL	Hiway Demon	J. C. Houghton
	G–MBFM	Hiway Hang Glider	G. P. Kimmons
	G–MBFN	Hiway Skytrike II	R. E. Todd
	G–MBFO	Eipper Quicksilver MX	M. L. Desoutter
	G–MBFP	Hiway Scorpion	J. G. Beesley
	G–MBFR	Eagle Microlight	W. G. Bradley
	G–MBFS	Electro Flyer Eagle	R. Fox
	G–MBFT	—	—
	G–MCDS	Cessna 210N	Merseyside Car Delivery (G–BHNB)
	G–MCEO	Beech A200 Super King Air	Colt Car Co. Ltd. (G–BILY)
	G–MDAS	PA-31-310 Navajo	Milford Haven Dry Dock Co. Ltd. (G–BCJZ)
	G–MDRB	PA-31-350 Navajo Chieftain	Iceni Aviation Ltd.
	G–MEDI	Beech C90 King Air	Airmore Aviation Ltd.
	G–MERI	PA-28-181 Archer II	Electronic Associates
	G–META	Bell 222	The Metropolitan Police
	G–METB	Bell 222	The Metropolitan Police
	G–MFEU	H.S.125 Srs. 600B	Clartacrest Ltd.
	G–MHGI	Cessna 414A	Keighley Finance Ltd. (G–BHKK)
	G–MICK	Cessna F.172N	J. W. Barwell
	G–MICV	BN-2B Islander	Pilatus BN Ltd. (G–BIXC)
	G–MIKE	Hornet Gyroplane	M. H. J. Goldring
	G–MILK	SOCATA TB.10 Tobago	Accent Aviation Ltd.
	G–MINI	Currie Wot	D. Collinson
	G–MISS	Taylor JT.2 Titch	A. Brenan
	G–MIST	Cessna T.210K	Allzones Travel Ltd. (G–AYGM)
	G–MKAY	Cessna 172N	Malkay Automobiles Ltd.
	G–MKEE	EAA Acro Sport	G. M. McKee
	G–MLAS	Cessna 182E	Mark Luton Aviation Services
	G–MOBL	EMB-110P2 Bandeirante	Fairflight Ltd./Air Ecosse
	G–MOGG	Cessna F.172N	R. M. W. & B. N. C. Mogg (G–BHDY)
	G–MOLY	PA-23 Apache 160	Steer Aviation Ltd. (G–APFV)
	G–MONA	M.S.880B Rallye Club	K. S. Pattinson (G–AWJK)
	G–MONO	Taylor JT.1 Monoplane	A. J. Holmes
	G–MORR	AS.350B Ecureuil	Colt Car Co. Ltd. (G–BHIU)
	G–MOSI	D.H.98 Mosquito 35	D. W. Arnold (G–ASKA)
	G–MOTH	D.H.82A Tiger Moth	M. C. Russell
	G–MOVE	Aerostar 601P	Knight Programming Support Ltd.
	G–MOZY	D.H.98 (replica)	J. Beck & G. L. Kemp
	G–MPWI	Robin HR.100/210	MPW Aviation Ltd.
	G–MSDS	Cessna 404	Executive Express
	G–MXIV	V.S. Spitfire FR.XIV	M. & K. Wickenden
	G–NASH	AA-5A Cheetah	T. R. Bamber
	G–NATT	R. Commander 114A	Tembo Music Ltd.
	G–NBSI	Cameron N-77 balloon	Nottingham Building Soc.
	G–NDNI	NDN-1 Firecracker	Norman Marsh Aircraft Ltd.
	G–NELL	R. Commander 112A	Arcdeal Ltd.
	G–NEUS	Brugger MB.2 Colibri	G. S. Smeaton
	G–NEVA	PA-31-310 Navajo	Air Continental Securities Ltd. (G–BGVO/G–AZAC)
	G–NEWR	PA-31-350 Navajo Chieftain	Burnthills Plant Hire Ltd.
	G–NEWS	Bell 206B JetRanger 3	Peter Press Ltd.
	G–NEWU	Partenavia P.68C	Biograft Private Clinic Ltd. (G–BHJX)
	G–NHVH	Maule M5-235C Lunar Rocket	Commercial Go-Karts Ltd.

Reg.	Type	Owner or Operator	Notes
G–NICK	PA-19 Super Cub 95	J. G. O'Donnell & I. Woolacott	
G–NILE	Colt 77A balloon	I. J. McDonnell & A. Gray	
G–NITE	PA-31-350 Navajo Chieftain	WT Shipping Group Ltd.	
G–NIUS	Cessna F.172N	Horizon Lighting Products Ltd.	
G–NJAG	Cessna 207	G. H. Nolan Ltd.	
G–NMAN	PA-31 Turbo Navajo	Machine Music Ltd. (G–AXDD)	
G–NNAC	PA-18 Super Cub 135	Norwich & Norfolk Aero Club Ltd.	
G–NODE	AA-5B Tiger	Curd & Green Ltd.	
G–NOEL	AB-206B JetRanger 2	N. Edmunds & P. R. McEnhill (G–BCWN)	
G–NOME	Baby Great Lakes	J. B. Scott & ptnrs.	
G–NORC	Cessna 425	Norcross Ltd.	
G–NORD	Nord NC.854	R. G. E. Simpson & A. M. Alexander	
G–NOVA	Cessna T.337H	C. C. Deane	
G–NRDC	NDN-6 Fieldmaster	NDN Aeroculture Ltd.	
G–NUTS	Cameron SS balloon	The Balloon Stable Ltd.	
G–NWBP	Thunder Ax7-77Z balloon	Lighter-Than-Air Ltd.	
G–OABI	Cessna 421C	ABI Caravans Ltd.	
G–OAHB	T-33 Mk. 3 Silver Star	—	
G–OAIR	EMB-110P1 Bandeirante	Air UK	
G–OAKS	Cessna 421C	Barratt Developments Ltd.	
G–OAMH	Agusta 109A	Willowbrook International Ltd.	
G–OAST	Cessna T.182RG	Airwork Services Ltd.	
G–OATS	PA-38-112 Tomahawk	B.S. Transport Ltd.	
G–OBAE	H.S.125 Srs. 700B	British Aerospace	
G–OBAT	Cessna F.152	Renco Aviation Ltd.	
G–OBCA	Cessna 421C	British Car Auctions Ltd.	
G–OBEY	PA-23 Aztec 250	Keller Bryant & Co. Ltd. (G–BAAJ)	
G–OBIA	EMB-110P1 Bandeirante	Air UK	
G–OBLE	C.A.S.A. 1.131 Jungmann	A.J.D. Securities Ltd.	
G–OBMW	AA-5 Traveler	Fretcourt Ltd. (G–BDPV)	
G–OCAL	Partenavia P.68B	Dowson Car Sales Ltd. (G–BCMY)	
G–OCAT	Eiri PIK-20E	D. S. Innes	
G–OCPC	Cessna FA.152	Western Isles Aviation	
G–OCUB	Piper J-3C-90 Cub	W. Savin & B. Ell	
G–ODAY	Cameron N-56 balloon	C. O. Day (Estate Agents)	
G–ODEL	Falconar F-II-3	J. R. D. Bygraves	
G–ODON	AA-5B Tiger	Moynihan Motor Engineering Ltd.	
G–OEZE	Rutan Vari-Eze	S. Stride & ptnrs.	
G–OFAR	Cessna 402C	Wm. Leach (Builders) Ltd.	
G–OFCM	Cessna F.172L	W. B. Garnham & P. J. Woodland (G–AZUN)	
G–OFED	Enstrom F-280C-UK-2	Spooner Aviation Ltd.	
G–OFLY	Cessna 210L	Northair Aviation Ltd.	
G–OFRL	Cessna 414A	Flight Refuelling Ltd.	
G–OGDN	Beech A200 Super King Air	A. Ogden & Sons Ltd.	
G–OGOJ	AA-5A Cheetah	Publishing Innovations Leasing Ltd.	
G–OHCA	SC.5 Belfast	HeavyLift Cargo Airlines Ltd.	
G–OHTL	Sikorsky S-76	Air Hanson Ltd.	
G–OILS	Cessna T.210L	Machine Music Ltd. (G–BCZP)	
G–OIML	AB-206B JetRanger 3	IML Group of Companies	
G–OIOO	PA-23 Aztec 250	A. A. Kelly (G–AVLV)	
G–OJCB	AB-206B JetRanger 2	J. C. Bamford Excavators Ltd.	
G–OJCW	PA-32RT-300 Lance II	J. C. Walsh Contractors & Plant Hire Ltd.	
G–OJEA	D.H.C.-6 Twin Otter 310	Jersey European Airways	
G–OJEE	Bede BD-4	G. Hodges	
G–OJON	Taylor JT.2 Titch	J. H. Fell	
G–OJOY	H.S.125 Srs. 700B	British Aerospace, Hatfield (G–BGGS)	
G–OJVH	Cessna F.150H	Westshells Ltd. (G–AWJZ)	
G–OKAY	Pitts S-1E Special	Sky Fever (Aviation Enterprises)	
G–OLDI	Ayres S2R-T15/500 Thrush Commander	Farm Aviation Services Ltd.	
G–OLDY	Luton LA-5 Major	M. P. & A. P. Sargent	
G–OLEE	Cessna F.152	Warwickshire Aero Club	
G–OLEN	Cessna 425	Northair Aviation Ltd.	
G–OLLI	Cameron O-31 SS balloon	Robertson Foods Ltd. *Golly III*	
G–OLLY	PA-31-350 Navajo Chieftain	Robertson Foods Ltd. (G–BCES)	
G–OLVR	FRED Srs. 2	A. R. Oliver	
G–OMET	Beech C90 King Air	Comet Radiovision Services Ltd. (G–COTE/G–BBKN)	
G–OMHC	PA-28RT-201 Arrow IV	M. H. Cundley	

Notes	Reg.	Type	Owner or Operator
	G-ONPN	H.S.125 Srs. 1B	Satellite Ltd. (G-BAXG)
	G-OODE	SNCAN SV-4A	V. S. E. Norman (G-AZNN)
	G-OODI	Pitts S-1D Special	R. N. Goode (G-BBBU)
	G-OODO	Stephens Akro	R. N. Goode
	G-OODY	PA-28R Cherokee Arrow 200	J. Traynor
	G-OOFY	Rollason Beta	G. Staples
	G-OOSE	Rutan Vari-Eze	J. A. Towers
	G-OPEL	Cessna F.172G	Concord Motor Services (Stansted) Ltd.
	G-ORAV	Cessna 337D	Rogers Aviation Sales Ltd. (G-AXGJ)
	G-ORAY	Cessna F.182Q-II	Ray Hoult (Land Drainage) Ltd. (G-BHDN)
	G-ORMC	Beech A200 Super King Air	RMC Group Services Ltd. (G-BEST)
	G-OROY	Partenavia P.68B	Astra Design Consultants Ltd. (G-BFSU)
	G-OSHH	Cessna 404	Northair Aviation Ltd.
	G-OSKY	Cessna 172M	Bevan Lynch Aviation Ltd.
	G-OSLA	Boeing 737-2U4ADV	Britannia Airways Ltd. *Sir Geoffrey de Havilland*
	G-OTOM	Cessna FR.172J	J. T. L. Jones
	G-OTRG	Cessna TR.182RG	G. F. Holdings (Contractors) Ltd.
	G-OTUX	PA-28R-201T Turbo Arrow IV	M. A. M. Quadrini
	G-OULD	Gould Mk. I balloon	C. A. Gould
	G-OVFR	Cessna F.172N	M. Harrison & Co. (Leeds) Ltd.
	G-OVMC	Cessna F.152 II	Staverton Flying Services Ltd.
	G-OWAC	Cessna F.152	Warwickshire Aero Club (G-BHEB)
	G-OWAK	Cessna F.152	Warwickshire Aero Club (G-BHEA)
	G-OWEN	K & S Jungster	R. C. Owen
	G-OWJM	AB-206B JetRanger 3	J. M. Gow (G-BHXV)
	G-PACE	Robin R.1180T	Millicron Instruments Ltd.
	G-PACY	Rutan Vari-Eze	E. Pace
	G-PADY	R. Commander 114	Wyndley Nurseries Ltd.
	G-PAGE	Cessna F.150L	Page Vehicle Hire (Strumpshaw) Ltd.
	G-PALS	Enstrom F-280C Shark	Southern Air
	G-PARA	Cessna 207	B. Goodwin
	G-PARI	Cessna 172RG Cutlass	Frank Slowey Ltd.
	G-PARS	Evans VP-2	A. Parsfield
	G-PATT	Cessna 404 Titan	Casair Aviation Sales Ltd. (G-BHGL)
	G-PCUB	PA-18 Super Cub 135 (L-21B)	M. J. Wilson
	G-PDON	WMB.2 Windtracker balloon	P. Donnellan
	G-PEET	Cessna 401A	Forest Aviation Ltd.
	G-PENI	Hughes 369D	Yewlands Engineering Ltd.
	G-PENN	AA-5B Tiger	Gulfstream School of Flying (Denham) Ltd.
	G-PENY	Sopwith LC-1T Triplane	J. S. Penny
	G-PERR	Cameron 60 bottle balloon	The Balloon Stable Ltd.
	G-PETE	Cessna 340A-II	Northair Aviation Ltd.
	G-PFAA	EAA Model P biplane	P. E. Barker
	G-PFAB	FRED Srs. 2	P. E. Barker
	G-PFAC	FRED Srs. 2	R. Bennett
	G-PFAD	Wittman W.8 Tailwind	M. R. Stamp
	G-PFAE	Taylor JT.1 Monoplane	G. Johnson
	G-PFAF	FRED Srs. 2	K. Fern
	G-PFAG	Evans VP-1	N. S. Giles-Townsend
	G-PFAH	Evans VP-1	J. A. Scott
	G-PFAI	Clutton EC.2 Easy Too	E. Clutton & A. Tabenor
	G-PFAK	Kendal Mayfly	R. Y. Kendal
	G-PFAL	FRED Srs. 2	R. A. Abrahams
	G-PFAM	FRED Srs. 2	W. C. Rigby
	G-PFAN	Avro 558 (replica)	N. P. Harrison
	G-PFAO	Evans VP-1	P. W. Price
	G-PFAP	Currie Wot	P. G. Abbey
	G-PFAR	Isaacs Fury II	C. J. Repik
	G-PFAS	GY-20 Minicab	J. Sproston & F. W. Speed
	G-PFAT	Monnet Sonerai II	H. B. Carter
	G-PFAU	Evans VP-2	D. E. Peace
	G-PFAV	D.31 Turbulent	B. A. Luckins
	G-PFAW	Evans VP-1	R. F. Shingler
	G-PFAX	FRED Srs. 2	A. J. Dunston
	G-PFAY	EAA Biplane	A. K. Lang & A. L. Young
	G-PFAZ	Evans VP-1	B. Kylo
	G-PHIL	Hornet Gyroplane	A. J. Philpotts
	G-PICS	Cessna 182F	Astral Aerial Surveys Ltd. (G-ASHO)

G–PFAK (Top) Kendal Mayfly

G–RHCN (Centre) Cessna FR. 182RG

G-SEAR (Bottom) Pazmany PL.4

Notes	Reg.	Type	Owner or Operator
	G–PIED	PA-23 Aztec 250	Air London (Executive Travel) Ltd.
	G–PIES	Thunder Ax7-77Z balloon	Porkfarms Ltd.
	G–PIGN	Bolmet Paloma Mk. I	T. P. Metson & J. A. Bollen
	G–PINK	Cameron N-77 balloon	Leaveglen Ltd.
	G–PIPE	Cameron N-56 SS balloon	Carreras Rothmans Ltd.
	G–PISA	Thunder Ax7-77 balloon	Thunder Balloons Ltd.
	G–PLAN	Cessna F.150L	S. S. Padam
	G–PLAY	Robin R.2100A	Miss A. C. Peacock
	G–PLEV	Cessna 340	KJ Bill Aviation Ltd.
	G–PLIV	Pazmany PL.4	B. P. North
	G–PLUS	PA-34-200T-2 Seneca	C. G. Strasser
	G–PMCN	Monnet Sonerai II	P. J. McNamee
	G–PNUT	Cameron Special Shape balloon	Balloon Stable Ltd.
	G–POKE	Pitts S-1E Special	D. C. Purley
	G–POLO	PA-31-350 Navajo Chieftain	Rowntree Mackintosh Ltd.
	G–POLY	Cameron N-77 balloon	A. J. Bingley Ltd. *Polywallets*
	G–POOH	Piper J-3C-65 Cub	P. Robinson
	G–POPE	Eiri PIK-20E-1	J. T. Pope
	G–PORR	AS.350B Ecureuil	Colt Car Co. Ltd.
	G–POST	EMB-110PI Bandeirante	Fairflight Ltd./Air Ecosse
	G–POWA	PA-24 Comanche 400	G. F. Miller
	G–PPHJ	Cessna F.172P	P. P. J. Howard-Johnston
	G–PPLI	Pazmany PL.1	G. Anderson
	G–PRAG	Brugger MB.2 Colibri	P. Russell
	G–PRES	Cessna 441 Conquest	Press Aviation Ltd. (G–BHFX)
	G–PRIX	Cessna 414A	Group Lotus Car Co. Ltd.
	G–PSID	P-51D Mustang	Fairoaks Aviation Services Ltd.
	G–PSPS	Thunder Colt AS-80 airship	Lighter-Than-Air Ltd.
	G–PTWO	Pilatus P2-05	B. J. S. Grey
	G–PUBS	Colt 56 SS balloon	Intervarsity Balloon Club *Puffin*
	G–PVAF	PA-44-180 Seminole	Pincus Vidler Arthur Fitzgerald Ltd.
	G–PVAM	Port Victoria 7 Grain Kitten	A. J. Manning
	G–PYRO	Cameron N-65 balloon	Masts Engineering Ltd.
	G–RACA	P.57 Sea Prince T.1	Atlantic & Caribbean Aviation Ltd.
	G–RACE	Aerostar 601P	Aerotime Ltd.
	G–RAFC	Robin R.2112	RAF Cranwell Flying Club
	G–RAFE	Thunder Ax7-77 balloon	Thunder Balloons Ltd. *Billboard II*
	G–RAIN	Maule M5-235C Lunar Rocket	Velcourt (East) Ltd.
	G–RAMS	PA-32R-301 Saratoga SP	Peacock & Archer Ltd.
	G–RAND	Rand KR-2	R. L. Wharmby
	G–RARE	Thunder Ax4 SS balloon	International Distillers & Vintners Ltd.
	G–RASC	Evans VP-2	R. A. Codling
	G–RAYS	Zenair CH.250	R. E. Delves
	G–RBBE	Cessna 421C	Ruston Bucyrus Ltd.
	G–RBIN	Robin DR.400/2+2	Headcorn Flying School Ltd.
	G–RBLA	D.H.C.-6 Twin Otter 310	Loganair Ltd.
	G–RCCL	Beech C90 King Air	Reckitt & Colman Products Ltd.
	G–RDON	WMB.2 Windtracker balloon	P. J. Donnellan (G–BICH)
	G–REAT	GA-7 Cougar	E. C. Heathcote & D. E. Nixon
	G–REEK	AA-5A Cheetah	JT Aviation Ltd.
	G–REES	Jodel D.140C	J. D. Rees
	G–REIS	PA-28R-201T Turbo Arrow III	H. Reis (Hard Chrome) Ltd.
	G–RETA	C.A.S.A. 1.131 Jungmann	C. Maron
	G–REXS	PA-28-181 Archer II	R. W. Willoughby
	G–RHCN	Cessna FR.182RG	R. H. C. Neville
	G–RHFI	Alexander Todd Skybolt	RHF (Estates) Ltd.
	G–RHHT	PA-32RT-300 Lance II	H.T. Air Freight Ltd.
	G–RIDE	Stephens Akro	R. Mitchell
	G–RIGS	Aerostar 601P	Rigs Design Services Ltd.
	G–RILL	Cessna 421C	Maxwell Restaurants Ltd. (G–BGZM)
	G–RILY	Monnet Sonerai II	K. D. Riley
	G–RIND	Cessna 335	ATA Grinding Processes
	G–RIST	Cessna 310R-II	Velcourt (East) Ltd. (G–DATS)
	G–RKSF	Pitts S-2A Special	Anvil Aviation Ltd.
	G–RLAY	EMB-110PI Bandeirante	Genair
	G–RMAM	Musselwhite MAM.1	M. A. Musselwhite
	G–ROAN	Boeing B.75N-1 Stearman	R. & A. Windley
	G–ROBN	Robin R.1180T	F. J. Franklin
	G–RODI	Isaacs Fury	J. R. C. Morgan
	G–ROLF	PA-28R-301 Saratoga SP	R. W. Burchardt
	G–ROLL	Pitts S-2A Special	G. Lynn

INTO BIRMINGHAM AIRPORT SOON AND SEE THE BEST SELECTION OF AIRBANDS IN THE U.K.

Or send 25p in stamps for our fully illustrated catalogue. Prices from £6.95 to over £200. We must have one to suit you! Plus our Exclusive DR600 Fully Portable scanner with a manual tune coupled to an *L.C.D. FREQUENCY READOUT*. Send a 14p stamp for separate data sheet. SEE the Signal range including the R517, R528 pocket scanner and the R512 8-channel scanner, also the Regency D.F.S. still one of the better synthesized Airbands. Callers welcome, friendly Personal service plus really helpful advice.

Access & Barclaycard accepted. Comprehensive frequency booklet 50p from:

HANGAR ROAD OFFICES, BIRMINGHAM AIRPORT
WEST MIDLANDS B26 3QJ
Tel: 021-742 8704

Pocket Sharp **£13.95** p&p extra
Saxon Concorde **£6.95** p&p extra

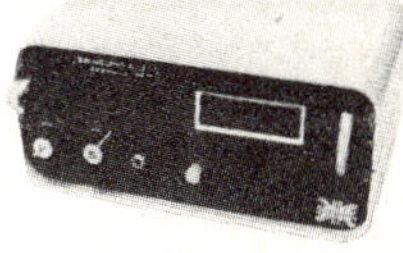

DR600 Scanning Airband Monitor
£189.00 Complete p&p extra

Air-Strip

Whether your interests are civil or military, Air-Strip, the journal of M.C.A.S., is the one for you. Air-Strip averages 36 pages per month and includes regular features covering Midland Civil Aviation News, Military Magazine, US Military News, Airline News, New Registrations, Airways, Ballooning, Gliding, Historical features, photographs, Society News etc.

Midland Civil Aviation News provides the most comprehensive news and movements service for the Midlands region.

Military Magazine and U.S. Military News cover News and movements from the U.K. and Europe and also regular features such as new orders, production lists etc.

M.C.A.S. hold regular meetings with guest speakers, slides and films in both the East and West Midlands, trips are arranged to Airshows, and an active Sales department is maintained, covering major shows and a regular Sunday stand at Birmingham Airport.

Subscription for 1982 is just **£5.90** — *can you afford not to join?*

Remittances to: Hon Registrar, M.C.A.S., 113 Ferndown Road, Solihull, West Midlands B91 2AX. Please quote reference 'CAM 82'

Midland Counties Aviation Society

Reg.	Type	Owner or Operator	Notes
G–RONW	FRED Srs. 2	P. J. D. Granow	
G–ROOK	Cessna F.172P	P. H. Johnston	
G–ROOM	Short SD3-60	Short Bros Ltd. (G–BSBL)	
G–RORO	Cessna 337B	Ronageny (Shipping) Ltd. (G–AVIX)	
G–ROSE	Evans VP-1	W. K. Rose	
G–ROSS	Practavia Pilot Sprite	F. M. T. Ross	
G–ROUS	PA-34-200T-2 Seneca	Linskill Air Charter Ltd.	
G–ROWL	AA-5B Tiger	K. E. Lundquist	
G–ROWS	PA-28-161 Warrior II	Robdow Productions Ltd.	
G–ROYL	Taylor JT.1 Monoplane	R. L. Wharmby	
G–ROYS	D.H.C.1 Chipmunk T.10	R. W. & S. Pullan	
G–RPAH	Rutan Vari-Eze	B. Hanson	
G–RRRR	Privateer Motor Glider	R. F. Selby	
G–RTHL	Leivers Special	R. Leivers	
G–RUTH	Cessna TR.182	W. H. & J. Rogers Group Ltd. (G–BJBC)	
G–RUIA	Cessna F.172M	Delamere & Norley Finance Ltd.	
G–RUMN	AA-1A Trainer	Reafly Ltd.	
G–RUSS	Cessna 172N	Leisure Lease	
G–SAAB	R. Commander 112TC	Continental Cars (Stansted) Ltd. (G–BEFS)	
G–SAAS	Ayres S2R-T34 Thrush Commander	Shoreham Flight Simulation	
G–SABA	PA-28R-201T Turbo Arrow III	Gill Aviation Ltd. (G–BFEN)	
G–SAFE	Cameron N-77 balloon	Derbyshire Building Soc.	
G–SAHI	Trago Mills SAH-1	Trago Mills Ltd.	
G–SAIL	Boeing 707-323C	Tradewinds Ltd.	
G–SALA	PA-32-300 Cherokee Six	Rodney Saunders Associates	
G–SALL	Cessna F.150L	Citation Flying Services Ltd.	
G–SAMS	M.S.880B Rallye Club	H. F. Hambling	
G–SARA	PA-28-181 Archer II	R. H. Ford	
G–SARO	Saro Skeeter Mk.12	F. F. Chamberlain	
G–SATC	Cessna F.150L	SAF Ltd.	
G–SATO	PA-23 Aztec 250	Colt Car Co. Ltd. (G–BCXP)	
G–SAVE	PA-31-350 Navajo Chieftain	Jetsave Ltd.	
G–SBRV	BRV Special	B. R. Vickers	
G–SCHH	BAe.146-100	British Aerospace Ltd. (G–BIAG)	
G–SCOT	PA-31-350 Navajo Chieftain	British Caledonian Airways	
G–SCUB	PA-18-135 Super Cub (542447)	N. D. Needham Farms	
G–SEAR	Pazmany PL.4	A. J. Sear	
G–SEED	Piper J-3C-65 Cub	J. H. Seed	
G–SEJW	PA-28-161 Warrior II	Grosvenor Homes Ltd.	
G–SEWL	PA-28-151 Warrior	A. R. Sewell & Sons	
G–SEXY	AA-1 Yankee	W. Davies (G–AYLM)	
G–SHAW	PA-30 Twin Comanche 160	Micro Metalsmiths Ltd.	
G–SHEL	Cameron O-56 balloon	The Shell Company of Hong Kong Ltd.	
G–SHIP	PA-23 Aztec 250	Shetland Line Air Services Ltd.	
G–SHOE	Cessna 421C-II	Shuimpex Services Ltd. (G–BHGD)	
G–SHOK	Cessna 421C-II	Armstrong Aviation Ltd.	
G–SHOW	M.S.733 Alcyon	L. M. Walton	
G–SIGN	PA-39 Twin Comanche C/R	K. W. Hawes (Electrical) Ltd.	
G–SILK	Aerostar 601P	R. Yorke	
G–SILV	Cessna 340A	Superprime Ltd.	
G–SIME	J/IN Alpha	J. T. Sime (G–AHHP)	
G–SIXA	Douglas DC-6B	Bowden Grange Enterprise Ltd. (G–ARXZ)	
G–SJAB	PA-39 Twin Comanche 160 C/R	S. J. Anderson-Brown (G–AYWZ)	
G–SKYE	Cessna TU.206G	RAF Sport Parachute Association	
G–SKYH	Cessna 172N	Ascombe Distributors	
G–SKYM	Cessna F.337E	A. P. Mothew & B. J. Ixer (G–AYHW)	
G–SLEA	Mudry/CAARP CAP.10B	C. J. Else & Co. Ltd.	
G–SLIK	Taylor JT.2 Titch	J. Jennings	
G–SMHK	Cameron D-38 airship	San Miguel Brewery Ltd.	
G–SMIT	Messerschmitt Bf.109G	Fairoaks Aviations Services Ltd.	
G–SMJJ	Cessna 414A	Gull Air Ltd.	
G–SNIP	Cessna F.172H	H. V. Jones (G–AXSI)	
G–SNOW	Cameron V-77 balloon	M. J. Snow	
G–SOAR	Eiri PIK-20E	C. P. Witter Ltd.	
G–SOLO	Pitts S-2S Special	Rothmans International Ltd.	
G–SOLY	Westland Bell 47G-3B1 (Soloy)	Heliwork Ltd.	

Notes	Reg.	Type	Owner or Operator
	G–SONA	SOCATA TB.10 Tobago	Sonardyne Ltd. (G–BIBI)
	G–SONG	Beech A200 Super King Air	Tembo Records Ltd. (G–BKTI)
	G–SPEY	AB-206B JetRanger 3	Craggan Chaise (Scotland) Management Ltd. (G–BIGO)
	G–SPIN	Pitts S-2A Special	R. N. Goode
	G–SPIT	V.S. Spitfire XIV (MV293)	D. W. Arnold (G–BGHB)
	G–SPOP	Thunder Ax7-77 balloon	Thunder Balloons Ltd.
	G–SPOT	Partenavia P.68B Observer	Twin Flight Ltd. (G–BCDK) (I)
	G–SPTS	Beech C90 King Air	Seabourne Aviation Ltd. (G–BHAP)
	G–SPUD	F.27 Friendship Mk.100	Air UK
	G–SSBS	Colting Ax77 balloon	Lighter-Than-Air Ltd.
	G–SSCH	BAe.146-100	British Aerospace Ltd. (G–BIAF)
	G–SSHH	BAe.146-100	British Aerospace Ltd. (G–BIAE)
	G–SSSH	BAe.146-100	British Aerospace Ltd. (G–BIAD)
	G–STAN	F.27 Friendship Mk.200	Air UK
	G–STAT	Cessna U.206F	Corbett Farms Ltd.
	G–STIO	ST.10 Diplomate	Rogers Autos Ltd.
	G–STOL	M.S.894A Rallye Minerva	Cole Cutters Ltd.
	G–STUD	D.H.C.-6 Twin Otter 310	Fairflight Ltd./Air Ecosse
	G–SUES	AT-6D Harvard III (133854)	P. W. Leaney Automatic Transmissions Ltd.
	G–SUPA	PA-18-150 Super Cub	Yorkshire Gliding Club (Pty) Ltd.
	G–SUZY	Taylor JT.1 Monoplane	S. A. Kanick
	G–SVHA	Partenavia P.68B	NPD Aviation Ltd.
	G–SWOT	Currie Super Wot	S. T. A. Albu
	G–TACA	P.57 Sea Prince T.1	Atlantic & Caribbean Aviation Ltd.
	G–TACE	H.S.125 Srs. 403B	Aeronautical Educational Trust Ltd. (G–AYIZ)
	G–TALY	AB-206B JetRanger 3	J. N. C. James
	G–TAMY	Cessna 421B	Abbergail Ltd.
	G–TATT	GY-20 Minicab	L. Tattershall
	G–TAXI	PA-23 Aztec 250	Northern Executive Aviation Ltd.
	G–TAXY	PA-31 Navajo	Air Continental Securities Ltd.
	G–TBCA	Bell 206L LongRanger	British Car Auctions Ltd. (G–BFAL)
	G–TCAT	G.164D Ag-Cat	Miller Aerial Spraying Ltd.
	G–TDAA	Cessna U.206G	Edward & Susan Dexter Ltd.
	G–TEAC	AT-6C Harvard IIA (MC280)	Trans Europe Ltd.
	G–TEAM	Cessna 414A	Northair Aviation Ltd. (G–BHJT)
	G–TEFC	PA-28 Cherokee 140	Thames Estuary Flying Club
	G–TFCI	Cessna FA.152	Tayside Aviation Ltd.
	G–THAM	Cessna F.182R	German Tourist Facilities Ltd.
	G–THEA	Boeing E75 Stearman	L. M. Walton
	G–THOM	Thunder Ax6-56 balloon	Thunder Balloons Ltd.
	G–THOR	Thunder Ax8-105 balloon	N. C. Faithfull *Turncoat*
	G–THSL	PA-28R-201 Arrow II	Thistle Metallics Ltd.
	G–TIME	Aerostar 601P	Marlborough Fine Art (London) Ltd.
	G–TIMK	PA-28-181 Archer II	T. Kilroe & Sons Ltd.
	G–TINA	SOCATA TB.10 Tobago	A. Lister
	G–TJCB	H.S.125 Srs. 700B	J. C. Bamford (Excavators) Ltd.
	G–TKHM	AB-206B JetRanger 3	T. Kilroe & Sons Ltd. (G–MKAN/G–DOUG)
	G–TLOL	Cessna 421C	Littlewoods Organisation Ltd.
	G–TOBY	Cessna 172B	J. A. Kelman (G–ARCM)
	G–TOMS	PA-38-112 Tomahawk	Channel Aviation Ltd.
	G–TONI	Cessna 421C	Coventry Aviation Ltd.
	G–TOUR	Robin R.2112	Finncharter Services
	G–TPTR	AB-206B JetRanger 3	Tolemans Holding Co. Ltd. (G–LOCK)
	G–TREV	Saffery S.330 balloon	T. W. Gurd
	G–TRIX	V.S. Spitfire T.IX	S. Atkins
	G–TROT	PA-31-350 Navajo Chieftain	Trottair Airways Ltd.
	G–TSIX	AT-6C Harvard IIA	D. Taylor
	G–TTAM	Taylor JT.2 Titch	A. J. Manning
	G–TTWO	Colt 56A balloon	D. M. Winder
	G–TUKE	Robin DR.400/160	Tukair
	G–TURB	D.31 Turbulent	P. S. E. Clifton
	G–TWEL	PA-28-181 Archer II	T.W. Electrical Ltd.
	G–TWIN	PA-44-180 Seminole	J. D. Apthorp
	G–TYPE	PA-28RT-201T Turbo Arrow IV	T. F. Guinness & B. Demsey (G–BIEA)
	G–TYRE	Cessna F.172M	Watts Aviation Ltd.
	G–UBHL	Beech A200 Super King Air	United Biscuits (UK) Ltd.
	G–UCPA	Eiri PIK-20E	B. J. Willson

Reg.	Type	Owner or Operator	Notes
G–UESS	Cessna 500 Citation	Osiwel Ltd.	
G–UIDE	Jodel D.120	S. T. Gilbert	
G–UPVC	Agusta A.109A	Anglian Double Glazing Co. Ltd.	
G–USTY	FRED Srs. 2	S. Styles	
G–VAGA	PA-15 Vagabond	R. Harris & I. S. Hodge	
G–VALE	AT-6C Harvard IIA (8810677)	Kayvale Finance Ltd. (G–RBAC)	
G–VAMP	Thunder Ax6-56 balloon	Thunder Balloons Ltd. *Vamp*	
G–VAUN	Cessna 340	Vaughan Associates Ltd.	
G–VEZE	Rutan Vari-Eze	P. J. Henderson	
G–VIKE	Bellanca 1730A Viking	MLP Aviation Ltd.	
G–VITE	Robin R.1180	Trans Global Aviation Supply Co. Ltd.	
G–VIVA	Thunder Ax6-56 balloon	G. Edwards *Dopey*	
G–VIZZ	Sportavia RS.180 Sportsman	Executive Air Sport Ltd.	
G–VMDE	Cessna P.210N	V. S. Evans & Horne & Sutton Ltd.	
G–VPTO	Evans VP-2	J. Cater	
G–VRES	Beech A200 Super King Air	Vernair Transport Services	
G–VTII	D.H.115 Vampire T.11 (WZ507)	J. Turnbull & ptnrs.	
G–VTOL	H.S. Harrier T.52	British Aerospace	
G–VWGB	Cessna 404 Titan	Volkswagon (GB) Ltd.	
G–VWSE	Cessna 404 Titan	Northair Aviation Ltd.	
G–WAAC	Cameron N-56 balloon	Zebedee Balloon Group	
G–WAGY	Cessna F.172N	J. B. Wagstaff	
G–WARD	Taylor JT.1 Monoplane	G. & G. D. Ward	
G–WARM	Bell 206L-1 LongRanger	Warmco (Manchester) Ltd.	
G–WASP	Brantly B.2B	Wasp Helicopter Hire Ltd. (G–ASXE)	
G–WEST	Agusta A.109A	Westland Helicopters Ltd.	
G–WETI	Cameron N-31 balloon	J. M. Albery *Weti*	
G–WGHB	T-33 Mk.3 Silver Star	—	
G–WHIT	Westland Bell 47G-3B1	C. G. Whittaker	
G–WHIZ	Pitts S-1 Special	K. M. McLeod	
G–WHIZ	*V.732 Viscount (fuselage only)	S. Wales Museum, Rhoose (G–ANRS)	
G–WICH	FRED Srs. 2	R. H. Hearn	
G–WICK	Partenavia P.68B	European Ferries Ltd. (G–BGFZ)	
G–WILL	AB-206B JetRanger 3	Anglian Double Glazing (Kent) Ltd.	
G–WIND	Boeing 707-323C	Tradewinds Ltd.	
G–WING	Cessna 404 Titan	Euroair Transport Ltd.	
G–WITT	PA-31P Navajo	C. E. Whittaker (G–BBRL)	
G–WIXY	Mudry/CAARP CAP.10B	G. Tanner & P. O. Wicks Ltd.	
G–WIZZ	AB-206B JetRanger 2	Wm. Monks (Builders Merchants) Ltd.	
G–WJMN	R. Commander 114	Shoreham Flight Simulation Ltd.	
G–WOOD	Beech 95-B55 Baron	Woods Management Services Ltd. (G–AYID)	
G–WOLF	PA-28 Cherokee 140	W. of Scotland Flying Club	
G–WOLL	G.164A Ag-Cat	Norfolk Aerial Spraying Ltd. (G–AYTM)	
G–WOSP	Bell 206B JetRanger 3	Burnthills Aviation Ltd.	
G–WPUI	Cessna P.172D	Oldham & Crowther (Promotions) Ltd. (G–AXPI)	
G–WRAY	PA-32RT-300T Turbo Lance II	Essex Insulation Ltd.	
G–WREN	Pitts S-2A Special	P. Meeson	
G–WSSC	PA-31-350 Navajo Chieftain	Spacegrand Ltd.	
G–WSSL	PA-31-350 Navajo Chieftain	G. W. Sparrow & Sons Ltd.	
G–WTVA	Cessna 404 Titan	Executive Express Ltd.	
G–WTVB	Cessna 404 Titan	Northair Aviation Ltd.	
G–WTVC	Cessna 404 Titan	Hay & Co. (Lerwick) Ltd.	
G–WTVE	Cessna 404 Titan	Executive Express Ltd.	
G–WULF	Focke-Wulf Fw.190 (04)	SBV Aeroservices Ltd. & ptnrs.	
G–WWII	V.S. Spitfire 18 (SM832)	D. W. Arnold & ptnrs.	
G–XCUB	PA-18-150 Super Cub	Unicol Engineering Ltd.	
G–XING	EMB-121A Xingu	C.S.E. Aviation Ltd.	
G–XMAS	PA-32RT-300 Lance II	A. G. Chrismas	
G–YIII	Cessna F.150L	Sherburn Aero Club Ltd.	
G–YKIV	Cessna F.150L	Sherburn Aero Club Ltd.	
G–YORK	Cessna F.172M	Sherburn Aero Club Ltd.	
G–YPSY	Andreasson BA-4B	J. Thomas	
G–YROS	Benson B.80-D	J. M. Montgomerie	
G–YTWO	Cessna F.172M	Sherburn Aero Club Ltd.	
G–YULL	PA-28 Cherokee 180E	Lansdowne Chemical Co. (G–BEAJ)	
G–ZEAL	Learjet 35A	C.S.E. Aviation Ltd.	

Notes	Reg.	Type	Owner or Operator
	G–ZEIZ	Learjet 36A	Royco Homes Ltd.
	G–ZEPP	Cameron D-96 balloon	Leaveglen Ltd.
	G–ZERO	AA-5B Tiger	Service Photography & Display Ltd.
	G–ZING	Learjet 35A	C.S.E. Aviation Ltd.
	G–ZIPP	Cessna 310Q	Citation Flying Services Ltd. (G–BAYU)
	G–ZLIN	Z.326 Trener Master	Systems Designers Aviation Ltd. (G–BBCR)
	G–ZONE	Learjet 35A	Jointair Ltd.
	G–ZSOL	Zlin 250L	R. N. Goode
	G–ZUMP	Cameron N-77 balloon	M. J. Allen *Gazump*
	G–ZZZZ	Point Maker Mk.I balloon	M. J. Wakelin

MILITARY TO CIVIL CROSS-REFERENCE

Serial carried	*Civil identity*	*Serial carried*	*Civil identity*
04 (Luftwaffe)	G-WULF	T5879	G-AXBW
14 (USSR)	G-AYAK	T6818	G-ANKT
45 (Aeronavale)	G-BHFG	T7187	G-AOBX
75	G-AFDX	T7281	G-ARTL
168	G-BFDE	T7404	G-ANMV
282 (Swiss A.F.)	G-BBMI	T7909	G-ANON
385 (RCAF)	G-BGPB	T9707	G-AKKR
422	G-AVJO	T9738	G-AKAT
2345	G-ATVP	V3388	G-AHTW
2807 (VE-111 USN)	G-BHTH	V9281 (RU-M)	G-BCWL
3066	G-AETA	V9441 (AR-A)	G-AZWT
5964	G-BFVH	Z2033	G-ASTL
8449M	G-ASWJ	Z7197	G-AKZN
18393 (C.A.F.)	G-BCYK	AB910 (XT-M)	G-AISU
133854	G-SUES	AP507	G-ACWP
133867	G-ELLY	AR213 (QG-A)	G-AIST
329417 (USAAF)	G-BDHK	AR501 (NN-A)	G-AWII
329601 (USAAF)	G-AXHR	BB814	G-AFWI
461748	G-BHDK	BS676 (K-U)	G-KUKU
480308	G-BDCD	DE208	G-AGYU
485784 (YB-E)	G-BEDF	DE363	G-ANFC
542447	G-SCUB	DE623	G-ANFI
8810677	G-VALE	DE992	G-AXXV
A16-199 (RAAF)	G-BEOX	DF130	G-BACK
A8226	G-BIDW	DF155	G-ANFV
B1807	G-EAVX	DF198	G-BBRB
B7270	G-BFCZ	DG590	G-ADMW
C1701	G-AWYY	EM903	G-APBI
D8096	G-AEPH	EV851	G-AJPI
E449	G-EBKN	FE992	G-BDAM
E3404	G-ADEV	FT229	G-AZKI
F904	G-EBIA	FT323	G-AZSC
F938	G-EBIC	FT391	G-AZBN
F939	G-EBIB	FX301 (FD-NQ)	G-JUDI
F1425	G-BEFR	HB751	G-BCBL
F8010	G-BDWJ	LB312	G-AHXE
F8614	G-AWAU	LS326	G-AJVH
G-29-1 (Class B)	G-APRJ	LZ766	G-ALCK
G-48-1 (Class B)	G-ALSX	MC280	G-TEAC
H2311	G-ABAA	MD497	G-ANLW
J9941	G-ABMR	MH434 (AC-S)	G-ASJV
K123	G-EACN	MP425	G-AITB
K1786	G-AFTA	MT360	G-AKWT
K2050	G-ASCM	MV293	G-SPIT
K2572	G-AOZH	MW100	G-AGNV
K3215	G-AHSA	NF875	G-AGTM
K4235	G-AHMJ	NJ695	G-AJXV
L2301	G-AIZG	NJ703	G-AKPI
L8032	G-AMRK	NM140	G-APGL
N220	G-BDFF	NP181	G-AOAR
N1854	G-AIBE	NP184	G-ANYP
N3788	G-AKPF	NP303	G-ANZJ
N4877 (VX-F)	G-AMDA	NX611	G-ASXX
N5054	G-AWSA	PG617	G-AYVY
N5180	G-EBKY	PG651	G-AYUX
N5182	G-APUP	PZ865	G-AMAU
N5430	G-BHEW	RG333	G-AIEK
N6848	G-BALX	RG333	G-AKEZ
N6985	G-AHMN	RH378	G-AJOE
N9191	G-ALND	RL962	G-AHED
N9238	G-ANEL	RM221	G-ANXR
N9389	G-ANJA	RM619 (AP-D)	G-ALGT
P3308 (UP-A)	G-AWLW	RR299 (HT-E)	G-ASKH
P6382	G-AJDR	RS709	G-ASKA
P7350	G-AWIJ	SM832	G-WWII
R1914	G-AHUJ	SM969	G-BRAF
R4907	G-ANCS	TA634	G-AWJV
R4959	G-ARAZ	TA719	G-ASKC
R7524	G-AIWA	TS423	G-DAKS
S1287	G-BEYB	VL348	G-AVVO
T5424	G-AJOA	VM360	G-APHV
T5493	G-ANEF	VR249	G-APIY
T5854	G-ANKK	VS356	G-AOLU

Serial carried	Civil identity	Serial carried	Civil identity
VS610	G–AOKL	XB733	G–ATBF
VX302 (77-M)	G–BCOV	XF690	G–BGKA
VZ728	G–AGOS	XF785	G–ALBN
WA576	G–ALSS	XF877	G–AWVF
WB763	G–BBMR	XG452	G–BRMB
WD413	G–BFIR	XG547	G–HAPR
WG307	G–BCYJ	XJ389	G–AJJP
WG316	G–BCAH	XK417	G–AVXY
WG348	G–BBMV	XK655	G–AMXA
WG719	G–BRMA	XL717	G–AOXG
WJ897	G–BDFT	XM529	G–BDNS
WJ945	G–BEDV	XM553	G–AWSV
WL626	G–BHDD	XM556	G–HELI
WP321	G–BRFC	XM685	G–AYZJ
WP790	G–BBNC	XP355	G–BEBC
WP808	G–BDEU	XR240	G–BDFH
WP857	G–BDRJ	XR241	G–AXRR
WP977	G–BHRD	XR269	G–BDXY
WV493	G–BDYG	XR279	G–BWKK
WV494	G–BGSB	F+IS (Luftwaffe)	G–BIRW
WV783	G–ALSP	AT+JX (Luftwaffe)	G–ATJX
WZ507	G–VTII	6J+PR (Luftwaffe)	G–AWHB
WZ672	G–BDER	7A+WN (Luftwaffe)	G–AZMH
WZ711	G–AVHT	N8+AA (Luftwaffe)	G–BFHD
WZ868	G–BCIW	N9+AA (Luftwaffe)	G–BECL

S1287 Fairey Flycatcher replica, civil identity G–BEYB

OVERSEAS AIRLINER REGISTRATIONS

(Aircraft included in this section are those most likely to be seen at UK airports on scheduled or charter services.)

A40 (Oman)

Reg.	*Type*	*Owner or Operator*	*Notes*
A40–TV	L.1011-385 TriStar 100	Gulf Air	
A40–TW	L-1011-385 TriStar 100	Gulf Air	
A40–TX	L-1011-385 TriStar 100	Gulf Air	
A40–TY	L-1011-385 TriStar 100	Gulf Air	
A40–TZ	L-1011-385 TriStar 100	Gulf Air	

NOTE: Gulf Air also operate TriStar 200s N92TA, N92TB & N92TC.

AP (Pakistan)

Reg.	*Type*	*Owner or Operator*	*Notes*
AP–ATQ	Boeing 720-051B	Pakistan International Airlines	
AP–AWU	Boeing 707-373C	Pakistan International Airlines	
AP–AWY	Boeing 707-340C	Pakistan International Airlines	
AP–AXA	Boeing 707-340C	Pakistan International Airlines	
AP–AXC	Douglas DC-10-30	Pakistan International Airlines	
AP–AXD	Douglas DC-10-30	Pakistan International Airlines	
AP–AXG	Boeing 707-340C	Pakistan International Airlines	
AP–AXK	Boeing 720-047B	Pakistan International Airlines	
AP–AXL	Boeing 720-047B	Pakistan International Airlines	
AP–AXM	Boeing 720-047B	Pakistan International Airlines	
AP–AYM	Douglas DC-10-30	Pakistan International Airlines	
AP–AYV	Boeing 747-282B	Pakistan International Airlines	
AP–AYW	Boeing 747-282B	Pakistan International Airlines	
AP–AZP	Boeing 720-030B	Pakistan International Airlines	
AP–AZW	Boeing 707-351B	Pakistan International Airlines	
AP–BAA	Boeing 707-351B	Pakistan International Airlines	
AP–BAF	Boeing 720-047B	Pakistan International Airlines	
AP–BAK	Boeing 747-240B	Pakistan International Airlines	
AP–BAT	Boeing 747-240B	Pakistan International Airlines	

B (China)

Reg.	*Type*	*Owner or Operator*	*Notes*
B–2402	Boeing 707-3J6B	CAAC	
B–2404	Boeing 707-3J6B	CAAC	
B–2406	Boeing 707-3J6B	CAAC	
B–2408	Boeing 707-3J6B	CAAC	
B–2410	Boeing 707-3J6C	CAAC	
B–2412	Boeing 707-3J6C	CAAC	
B–2414	Boeing 707-3J6C	CAAC	
B–2416	Boeing 707-3J6C	CAAC	
B–2418	Boeing 707-3J6C	CAAC	
B–2420	Boeing 707-3J6C	CAAC	
B–2442	Boeing 747SP-J6	CAAC	
B–2444	Boeing 747SP-J6	CAAC	

NOTE: CAAC also operate Boeing 747SP-J6 N1304E.

C–F and C–G (Canada)

Notes	*Reg.*	*Type*	*Owner or Operator*
	C–FCPL	Douglas DC-8-63 (805)	CP Air *Empress of Manitoba*
	C–FCPM	Douglas DC-8-53 (607)	CP Air *Empress of Lisbon*
	C–FCPO	Douglas DC-8-63 (801)	CP Air *Empress of Quebec*
	C–FCPP	Douglas DC-8-63 (802)	CP Air *Empress of Alberta*
	C–FCPQ	Douglas DC-8-63 (803)	CP Air *Empress of Ontario*
	C–FCPS	Douglas DC-8-63 (804)	CP Air *Empress of Sydney*
	C–FCRA	Boeing 747-217B (741)	CP Air *Empress of Japan*
	C–FCRB	Boeing 747-217B (742)	CP Air *Empress of Canada*
	C–FCRD	Boeing 747-217B (743)	CP Air *Empress of Australia*
	C–FCRE	Boeing 747-217B (744)	CP Air *Empress of Italy*
	C–FDJC	Boeing 747-1D1 (399)	Wardair Canada *Phil Garratt*
	C–FFUN	Boeing 747-1D1 (398)	Wardair Canada *Romeo Vachon*
	C–FNWF	L-100-30 Hercules	North West Territorial Airways
	C–FNWY	L-100-30 Hercules	North West Territorial Airways
	C–FPWK	L-100-20 Hercules	North West Territorial Airways
	C–FPWN	L-100-20 Hercules (383)	Pacific Western
	C–FPWR	L-100-20 Hercules (385)	Pacific Western
	C–FTIK	Douglas DC-8-63F (867)	Air Canada
	C–FTIL	Douglas DC-8-63 (868)	Air Canada
	C–FTIM	Douglas DC-8-63 (869)	Air Canada
	C–FTIN	Douglas DC-8-63 (870)	Air Canada
	C–FTIO	Douglas DC-8-63AF (871)	Air Canada
	C–FTIP	Douglas DC-8-63 (872)	Air Canada
	C–FTIQ	Douglas DC-8-63AF (873)	Air Canada
	C–FTIR	Douglas DC-8-63 (874)	Air Canada
	C–FTIS	Douglas DC-8-63 (875)	Air Canada
	C–FTIU	Douglas DC-8-63 (876)	Air Canada
	C–FTIV	Douglas DC-8-63 (877)	Air Canada
	C–FTIX	Douglas DC-8-63 (879)	Air Canada
	C–FTJL	Douglas DC-8-54F (812)	Air Canada
	C–FTJO	Douglas DC-8-54F (815)	Air Canada
	C–FTJP	Douglas DC-8-54F (816)	Air Canada
	C–FTJQ	Douglas DC-8-54F (817)	Air Canada
	C–FTJR	Douglas DC-8-54F (818)	Air Canada
	C–FTJS	Douglas DC-8-54F (819)	Air Canada
	C–FTJT	Douglas DC-8-61 (860)	Air Canada
	C–FTJU	Douglas DC-8-61 (861)	Air Canada
	C–FTJV	Douglas DC-8-61 (862)	Air Canada
	C–FTJW	Douglas DC-8-61 (863)	Air Canada
	C–FTJX	Douglas DC-8-61 (864)	Air Canada
	C–FTJY	Douglas DC-8-61 (865)	Air Canada
	C–FTJZ	Douglas DC-8-61 (866)	Air Canada
	C–FTNI	L.1011-385 TriStar 100 (509)	Air Canada
	C–FTNJ	L.1011-385 TriStar 100 (510)	Air Canada
	C–FTNK	L.1011-385 TriStar 100 (511)	Air Canada
	C–FTNL	L.1011-385 TriStar 100 (512)	Air Canada
	C–FTOA	Boeing 747-133 (301)	Air Canada
	C–FTOB	Boeing 747-133 (302)	Air Canada
	C–FTOC	Boeing 747-133 (303)	Air Canada
	C–FTOD	Boeing 747-133 (304)	Air Canada
	C–FTOE	Boeing 747-133 (305)	Air Canada
	C–GAGA	Boeing 747-233B (306)	Air Canada
	C–GAGB	Boeing 747-233B (307)	Air Canada
	C–GAGF	L.1011-385 TriStar 500 (551)	Air Canada
	C–GAGG	L.1011-385 TriStar 500 (552)	Air Canada
	C–GAGH	L.1011-385 TriStar 500 (553)	Air Canada
	C–GAGI	L.1011-385 TriStar 500 (554)	Air Canada
	C–GAGJ	L.1011-385 TriStar 500 (555)	Air Canada
	C–GAGK	L.1011-385 TriStar 500 (556)	Air Canada
	C–GCPC	Douglas DC-10-30 (901)	CP Air *Empress of Amsterdam*

Reg.	Type	Owner or Operator	Notes
C-GCPD	Douglas DC-10-30 (902)	CP Air *Empress of Sydney*	
C-GCPE	Douglas DC-10-30 (903)	CP Air *Empress of Rome*	
C-GCPF	Douglas DC-10-30 (904)	CP Air *Empress of Buenos Aires*	
C-GCPG	Douglas DC-10-30 (905)	CP Air *Empress of Fiji*	
C-GCPH	Douglas DC-10-30 (906)	CP Air *Empress of Lima*	
C-GFHX	Douglas DC-10-30	Wardair Canada	
C-GHPW	L-100-30 Hercules (387)	Pacific Western	
C-GXRA	Boeing 747-211B (397)	Wardair Canada *Herbert Hollick Kenyon*	
C-GXRB	Douglas DC-10-30 (101)	Wardair Canada *C. H. Punch Dickens*	
C-GXRC	Douglas DC-10-30 (102)	Wardair Canada *W. R. Wop May*	
C-GXRD	Boeing 747-211B (396)	Wardair Canada *H. A. Doc Oakes*	

NOTE: Airline fleet number carried on aircraft is shown in parenthesis.

CCCP (Russia)

All aircraft listed are operated by Aeroflot. The registrations are prefixed by CCCP in each case.

Reg.	Type	Notes
65001	Tu-134A	
65004	Tu-134A	
65005	Tu-134A	
65008	Tu-134A	
65009	Tu-134A	
65010	Tu-134A	
65011	Tu-134A	
65013	Tu-134A	
65014	Tu-134A	
65015	Tu-134A	
65017	Tu-134A	
65018	Tu-134A	
65020	Tu-134A	
65021	Tu-134A	
65023	Tu-134A	
65024	Tu-134A	
65025	Tu-134A	
65027	Tu-134A	
65028	Tu-134A	
65030	Tu-134A	
65031	Tu-134A	
65032	Tu-134A	
65034	Tu-134A	
65035	Tu-134A	
65036	Tu-134A	
65037	Tu-134A	
65038	Tu-134A	
65039	Tu-134A	
65040	Tu-134A	
65042	Tu-134A	
65043	Tu-134A	
65044	Tu-134A	
65045	Tu-134A	
65046	Tu-134A	
65047	Tu-134A	
65048	Tu-134A	
65049	Tu-134A	
65050	Tu-134A	
65051	Tu-134A	
65052	Tu-134A	
65053	Tu-134A	
65055	Tu-134A	
65056	Tu-134A	
65058	Tu-134A	
65059	Tu-134A	
65060	Tu-134A	
65061	Tu-134A	
65062	Tu-134A	
65065	Tu-134A	
65066	Tu-134A	
65069	Tu-134A	
65070	Tu-134A	
65071	Tu-134A	
65072	Tu-134A	
65074	Tu-134A	
65075	Tu-134A	
65076	Tu-134A	
65077	Tu-134A	
65080	Tu-134A	
65082	Tu-134A	
65083	Tu-134A	
65085	Tu-134A	
65086	Tu-134A	
65088	Tu-134A	
65089	Tu-134A	
65090	Tu-134A	
65095	Tu-134A	
65096	Tu-134A	
65098	Tu-134A	
65100	Tu-134A	
65101	Tu-134A	
65105	Tu-134A	
65107	Tu-134A	
65109	Tu-134A	
65112	Tu-134A	
65113	Tu-134A	
65114	Tu-134A	
65118	Tu-134A	
65119	Tu-134A	
65120	Tu-134A	
65132	Tu-134A	
65134	Tu-134A	
65135	Tu-134A	
65136	Tu-134A	
65140	Tu-134A	
65600	Tu-134	
65601	Tu-134	
65602	Tu-134	
65603	Tu-134	
65604	Tu-134	
65605	Tu-134	
65606	Tu-134	

Notes	Reg.	Type	Notes	Reg.	Type
	65607	Tu-134		65687	Tu-134A
	65608	Tu-134		65689	Tu-134A
	65609	Tu-134		65690	Tu-134A
	65610	Tu-134		65691	Tu-134A
	65611	Tu-134		65692	Tu-134A
	65612	Tu-134		65694	Tu-134A
	65613	Tu-134		65696	Tu-134A
	65614	Tu-134		65697	Tu-134A
	65615	Tu-134		65705	Tu-134A
	65616	Tu-134		65706	Tu-134A
	65617	Tu-134		65711	Tu-134A
	65618	Tu-134		65713	Tu-134A
	65619	Tu-134		65714	Tu-134A
	65620	Tu-134		65717	Tu-134A
	65621	Tu-134		65718	Tu-134A
	65622	Tu-134		65727	Tu-134A
	65623	Tu-134		65728	Tu-134A
	65624	Tu-134A		65729	Tu-134A
	65625	Tu-134		65730	Tu-134A
	65626	Tu-134A		65731	Tu-134A
	65627	Tu-134		65732	Tu-134A
	65628	Tu-134		65733	Tu-134A
	65629	Tu-134		65734	Tu-134A
	65630	Tu-134		65735	Tu-134A
	65631	Tu-134		65739	Tu-134A
	65632	Tu-134		65741	Tu-134A
	65633	Tu-134		65742	Tu-134A
	65634	Tu-134		65743	Tu-134A
	65635	Tu-134		65744	Tu-134A
	65636	Tu-134		65745	Tu-134A
	65637	Tu-134		65746	Tu-134A
	65638	Tu-134		65747	Tu-134A
	65639	Tu-134		65748	Tu-134A
	65640	Tu-134		65749	Tu-134A
	65641	Tu-134		65753	Tu-134A
	65642	Tu-134		65757	Tu-134A
	65643	Tu-134		65761	Tu-134A
	65644	Tu-134A		65762	Tu-134A
	65645	Tu-134A		65764	Tu-134A
	65646	Tu-134A		65765	Tu-134A
	65647	Tu-134A		65766	Tu-134A
	65648	Tu-134A		65769	Tu-134A
	65649	Tu-134A		65770	Tu-134A
	65650	Tu-134A		65771	Tu-134A
	65651	Tu-134A		65777	Tu-134A
	65652	Tu-134A		65780	Tu-134A
	65653	Tu-134A		65781	Tu-134A
	65654	Tu-134A		65782	Tu-134A
	65655	Tu-134A		65783	Tu-134A
	65656	Tu-134A		65784	Tu-134A
	65657	Tu-134A		65785	Tu-134A
	65658	Tu-134A		65789	Tu-134A
	65659	Tu-134A		65791	Tu-134A
	65660	Tu-134A		65792	Tu-134A
	65661	Tu-134A		65794	Tu-134A
	65662	Tu-134A		65801	Tu-134A
	65663	Tu-134A		65802	Tu-134A
	65664	Tu-134A		65804	Tu-134A
	65665	Tu-134A		65806	Tu-134A
	65666	Tu-134A		65810	Tu-134A
	65667	Tu-134A		65812	Tu-134A
	65668	Tu-134A		65815	Tu-134A
	65669	Tu-134A		65817	Tu-134A
	65670	Tu-134A		65818	Tu-134A
	65671	Tu-134A		65820	Tu-134A
	65672	Tu-134A		65821	Tu-134A
	65673	Tu-134A		65822	Tu-134A
	65674	Tu-134A		65823	Tu-134A
	65675	Tu-134A		65825	Tu-134A
	65676	Tu-134A		65828	Tu-134A
	65677	Tu-134A		65829	Tu-134A
	65678	Tu-134A		65830	Tu-134A
	65679	Tu-134A		65831	Tu-134A
	65680	Tu-134A		65832	Tu-134A
	65683	Tu-134A		65833	Tu-134A

Reg.	Type	Notes
65834	Tu-134A	
65837	Tu-134A	
65839	Tu-134A	
65840	Tu-134A	
65841	Tu-134A	
65843	Tu-134A	
65844	Tu-134A	
65845	Tu-134A	
65848	Tu-134A	
65851	Tu-134A	
65852	Tu-134A	
65853	Tu-134A	
65854	Tu-134A	
65857	Tu-134A	
65861	Tu-134A	
65862	Tu-134A	
65863	Tu-134A	
65864	Tu-134A	
65865	Tu-134A	
65866	Tu-134A	
65867	Tu-134A	
65868	Tu-134A	
65869	Tu-134A	
65870	Tu-134A	
65871	Tu-134A	
65872	Tu-134A	
65873	Tu-134A	
65874	Tu-134A	
65877	Tu-134A	
65878	Tu-134A	
65879	Tu-134A	
65880	Tu-134A	
65881	Tu-134A	
65882	Tu-134A	
65883	Tu-134A	
65884	Tu-134A	
65886	Tu-134A	
65888	Tu-134A	
65890	Tu-134A	
65891	Tu-134A	
65892	Tu-134A	
65893	Tu-134A	
65894	Tu-134A	
65895	Tu-134A	
65898	Tu-134A	
65899	Tu-134A	
65903	Tu-134A	
65950	Tu-134A	
65951	Tu-134A	
65952	Tu-134A	
65953	Tu-134A	
65954	Tu-134A	
65955	Tu-134A	
65957	Tu-134A	
65960	Tu-134A	
65961	Tu-134A	
65962	Tu-134A	
65963	Tu-134A	
65964	Tu-134A	
65965	Tu-134A	
65967	Tu-134A	
65969	Tu-134A	
65970	Tu-134A	
65971	Tu-134A	
65972	Tu-134A	
65973	Tu-134A	
65974	Tu-134A	
65975	Tu-134A	
65976	Tu-134A	
85000	Tu-154	
85005	Tu-154	
85007	Tu-154	
85008	Tu-154	
85009	Tu-154	
85010	Tu-154	
85011	Tu-154	
85012	Tu-154	
85013	Tu-154	
85014	Tu-154	
85016	Tu-154	
85017	Tu-154	
85018	Tu-154	
85019	Tu-154	
85021	Tu-154	
85022	Tu-154	
85024	Tu-154	
85025	Tu-154	
85028	Tu-154	
85029	Tu-154	
85030	Tu-154	
85031	Tu-154	
85032	Tu-154	
85033	Tu-154	
85034	Tu-154	
85035	Tu-154	
85037	Tu-154	
85038	Tu-154	
85039	Tu-154	
85040	Tu-154	
85041	Tu-154	
85042	Tu-154	
85043	Tu-154	
85044	Tu-154	
85047	Tu-154	
85048	Tu-154	
85050	Tu-154	
85051	Tu-154	
85052	Tu-154	
85053	Tu-154	
85054	Tu-154	
85055	Tu-154	
85057	Tu-154	
85058	Tu-154A	
85060	Tu-154A	
85061	Tu-154A	
85062	Tu-154A	
85063	Tu-154A	
85064	Tu-154A	
85065	Tu-154A	
85066	Tu-154A	
85067	Tu-154A	
85068	Tu-154A	
85069	Tu-154A	
85070	Tu-154A	
85071	Tu-154A	
85072	Tu-154A	
85074	Tu-154A	
85075	Tu-154B	
85076	Tu-154A	
85080	Tu-154A	
85081	Tu-154A	
85082	Tu-154A	
85083	Tu-154A	
85084	Tu-154A	
85085	Tu-154A	
85086	Tu-154A	
85088	Tu-154A	
85090	Tu-154A	
85091	Tu-154A	
85092	Tu-154A	
85093	Tu-154A	
85094	Tu-154A	
85095	Tu-154A	
85096	Tu-154A	
85097	Tu-154A	
85098	Tu-154A	
85099	Tu-154A	
85100	Tu-154A	
85101	Tu-154A	
85102	Tu-154A	

Notes	Reg.	Type
	85103	Tu-154A
	85104	Tu-154A
	85105	Tu-154A
	85106	Tu-154B
	85107	Tu-154A
	85108	Tu-154A
	85109	Tu-154A
	85111	Tu-154A
	85112	Tu-154A
	85113	Tu-154A
	85114	Tu-154A
	85115	Tu-154A
	85116	Tu-154A
	85117	Tu-154A
	85118	Tu-154B
	85119	Tu-154A
	85121	Tu-154B
	85122	Tu-154B
	85123	Tu-154B
	85124	Tu-154B
	85128	Tu-154B
	85129	Tu-154B
	85130	Tu-154B
	85131	Tu-154B
	85134	Tu-154B
	85136	Tu-154B
	85137	Tu-154B
	85138	Tu-154B
	85139	Tu-154B
	85140	Tu-154B
	85141	Tu-154B
	85142	Tu-154B
	85145	Tu-154B
	85146	Tu-154B
	85147	Tu-154B
	85148	Tu-154B
	85149	Tu-154B
	85150	Tu-154B
	85152	Tu-154B
	85153	Tu-154B
	85154	Tu-154B
	85155	Tu-154B
	85156	Tu-154B
	85157	Tu-154B
	85160	Tu-154B
	85162	Tu-154B
	85163	Tu-154B
	85164	Tu-154B
	85165	Tu-154B
	85166	Tu-154B
	85167	Tu-154B
	85168	Tu-154B
	85169	Tu-154B
	85170	Tu-154B
	85171	Tu-154B
	85172	Tu-154B
	85174	Tu-154B
	85175	Tu-154B
	85176	Tu-154B
	85177	Tu-154B
	85178	Tu-154B
	85179	Tu-154B
	85180	Tu-154B
	85181	Tu-154B
	85182	Tu-154B
	85184	Tu-154B
	85185	Tu-154B
	85186	Tu-154B
	85187	Tu-154B
	85188	Tu-154B
	85189	Tu-154B
	85190	Tu-154B
	85192	Tu-154B
	85194	Tu-154B
	85196	Tu-154B

Notes	Reg.	Type
	85197	Tu-154B
	85198	Tu-154B
	85199	Tu-154B
	85200	Tu-154B
	85201	Tu-154B
	85202	Tu-154B
	85203	Tu-154B
	85204	Tu-154B
	85206	Tu-154B
	85207	Tu-154B
	85210	Tu-154B
	85211	Tu-154B
	85212	Tu-154B
	85214	Tu-154B
	85215	Tu-154B
	85216	Tu-154B
	85217	Tu-154B
	85218	Tu-154B
	85219	Tu-154B
	85220	Tu-154B
	85221	Tu-154B
	85222	Tu-154B
	85223	Tu-154B
	85228	Tu-154B
	85229	Tu-154B
	85230	Tu-154B
	85231	Tu-154B
	85232	Tu-154B
	85233	Tu-154B
	85234	Tu-154B
	85235	Tu-154B
	85236	Tu-154B
	85238	Tu-154B
	85240	Tu-154B
	85241	Tu-154B
	85242	Tu-154B
	85243	Tu-154B
	85244	Tu-154B
	85245	Tu-154B
	85247	Tu-154B
	85248	Tu-154B
	85249	Tu-154B
	85252	Tu-154B
	85253	Tu-154B
	85256	Tu-154B
	85257	Tu-154B
	85259	Tu-154B
	85260	Tu-154B
	85261	Tu-154B
	85263	Tu-154B
	85264	Tu-154B
	85265	Tu-154B
	85269	Tu-154B
	85271	Tu-154B
	85272	Tu-154B
	85274	Tu-154B
	85275	Tu-154B
	85276	Tu-154B
	85277	Tu-154B
	85278	Tu-154B
	85279	Tu-154B
	85280	Tu-154B
	85281	Tu-154B
	85282	Tu-154B
	85283	Tu-154B
	85284	Tu-154B
	85285	Tu-154B
	85286	Tu-154B
	85287	Tu-154B
	85288	Tu-154B
	85290	Tu-154B
	85295	Tu-154B
	85296	Tu-154B
	85298	Tu-154B
	85300	Tu-154B

Reg.	Type	Notes
85301	Tu-154B	
85302	Tu-154B	
85303	Tu-154B	
85304	Tu-154B	
85305	Tu-154B	
85306	Tu-154B	
85311	Tu-154B	
85313	Tu-154B	
85314	Tu-154B	
85316	Tu-154B	
85323	Tu-154B	
85328	Tu-154B	
85330	Tu-154B	
85331	Tu-154B	
85334	Tu-154B	
85335	Tu-154B	
85336	Tu-154B	
85337	Tu-154B	
85339	Tu-154B	
85346	Tu-154B	
85347	Tu-154B	
85349	Tu-154B	
85350	Tu-154B	
85353	Tu-154B	
85358	Tu-154B	
85359	Tu-154B	
85362	Tu-154B	
85363	Tu-154B	
85364	Tu-154B	
85365	Tu-154B	
85366	Tu-154B	
85367	Tu-154B	
85368	Tu-154B	
85372	Tu-154B	
85374	Tu-154B	
85375	Tu-154B	
85376	Tu-154B	
85377	Tu-154B	
85378	Tu-154B	
85379	Tu-154B	
85381	Tu-154B	
85382	Tu-154B	
85385	Tu-154B	
85390	Tu-154B	
85395	Tu-154B	
85396	Tu-154B	
85397	Tu-154B	
85398	Tu-154B	
85399	Tu-154B	
85400	Tu-154B	
85402	Tu-154B	
85407	Tu-154B	
85409	Tu-154B	
85410	Tu-154B	
85411	Tu-154B	
85412	Tu-154B	
85413	Tu-154B	
85414	Tu-154B	
85423	Tu-154B	
85424	Tu-154B	
85438	Tu-154B	
85441	Tu-154B	
85455	Tu-154B	
85460	Tu-154B	
85469	Tu-154B	
85476	Tu-154B	
85478	Tu-154B	
85479	Tu-154B	
85490	Tu-154B	
85491	Tu-154B	
85494	Tu-154B	
85495	Tu-154B	
85496	Tu-154B	
85497	Tu-154B	
85498	Tu-154B	

Reg.	Type	Notes
86450	IL-62	
86451	IL-62	
86452	IL-62M	
86453	IL-62M	
86454	IL-62M	
86455	IL-62M	
86456	IL-62M	
86457	IL-62M	
86458	IL-62M	
86459	IL-62M	
86460	IL-62	
86461	IL-62	
86462	IL-62M	
86463	IL-62M	
86464	IL-62M	
86465	IL-62M	
86469	IL-62M	
86470	IL-62M	
86471	IL-62M	
86472	IL-62M	
86473	IL-62M	
86474	IL-62M	
86475	IL-62M	
86476	IL-62M	
86477	IL-62M	
86478	IL-62M	
86479	IL-62M	
86480	IL-62M	
86481	IL-62M	
86482	IL-62M	
86483	IL-62M	
86484	IL-62M	
86485	IL-62M	
86486	IL-62M	
86487	IL-62M	
86488	IL-62M	
86489	IL-62M	
86490	IL-62M	
86491	IL-62M	
86497	IL-62M	
86498	IL-62M	
86499	IL-62M	
86500	IL-62M	
86501	IL-62M	
86502	IL-62M	
86504	IL-62M	
86507	IL-62M	
86509	IL-62M	
86510	IL-62M	
86511	IL-62M	
86512	IL-62M	
86513	IL-62M	
86517	IL-62M	
86518	IL-62M	
86605	IL-62	
86606	IL-62	
86607	IL-62M	
86608	IL-62	
86609	IL-62	
86610	IL-62	
86611	IL-62	
86612	IL-62	
86613	IL-62	
86614	IL-62M	
86615	IL-62	
86616	IL-62	
86617	IL-62	
86618	IL-62M	
86619	IL-62	
86620	IL-62M	
86621	IL-62M	
86622	IL-62M	
86623	IL-62M	
86624	IL-62	
86648	IL-62	

Notes	Reg.	Type
	86649	IL-62
	86650	IL-62
	86652	IL-62
	86653	IL-62
	86654	IL-62
	86655	IL-62
	86656	IL-62M
	86657	IL-62
	86658	IL-62M
	86659	IL-62
	86661	IL-62
	86662	IL-62
	86663	IL-62
	86664	IL-62
	86665	IL-62
	86666	IL-62
	86667	IL-62
	86668	IL-62
	86669	IL-62
	86670	IL-62
	86671	IL-62
	86672	IL-62
	86673	IL-62M
	86674	IL-62
	86675	IL-62
	86676	IL-62
	86677	IL-62
	86678	IL-62
	86679	IL-62
	86680	IL-62
	86681	IL-62
	86682	IL-62
	86683	IL-62
	86684	IL-62
	86685	IL-62
	86686	IL-62
	86687	IL-62
	86688	IL-62
	86689	IL-62
	86690	IL-62
	86691	IL-62
	86692	IL-62M
	86693	IL-62M
	86694	IL-62
	86695	IL-62
	86696	IL-62
	86697	IL-62
	86698	IL-62
	86699	IL-62
	86700	IL-62M
	86701	IL-62M
	86702	IL-62M
	86703	IL-62
	86704	IL-62
	86705	IL-62M

CN (Morocco)

Notes	Reg.	Type	Owner or Operator
	CN–CCF	Boeing 727-2B6	Royal Air Maroc *Fes*
	CN–CCG	Boeing 727-2B6	Royal Air Maroc *l'Oiseau de la Providence*
	CN–CCH	Boeing 727-2B6	Royal Air Maroc *Marrakesh*
	CN–CCW	Boeing 727-2B6	Royal Air Maroc *Agadir*
	CN–RMO	Boeing 727-2B6	Royal Air Maroc
	CN–RMP	Boeing 727-2B6	Royal Air Maroc
	CN–RMQ	Boeing 727-2B6	Royal Air Maroc
	CN–RMR	Boeing 727-2B6	Royal Air Maroc

CS (Portugal)

Notes	Reg.	Type	Owner or Operator
	CS–TBA	Boeing 707-382B	TAP—Air Portugal *Santa Cruz*
	CS–TBB	Boeing 707-382B	TAP—Air Portugal *Santa Maria*
	CS–TBC	Boeing 707-382B	TAP—Air Portugal *Cidade de Luanda*
	CS–TBD	Boeing 707-382B	TAP—Air Portugal *Cidade de Lourenço Marques*
	CS–TBE	Boeing 707-382B	TAP—Air Portugal *Pedro Alvares Cabral*
	CS–TBF	Boeing 707-382B	TAP—Air Portugal *Vasco da Gama*
	CS–TBG	Boeing 707-382B	TAP—Air Portugal *Fernao de Magalhaes*
	CS–TBH	Boeing 707-399C	TAP—Air Portugal *Pedro Nunes*
	CS–TBI	Boeing 707-399C	TAP—Air Portugal *D. João de Castro*
	CS–TBJ	Boeing 707-373C	TAP—Air Portugal *Lisboa*
	CS–TBK	Boeing 727-82	TAP—Air Portugal *Açores*
	CS–TBL	Boeing 727-82	TAP—Air Portugal *Madeira*
	CS–TBM	Boeing 727-82	TAP—Air Portugal *Algarve*
	CS–TBN	Boeing 727-82QC	TAP—Air Portugal *Cidade do Porto*

B–2442 (Top) Boeing 747SP-J6 of CAAC/*A. S. Wright*

CS–TBS (Centre) Boeing 727-282 of Air Portugal

F–BYCY (Bottom) S.E.210 Caravelle VI-N of Corse Air/*A. S. Wright*

Notes	Reg.	Type	Owner or Operator
	CS–TBO	Boeing 727-82QC	TAP—Air Portugal *Costa do Sol*
	CS–TBP	Boeing 727-82	TAP—Air Portugal *Cabo Verde*
	CS–TBQ	Boeing 727-172C	TAP—Air Portugal *Bissau*
	CS–TBS	Boeing 727-282	TAP—Air Portugal *Gago Coutinho*
	CS–TBT	Boeing 707-3F5C	TAP—Air Portugal *Humberto Delgado*
	CS–TBU	Boeing 707-3F5C	TAP—Air Portugal *Jaime Cortesão*
	CS–TBV	Boeing 727-155C	TAP—Air Portugal
	CS–TBW	Boeing 727-282	TAP—Air Portugal *Coimbra*
	CS–TBX	Boeing 727-282	TAP—Air Portugal *Faro*
	CS–TBY	Boeing 727-2B2	TAP—Air Portugal *Amadora*
	CS–TBZ	Boeing 727-2B2	TAP—Air Portugal
	CS–TEA	L.1011-385 TriStar 500	TAP—Air Portugal
	CS–TEB	L.1011-385 TriStar 500	TAP—Air Portugal
	CS–TEC	L.1011-385 TriStar 500	TAP—Air Portugal
	CS–TJA	Boeing 747-282B	TAP—Air Portugal *Portugal*
	CS–TJB	Boeing 747-282B	TAP—Air Portugal *Brasil*

CU (Cuba)

Notes	Reg.	Type	Owner or Operator
	CU–T1208	Ilyushin IL-62M	Cubana *Capt. Wifredo Perez*
	CU–T1209	Ilyushin IL-62M	Cubana
	CU–T1215	Ilyushin IL-62M	Cubana
	CU–T1216	Ilyushin IL-62M	Cubana
	CU–T1217	Ilyushin IL-62M	Cubana
	CU–T1218	Ilyushin IL-62M	Cubana
	CU–T1225	Ilyushin IL-62M	Cubana

D (German Federal Republic)

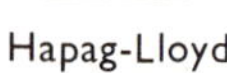
Hapag-Lloyd

Lufthansa

Condor Flugdienst

Notes	Reg.	Type	Owner or Operator
	D–AAST	S.E.210 Caravelle 10-R	Aero lloyd
	D–ABAK	S.E.210 Caravelle 10-R	Aero Lloyd
	D–ABAP	S.E.210 Caravelle 10-R	Special Air Transport
	D–ABAV	S.E.210 Caravelle 10-R	Special Air Transport
	D–ABAW	S.E.210 Caravelle 10-R	Special Air Transport
	D–ABBE	Boeing 737-230C	Lufthansa *City Jet Remscheid*
	D–ABCE	Boeing 737-230C	Lufthansa *City Jet Landshut*
	D–ABCI	Boeing 727-230	Lufthansa *Karlsruhe*
	D–ABDE	Boeing 737-230C	Lufthansa *City Jet Bamberg*
	D–ABDI	Boeing 727-230	Lufthansa *Lübeck*
	D–ABFA	Boeing 737-230	Lufthansa *Regensburg*
	D–ABFB	Boeing 737-230	Lufthansa *Flensburg*
	D–ABFC	Boeing 737-230	Lufthansa *Würzburg*
	D–ABFD	Boeing 737-230	Lufthansa *Bamberg*
	D–ABFE	Boeing 737-230C	Lufthansa *Trier*
	D–ABFF	Boeing 737-230	Lufthansa *Gelsenkirchen*
	D–ABFH	Boeing 737-230	Lufthansa *Pforzheim*
	D–ABFI	Boeing 727-230	Lufthansa *Münster*
	D–ABFK	Boeing 737-230	Lufthansa *Wuppertal*
	D–ABFL	Boeing 737-230	Lufthansa *Coburg*
	D–ABFM	Boeing 737-230	Lufthansa *Osnabrück*
	D–ABFN	Boeing 737-230	Lufthansa *Kempten*
	D–ABFP	Boeing 737-230	Lufthansa *Offenbach*
	D–ABFR	Boeing 737-230	Lufthansa *Solingen*
	D–ABFS	Boeing 737-230	Lufthansa *Oldenburg*
	D–ABFT	Boeing 737-230	Condor Flugdienst
	D–ABFU	Boeing 737-230	Lufthansa *Mülheim a.d.R*
	D–ABFW	Boeing 737-230	Lufthansa *Wolfsberg*

Reg.	Type	Owner or Operator	Notes
D-ABFX	Boeing 737-230	Lufthansa *Tübingen*	
D-ABFY	Boeing 737-230	Lufthansa *Göttingen*	
D-ABFZ	Boeing 737-230	Lufthansa *Wilhelmshaven*	
D-ABGE	Boeing 737-230C	Lufthansa *Erlangen*	
D-ABGI	Boeing 727-230	Lufthansa *Leverkusen*	
D-ABHA	Boeing 737-230	Lufthansa *Koblenz*	
D-ABHB	Boeing 737-230	Lufthansa *Gosler*	
D-ABHC	Boeing 737-230	Lufthansa *Friedrichshafen*	
D-ABHD	Boeing 737-230	Condor Flugdienst	
D-ABHE	Boeing 737-230C	Lufthansa *Darmstadt*	
D-ABHF	Boeing 737-230	Lufthansa *Heilbronn*	
D-ABHH	Boeing 737-230	Lufthansa *Marburg*	
D-ABHI	Boeing 727-230	Lufthansa *Mönchengladbach*	
D-ABHK	Boeing 737-230	Lufthansa *Bayreuth*	
D-ABHL	Boeing 737-230	Lufthansa *Worms*	
D-ABHM	Boeing 737-230	Lufthansa *Landsnut*	
D-ABHN	Boeing 737-230	Lufthansa *Trier*	
D-ABHP	Boeing 737-230	Lufthansa *Erlangen*	
D-ABHR	Boeing 737-230	Lufthansa *Darmstadt*	
D-ABHS	Boeing 737-230	Lufthansa *Remscheid*	
D-ABHT	Boeing 737-230	Condor Flugdienst	
D-ABHU	Boeing 737-230	Condor Flugdienst	
D-ABHW	Boeing 737-230	Condor Flugdienst	
D-ABHX	Boeing 737-230	Condor Flugdienst	
D-ABIL	Boeing 727-30	Condor Flugdienst	
D-ABIP	Boeing 727-30	Condor Flugdienst	
D-ABIQ	Boeing 727-30	Condor Flugdienst	
D-ABKA	Boeing 727-230	Lufthansa *Heidelberg*	
D-ABKB	Boeing 727-230	Lufthansa *Augsburg*	
D-ABKC	Boeing 727-230	Lufthansa *Braunschweig*	
D-ABKD	Boeing 727-230	Lufthansa *Freiburg*	
D-ABKE	Boeing 727-230	Lufthansa *Mannheim*	
D-ABKF	Boeing 727-230	Lufthansa *Saarbrücken*	
D-ABKG	Boeing 727-230	Lufthansa *Kassel*	
D-ABKH	Boeing 727-230	Lufthansa *Kiel*	
D-ABKI	Boeing 727-230	Lufthansa *Bremerhaven*	
D-ABKJ	Boeing 727-230	Lufthansa *Wiesbaden*	
D-ABKK	Boeing 727-230	Condor Flugdienst	
D-ABKL	Boeing 727-230	Condor Flugdienst	
D-ABKM	Boeing 727-230	Lufthansa *Hagen*	
D-ABKN	Boeing 727-230	Lufthansa *Ulm*	
D-ABKP	Boeing 727-230	Lufthansa *Krefeld*	
D-ABKQ	Boeing 727-230	Lufthansa *Mainz*	
D-ABKR	Boeing 727-230	Lufthansa *Bielefeld*	
D-ABKS	Boeing 727-230	Lufthansa *Oberhausen*	
D-ABKT	Boeing 727-230	Lufthansa *Aachen*	
D-ABLI	Boeing 727-230	Lufthansa *Ludwigshafen a.Rh.*	
D-ABMI	Boeing 727-230	Condor Flugdienst	
D-ABNI	Boeing 727-230	Condor Flugdienst	
D-ABPI	Boeing 727-230	Condor Flugdienst	
D-ABQI	Boeing 727-230	Lufthansa *Hildesheim*	
D-ABRI	Boeing 727-230	Lufthansa *Esslingen*	
D-ABSI	Boeing 727-230	Lufthansa *Hof*	
D-ABTI	Boeing 727-230	Condor Flugdienst	
D-ABUA	Boeing 707-330C	German Cargo	
D-ABUD	Boeing 707-330B	Lufthansa	
D-ABUE	Boeing 707-330C	German Cargo	
D-ABUF	Boeing 707-330B	Condor Flugdienst	
D-ABUH	Boeing 707-330B	Lufthansa	
D-ABUI	Boeing 707-330C	German Cargo	
D-ABUL	Boeing 707-330B	Lufthansa *Duisburg*	
D-ABUM	Boeing 707-330B	Lufthansa	
D-ABUO	Boeing 707-330C	German Cargo	
D-ABVI	Boeing 727-230	Condor Flugdienst	
D-ABWI	Boeing 727-230	Condor Flugdienst	
D-ABYJ	Boeing 747-230B	Lufthansa *Hessen*	
D-ABYK	Boeing 747-230B	Lufthansa *Rheinland-Pfalz*	
D-ABYL	Boeing 747-230B	Lufthansa *Saarland*	
D-ABYM	Boeing 747-230B	Lufthansa *Schleswig-Holstein*	
D-ABYN	Boeing 747-230B	Lufthansa *Baden-Wurttemberg*	
D-ABYO	Boeing 747-230F	Lufthansa	
D-ABYP	Boeing 747-230B	Lufthansa *Niedersachen*	

Notes	Reg.	Type	Owner or Operator
	D–ABYQ	Boeing 747-230B	Lufthansa *Bremen*
	D–ABYR	Boeing 747-230B	Lufthansa *Nordrhein-Westfalen*
	D–ABYS	Boeing 747-230B	Lufthansa *Bayern*
	D–ABYT	Boeing 747-230B	Lufthansa *Hamburg*
	D–ABYU	Boeing 747-230F	Lufthansa
	D–ABYW	Boeing 747-230B	Lufthansa
	D–ABYX	Boeing 747-230B	Lufthansa
	D–ABYY	Boeing 747-230B	Lufthansa
	D–ACVK	S.E.210 Caravelle 10-R	Aero Lloyd
	D–ADAO	Douglas DC-10-30	Lufthansa *Düsseldorf*
	D–ADBO	Douglas DC-10-30	Lufthansa *Berlin*
	D–ADCO	Douglas DC-10-30	Lufthansa *Frankfurt*
	D–ADDO	Douglas DC-10-30	Lufthansa *Hamburg*
	D–ADFO	Douglas DC-10-30	Lufthansa *München*
	D–ADGO	Douglas DC-10-30	Lufthansa *Bonn*
	D–ADHO	Douglas DC-10-30	Lufthansa *Hannover*
	D–ADJO	Douglas DC-10-30	Lufthansa *Essen*
	D–ADKO	Douglas DC-10-30	Lufthansa *Stuttgart*
	D–ADLO	Douglas DC-10-30	Lufthansa *Nurnberg*
	D–ADMO	Douglas DC-10-30	Lufthansa *Dortmund*
	D–ADPO	Douglas DC-10-30	Condor Flugdienst
	D–ADQO	Douglas DC-10-30	Condor Flugdienst
	D–ADSO	Douglas DC-10-30	Condor Flugdienst
	D–AERE	L.1011-385 TriStar 1	LTU
	D–AERI	L.1011-385 TriStar 1	LTU
	D–AERL	L.1011-385 TriStar 500	LTU
	D–AERM	L.1011-385 TriStar 1	LTU
	D–AERN	L.1011-385 TriStar 1	LTU
	D–AERT	L.1011-385 TriStar 500	LTU
	D–AERU	L.1011-385 TriStar 100	LTU
	D–AHLA	A.300B4 Airbus	Hapag-Lloyd
	D–AHLB	A.300B4 Airbus	Hapag-Lloyd
	D–AHLC	A.300B4 Airbus	Hapag-Lloyd
	D–AHLD	Boeing 737-2K5	Hapag-Lloyd
	D–AHLE	Boeing 737-2K5	Hapag-Lloyd
	D–AHLF	Boeing 737-2K5	Hapag-Lloyd
	D–AHLG	Boeing 737-2K5	Hapag-Lloyd
	D–AHLH	Boeing 737-2K5	Hapag-Lloyd
	D–AHLI	Boeing 737-2K5	Hapag-Lloyd
	D–AHLL	Boeing 727-81	Hapag-Lloyd
	D–AHLM	Boeing 727-81	Hapag-Lloyd
	D–AHLN	Boeing 727-81	Hapag-Lloyd
	D–AHLP	Boeing 727-14	Hapag-Lloyd
	D–AHLR	Boeing 727-89	Hapag-Lloyd
	D–AHLS	Boeing 727-89	Hapag-Lloyd
	D–AHLT	Boeing 727-2K5	Hapag-Lloyd
	D–AHLU	Boeing 727-2K5	Hapag-Lloyd
	D–AHLV	Boeing 727-2K5	Hapag-Lloyd
	D–AIAA	A.300B2 Airbus	Lufthansa *Garmisch-Partenkirchen*
	D–AIAB	A.300B2 Airbus	Lufthansa *Rüdesheim*
	D–AIAC	A.300B2 Airbus	Lufthansa *Lüneburg*
	D–AIAD	A.300B2 Airbus	Lufthansa *Westerland/Sylt*
	D–AIAE	A.300B2 Airbus	Lufthansa *Neustadt an der Weinstrasse*
	D–AIBA	A.300B4 Airbus	Lufthansa *Rothenburg ob der Tauber*
	D–AIBB	A.300B4 Airbus	Lufthansa *Freudenstadt/Schwarzwald*
	D–AIBC	A.300B4 Airbus	Lufthansa *Lindau-Bodenzee*
	D–AIBD	A.300B4 Airbus	Lufthansa *Erbach-Odenwald*
	D–AIBF	A.300B4 Airbus	Lufthansa *Kronberg im Taunus*
	D–AMAP	A.300B4 Airbus	Hapag-Lloyd
	D–AMAX	A.300B4 Airbus	Hapag-Lloyd *Maximilian*
	D–AMAY	A.300B4 Airbus	Hapag-Lloyd *Ludwig I*
	D–AMAZ	A.300B4 Airbus	Hapag-Lloyd
	D–A	Boeing 737-2K9	Bavaria Fluggesellschaft
	D–A	Boeing 737-2K9	Bavaria Fluggesellschaft
	D–A	Boeing 737-2K9	Bavaria Fluggesellschaft
	D–A	Boeing 737-2K9	Bavaria Fluggesellschaft
	D–A	L-100-30 Hercules	Wirtschaftsflug Rhein-Main
	D–BAKA	F.27 Friendship Mk. 100	WDL
	D–BOOM	F.27J Friendship	Wirtschaftsflug Rhein-Main

NOTE: Lufthansa is in the process of replacing its Boeing 737-130s with Series 230s.

DDR (German Democratic Republic)

Reg.	Type	Owner or Operator	Notes
DDR–SCB	Tupolev Tu-134	Interflug	
DDR–SCE	Tupolev Tu-134	Interflug	
DDR–SCF	Tupolev Tu-134	Interflug	
DDR–SCG	Tupolev Tu-134	Interflug	
DDR–SCH	Tupolev Tu-134	Interflug	
DDR–SCI	Tupolev Tu-134A	Interflug	
DDR–SCK	Tupolev Tu-134A	Interflug	
DDR–SCL	Tupolev Tu-134A	Interflug	
DDR–SCN	Tupolev Tu-134A	Interflug	
DDR–SCO	Tupolev Tu-134A	Interflug	
DDR–SCP	Tupolev Tu-134A	Interflug	
DDR–SCR	Tupolev Tu-134A	Interflug	
DDR–SCS	Tupolev Tu-134A	Interflug	
DDR–SCT	Tupolev Tu-134A	Interflug	
DDR–SCU	Tupolev Tu-134A	Interflug	
DDR–SCV	Tupolev Tu-134A	Interflug	
DDR–SCW	Tupolev Tu-134A	Interflug	
DDR–SCX	Tupolev Tu-134A	Interflug	
DDR–SCY	Tupolev Tu-134A	Interflug	
DDR–SCZ	Tupolev Tu-134	Interflug	
DDR–SDE	Tupolev Tu-134A	Interflug	
DDR–SDF	Tupolev Tu-134A	Interflug	
DDR–SDG	Tupolev Tu-134A	Interflug	
DDR–SDH	Tupolev Tu-134A	Interflug	
DDR–SDI	Tupolev Tu-134A	Interflug	
DDR–SDK	Tupolev Tu-134A	Interflug	
DDR–SDL	Tupolev Tu-134A	Interflug	
DDR–SDM	Tupolev Tu-134A	Interflug	
DDR–SDN	Tupolev Tu-134A	Interflug	
DDR–SDO	Tupolev Tu-134A	Interflug	
DDR–SDP	Tupolev Tu-134A	Interflug	
DDR–SEB	Ilyushin IL-62	Interflug	
DDR–SEC	Ilyushin IL-62	Interflug	
DDR–SEF	Ilyushin IL-62	Interflug	
DDR–SEG	Ilyushin IL-62	Interflug	
DDR–SEH	Ilyushin IL-62	Interflug	
DDR–SEI	Ilyushin IL-62	Interlug	
DDR–SEK	Ilyushin IL-62M	Interflug	
DDR–SEL	Ilyushin IL-62M	Interflug	

EC (Spain)

Transeuropa

Reg.	Type	Owner or Operator	Notes
EC–ASN	Douglas DC-8-52	Aviaco *Murillo*	
EC–AUM	Douglas DC-8-52	Aviaco *Zurbaran*	
EC–BIB	S.E.210 Caravelle 10-R	Transeuropa	
EC–BIG	Douglas DC-9-32	Iberia *Villa de Madrid*	
EC–BIH	Douglas DC-9-32	Iberia *Ciudad de Barcelona*	
EC–BIJ	Douglas DC-9-32	Iberia *Santa Cruz de Tenerife*	
EC–BIK	Douglas DC-9-32	Iberia *Las Palmas de Gran Canaria*	
EC–BIL	Douglas DC-9-32	Iberia *Ciudad de Zaragoza*	
EC–BIM	Douglas DC-9-32	Iberia *Ciudad de Santander*	
EC–BIN	Douglas DC-9-32	Iberia *Ciudad de Palma de Mallorca*	
EC–BIO	Douglas DC-9-32	Iberia *Villa de Bilbao*	
EC–BIP	Douglas DC-9-32	Aviaco *Santiago de Compostela*	
EC–BIQ	Douglas DC-9-32	Aviaco *Ciudad de Malaga*	

Notes	Reg.	Type	Owner or Operator
	EC-BIR	Douglas DC-9-32	Iberia *Ciudad de Valencia*
	EC-BIS	Douglas DC-9-32	Iberia *Ciudad de Alicante*
	EC-BIT	Douglas DC-9-32	Iberia *Ciudad de San Sebastian*
	EC-BIU	Douglas DC-9-32	Iberia *Ciudad de Oviedo*
	EC-BJD	CV-990A Coronado	Spantax
	EC-BMV	Douglas DC-8-55F	Aviaco *Pedro Berruguete*
	EC-BMY	Douglas DC-8-63CF	Aviaco *Playa del Puerto de la Cruz*
	EC-BMZ	Douglas DC-8-63CF	Iberia *Los Madrazo*
	EC-BPF	Douglas DC-9-32	Iberia *Ciudad de Almeria*
	EC-BPG	Douglas DC-9-32	Iberia *Ciudad de Vigo*
	EC-BPH	Douglas DC-9-32	Iberia *Ciudad de Gerona*
	EC-BQA	CV-990A Coronado	Spantax
	EC-BQQ	CV.990A Coronado	Spantax
	EC-BQS	Douglas DC-8-63	Aviaco
	EC-BQT	Douglas DC-9-32	Iberia *Villa de Murcia*
	EC-BQU	Douglas DC-9-32	Iberia *Ciudad de la Coruna*
	EC-BQV	Douglas DC-9-32	Iberia *Ciudad de Ibiza*
	EC-BQX	Douglas DC-9-32	Iberia *Ciudad de Valladolid*
	EC-BQY	Douglas DC-9-32	Iberia *Ciudad de Cordoba*
	EC-BQZ	Douglas DC-9-32	Iberia *Ciudad de Santa Cruz de la Palma*
	EC-BRQ	Boeing 747-256B	Iberia *Calderon de la Barca*
	EC-BRX	S.E.210 Caravelle 11-R	Transeuropa *Renacuajo I*
	EC-BSD	Douglas DC-8-63	Aviaco *Playa de las Canteras*
	EC-BSE	Douglas DC-8-63	Aviaco
	EC-BTE	CV-990A Coronado	Spantax
	EC-BYD	Douglas DC-9-32	Iberia *Ciudad de Arrecife de Lanzarote*
	EC-BYE	Douglas DC-9-32	Iberia *Ciudad de Mahon*
	EC-BYF	Douglas DC-9-32	Iberia *Ciudad de Granada*
	EC-BYG	Douglas DC-9-32	Iberia *Ciudad de Pamplona*
	EC-BYH	Douglas DC-9-32	Aviaco *Ciudad de Cadiz*
	EC-BYI	Douglas DC-9-32	Iberia *Ciudad de Vitoria*
	EC-BYJ	Douglas DC-9-32	Iberia *Ciudad de Salamanca*
	EC-BYK	Douglas DC-9-33RC	Iberia *Ciudad de Badajoz*
	EC-BYL	Douglas DC-9-33RC	Iberia *Ciudad de Albacete*
	EC-BYM	Douglas DC-9-33RC	Iberia *Ciudad de Cangas de Onis*
	EC-BYN	Douglas DC-9-33RC	Iberia *Ciudad de Caceres*
	EC-BZO	CV-990A Coronado	Spantax
	EC-BZP	CV-990A Coronado	Spantax
	EC–CAI	Boeing 727-256	Iberia *Castilla la Nueva*
	EC–CAJ	Boeing 727-256	Iberia *Cataluna*
	EC–CAK	Boeing 727-256	Iberia *Aragon*
	EC–CBA	Boeing 727-256	Iberia *Vascongadas*
	EC–CBB	Boeing 727-256	Iberia *Valencia*
	EC–CBC	Boeing 727-256	Iberia *Navarra*
	EC–CBD	Boeing 727-256	Iberia *Murcia*
	EC–CBE	Boeing 727-256	Iberia *Leon*
	EC–CBF	Boeing 727-256	Iberia *Gran Canaria*
	EC–CBG	Boeing 727-256	Iberia *Extremadura*
	EC–CBH	Boeing 727-256	Iberia *Galicia*
	EC–CBI	Boeing 727-256	Iberia *Asturias*
	EC–CBJ	Boeing 727-256	Iberia *Andalucia*
	EC–CBK	Boeing 727-256	Iberia *Baleares*
	EC–CBL	Boeing 727-256	Iberia *Tenerife*
	EC–CBM	Boeing 727-256	Iberia *Castilla la Vieja*
	EC–CBO	Douglas DC-10-30	Iberia *Costa del Sol*
	EC–CBP	Douglas DC-10-30	Iberia *Costa Dorada*
	EC–CCF	Douglas DC-8-61CF	Spantax
	EC–CCG	Douglas DC-8-61CF	Spantax
	EC–CCN	Douglas DC-8-33	T.A.E.
	EC–CEZ	Douglas DC-10-30	Iberia *Costa del Azahar*
	EC–CFA	Boeing 727-256	Iberia *Jerez Xeres Sherry*
	EC–CFB	Boeing 727-256	Iberia *Rioja*
	EC–CFC	Boeing 727-256	Iberia *Tarragona*
	EC–CFD	Boeing 727-256	Iberia *Montilla Moriles*
	EC–CFE	Boeing 727-256	Iberia *Penedes*
	EC–CFF	Boeing 727-256	Iberia *Valdepenas*
	EC–CFG	Boeing 727-256	Iberia *La Mancha*
	EC–CFH	Boeing 727-256	Iberia *Priorato*
	EC–CFI	Boeing 727-256	Iberia *Carinena*
	EC–CFJ	Boeing 727-256	Iberia *Jumilla*
	EC–CFK	Boeing 727-256	Iberia *Rivero*
	EC–CGN	Douglas DC-9-32	Aviaco *Martin Alonso Pinzon*
	EC–CGO	Douglas DC-9-32	Aviaco *Pedro Alonso Nino*
	EC–CGP	Douglas DC-9-32	Aviaco *Juan Sebastian Elcano*
	EC–CGQ	Douglas DC-9-32	Aviaco *Alonso de Ojeda*

Reg.	Type	Owner or Operator	Notes
EC–CGR	Douglas DC-9-32	Aviaco *Francisco de Orellana*	
EC–CGS	Douglas DC-9-32	Aviaco *Vasco Nunez de Balboa*	
EC–CGY	Douglas DC-9-14	Spantax	
EC–CGZ	Douglas DC-9-14	Spantax	
EC-CID	Boeing 727-256	Iberia *Malaga*	
EC-CIE	Boeing 727-256	Iberia *Esparragosa*	
EC-CIZ	S.E.210 Caravelle 10-R	Transeuropa	
EC-CLB	Douglas DC-10-30	Iberia *Costa Blanca*	
EC-CLD	Douglas DC-9-32	Aviaco *Hernando de Soto*	
EC-CLE	Douglas DC-9-32	Aviaco *Juan Ponce de Leon*	
EC-CMS	S.E.210 Caravelle 10-B	T.A.E.	
EC–CNF	CV-990A Coronado	Spantax	
EC–CNG	CV-990A Coronado	Spantax	
EC–CNH	CV-990A Coronado	Spantax	
EC–CPI	S.E.210 Caravelle 10-R	Transeuropa	
EC–CQM	Douglas DC-8-54F	Aviaco	
EC–CSJ	Douglas DC-10-30	Iberia *Costa de la Luz*	
EC–CSK	Douglas DC-10-30	Iberia *Cornisa Cantabrica*	
EC–CTR	Douglas DC-9-34CF	Aviaco *Hernan Cortes*	
EC–CTS	Douglas DC-9-34CF	Aviaco *Francisco Pizarro*	
EC–CTT	Douglas DC-9-34CF	Aviaco *Pedro de Valdivia*	
EC–CTU	Douglas DC-9-34CF	Aviaco *Pedro de Alvarado*	
EC–CYI	S.E.210 Caravelle 10-R	Transeuropa	
EC–CZE	Douglas DC-8-61	Spantax	
EC–DBE	Douglas DC-8-55F	Aviaco	
EC–DCC	Boeing 727-256	Iberia *Alvarino*	
EC–DCD	Boeing 727-256	Iberia *Chacoli*	
EC–DCE	Boeing 727-256	Iberia *Mentrida*	
EC–DCN	S.E.210 Caravelle 10-R	Transeuropa	
EC–DDU	Boeing 727-256	Iberia *Alhambra de Granada*	
EC–DDV	Boeing 727-256	Iberia *Acueducto de Segovia*	
EC–DDX	Boeing 727-256	Iberia *Monasterio de Poblet*	
EC–DDY	Boeing 727-256	Iberia *Cuevas de Altamira*	
EC–DDZ	Boeing 727-256	Iberia *Murallas de Avila*	
EC–DEA	Douglas DC-10-30	Iberia *Rias Gallegas*	
EC–DEG	Douglas DC-10-30CF	Spantax	
EC–DEM	Douglas DC-8-55F	Aviaco *Goya*	
EC–DFP	S.E.210 Caravelle 10-B	T.A.E.	
EC–DGB	Douglas DC-9-34	Aviaco *Castillo de Javier*	
EC–DGC	Douglas DC-9-34	Aviaco *Castillo de Sotomayor*	
EC–DGD	Douglas DC-9-34	Aviaco *Castillo de Arcos*	
EC–DGE	Douglas DC-9-34	Aviaco *Castillo de Bellver*	
EC–DHZ	Douglas DC-10-30	Iberia	
EC–DIA	Boeing 747-256B	Iberia	
EC–DIB	Boeing 747-256B	Iberia *Cervantes*	
EC–DIH	Douglas DC-8-55	T.A.E.	
EC-DIR	Douglas DC-9-14	Spantax	
EC–DLC	Boeing 747-256B	Iberia *Francisco de Quevedo*	
EC–DLD	Boeing 747-256B	Iberia *Lupe de Vega*	
EC–DLE	A.300B4 Airbus	Iberia *Donana*	
EC–DLF	A.300B4 Airbus	Iberia *Canadas del Teide*	
EC–DLG	A.300B4 Airbus	Iberia *Tablas de Daimiel*	
EC–DLH	A.300B4 Airbus	Iberia *Aigues Tortes*	
EC–DNQ	A.300B4 Airbus	Iberia *Islas Cies*	
EC–	A.300B4 Airbus	Iberia	
EC–	A.300B4 Airbus	Iberia	
EC–	A.300B4 Airbus	Iberia	
EC–	A.300B4 Airbus	Iberia	

EI (Republic of Ireland)

Including complete current Irish Civil Register

Notes	Reg.	Type	Owner or Operator
	EI–ADV	PA-12 Super Cruiser	R. E. Levis
	EI–AFF	*B.A. Swallow 2	J. McCarthy
	EI–AFK	*D.H.84 Dragon (EI–ABI)	Aer Lingus-Irish
	EI–AFN	*B.A. Swallow 2	J. McCarthy
	EI–AGB	*Miles M.38 Messenger 4	J. McLoughlin
	EI–AGD	*Taylorcraft Plus D	H. Wolf
	EI–AGE	*Miles M.38 Messenger 3	J. McLoughlin
	EI–AGJ	J/I Autocrat	W. G. Rafter
	EI–AHA	*D.H.82A Tiger Moth	J. H. Maher
	EI–AHR	*D.H.C.I Chipmunk 22	C. Lane
	EI–AKM	Piper J-3C-65 Cub	Setanta Flying Group
	EI–ALH	Taylorcraft Plus D	N. Reilly
	EI–ALP	Avro 643 Cadet	J. C. O'Loughlin
	EI–ALU	Avro 631 Cadet	M. P. Cahill
	EI–AMD	M.S.880B Rallye Club	Morane Aircraft Ltd.
	EI–AMK	J/I Autocrat	Irish Aero Club
	EI–AMO	J/IB Aiglet	R. Hassett
	EI–AMY	J/IN Alpha	Meath Flying Group Ltd.
	EI–AND	Cessna 175A	Jack Braithwaite (Ireland) Ltd.
	EI–ANE	BAC One-Eleven 208AL	Aer Lingus *St. Mel*
	EI–ANF	BAC One-Eleven 208AL	Aer Lingus *St. Malachy*
	EI–ANG	BAC One-Eleven 208AL	Aer Lingus *St. Declan*
	EI–ANH	BAC One-Eleven 208AL	Aer Lingus *St. Ronan*
	EI–ANT	Champion 7ECA Citabria	Setanta Flying Group
	EI–AOB	PA-28 Cherokee 140	Dr. H. J. R. Henderson
	EI–AOD	Cessna 182J Skylane	Oscar Delta Flying Training Co. Ltd.
	EI–AOK	Cessna F.172G	R. J. Cloughley & N. J. Simpson
	EI–AOO	Cessna 150E	R. Hassett
	EI–AOP	D.H.82A Tiger Moth	Dublin Tiger Group
	EI–AOS	Cessna 310B	Joyce Aviation Ltd.
	EI–APF	Cessna F.150F	Midland Flying Group Ltd.
	EI–APS	Schleicher ASK 14	G. W. Connolly & M. Slazenger
	EI–APT	Fokker D.VII/65 Replica	Blue Max Aviation Ltd.
	EI–APU	Fokker D.VII/65 Replica	Blue Max Aviation Ltd.
	EI–APV	Fokker D.VII/65 Replica	Blue Max Aviation Ltd.
	EI–APW	Fokker Dr.I Replica	Blue Max Aviation Ltd.
	EI–ARA	SE.5A Replica	Blue Max Aviation Ltd.
	EI–ARB	SE.5A Replica	Blue Max Aviation Ltd.
	EI–ARC	Pfalz D.III Replica	Blue Max Aviation Ltd.
	EI–ARD	Pfalz D.III Replica	Blue Max Aviation Ltd.
	EI–ARE	Stampe SV.4C	Blue Max Aviation Ltd.
	EI–ARF	Caudron C.277 Luciole	Blue Max Aviation Ltd.
	EI–ARH	Currie Wot/S.E.5 Replica	Blue Max Aviation Ltd.
	EI–ARI	Currie Wot/S.E.5 Replica	Blue Max Aviation Ltd.
	EI–ARJ	Currie Wot/S.E.5 Replica	Blue Max Aviation Ltd.
	EI–ARK	Currie Wot/S.E.5 Replica	Blue Max Aviation Ltd.
	EI–ARL	Currie Wot/S.E.5 Replica	Blue Max Aviation Ltd.
	EI–ARM	Currie Wot/S.E.5 Replica	Blue Max Aviation Ltd.
	EI–ARW	Jodel D.R.1050	John B. Healy
	EI–ASA	Boeing 737-248	Aer Lingus *St. Jarlath*
	EI–ASB	Boeing 737-248	Aer Lingus *St. Albert*
	EI–ASC	Boeing 737-248C	Aer Lingus *St. Macartan*
	EI–ASD	Boeing 737-248C	Aer Lingus *St. Ide*
	EI–ASE	Boeing 737-248C	Aer Lingus *St. Fachtna*
	EI–ASF	Boeing 737-248	Aer Lingus *St. Nathy*
	EI–ASG	Boeing 737-248	Aer Lingus *St. Cormack*
	EI–ASH	Boeing 737-248	Aer Lingus *St. Eugene*
	EI–ASI	Boeing 747-148	Aer Lingus *St. Columcille*
	EI–ASJ	Boeing 747-148	Aer Lingus *St. Patrick*
	EI–ASL	Boeing 737-248C	Aer Lingus *St. Cillian*
	EI–ASO	Boeing 707-349C	Aer Lingus *St. Canice*
	EI–AST	Cessna F.150H	Liberty Flying Club Group
	EI–ASV	PA-28R Cherokee Arrow 180	F. E. A. Biggar

Reg.	Type	Owner or Operator	Notes
EI-ATC	Cessna 310G	Iona National Airways	
EI-ATH	Cessna F.150J	Hibernian Flying Club	
EI-ATJ	B.121 Pup 1	Wexford Aero Club	
EI-ATK	PA-28 Cherokee 140	Mayo Flying Club Ltd.	
EI-ATS	M.S.880B Rallye Club	O. Bruton & G. Farrar	
EI-AUC	Cessna FA.150K Aerobat	Flying Fifteen Aero Club Ltd.	
EI-AUD	M.S.880B Rallye Club	Kilkenny Flying Club Ltd.	
EI-AUE	M.S.880B Rallye Club	Slievenamon Air Ltd.	
EI-AUG	M.S.894 Rallye Minerva 220	Weston Ltd.	
EI-AUM	J/1 Autocrat	J. G. Rafter	
EI-AUO	Cessna FA.150K Aerobat	Iona National Airways	
EI-AUP	M.S.880B Rallye Club	Limerick Flying Club	
EI-AUT	Forney F-1A Aircoupe	Joyce Aviation Ltd.	
EI-AUV	PA-23 Aztec 250	Shannon Executive Aviation	
EI-AUY	Morane-Saulnier M.S.502	Historical Aircraft Preservation Group	
EI-AVB	Aeronca 7AC Champion	G. G. Bracken	
EI-AVC	Cessna F.337F	Iona National Airways	
EI-AVG	Beech A.55 Baron	C. O'Neale	
EI-AVL	J/5F Aiglet	George E. Flood	
EI-AVM	Cessna F.150L	Victor Mike Flying Group	
EI-AVN	Hughes 369HM	Helicopter Maintenance Ltd.	
EI-AVU	Stampe SV.4C	S. P. O'Carroll	
EI-AWA	Bell 206B JetRanger 2	Helicopter Maintenance Ltd.	
EI-AWE	Cessna F.150M	Third Flight Group	
EI-AWH	Cessna 210J	Southern Air Ltd.	
EI-AWJ	M.S.893A Rallye Commodore	Dr. W. J. Phelan	
EI-AWM	BN-2A Islander	Aer Arann	
EI-AWP	D.H.82A Tiger Moth	A. Lyons	
EI-AWR	Malmo MFI-9 Junior	G. Fawcett	
EI-AWU	M.S.880B Rallye Club	Longford Aviation Ltd.	
EI-AWW	Cessna 414	T. Farrington	
EI-AYA	M.S.880B Rallye Club	Dundalk Aero Club Ltd.	
EI-AYB	GY-80 Horizon 180	Westwing Flying Group	
EI-AYD	AA-5 Traveler	Kenneth Walters	
EI-AYF	Cessna FRA.150L	Garda Flying Club	
EI-AYI	M.S.880B Rallye Club	Irish Air Training Group	
EI-AYK	Cessna F.172M	P. J. Meade	
EI-AYL	A.109 Airedale	J. Ronan	
EI-AYN	BN-2A Islander	Aer Arann	
EI-AYO	*Douglas DC-3A	Science Museum, Wroughton	
EI-AYR	Schleicher ASK-16	Kilkenny Airport Ltd.	
EI-AYS	PA-22 Colt 108	George Farrer & ptnrs.	
EI-AYT	M.S.894A Rallye Minerva	R. C. Cunningham	
EI-AYV	M.S.892A Rallye Commodore 150	P. Murtagh	
EI-AYY	Evans VP-1	Michael Donoghue	
EI-BAB	M.S.894E Rallye Minerva	J. Phelan	
EI-BAF	Thunder Ax6-56 balloon	W. G. Woollett	
EI-BAG	Cessna 172A	Cork Skydiving Club Ltd.	
EI-BAM	Bell 212	Irish Helicopters Ltd.	
EI-BAN	Cameron O-65 balloon	C. M. Alexander	
EI-BAO	Cessna F.172G	Kingdom Air Ltd.	
EI-BAR	Thunder Ax8-105 balloon	Fr. D. Reid	
EI-BAS	Cessna F.172M	F. Fahey & B. Fitzgerald	
EI-BAT	Cessna F.150M	20th Air Training Co. Ltd.	
EI-BAU	Stampe SV.4C	S. P. O'Carroll	
EI-BAV	PA-22 Colt 108	J. P. Montcalm	
EI-BAY	Cameron Ax8-84 balloon	F. N. Lewis	
EI-BBB	R. Commander 112	W. J. O'Connor	
EI-BBC	PA-28 Cherokee 180C	The Cherokee Group	
EI-BBD	Evans VP-1	Volksplane Group	
EI-BBE	7FC Tri-Traveler (tailwheel)	Aeronca Flying Group	
EI-BBG	M.S.880B Rallye Club	Weston Ltd.	
EI-BBH	B.175 Britannia 253F	Aer Turas Teo *City of Cork*	
EI-BBI	M.S.892 Rallye Commodore	Kilkenny Airport Ltd.	
EI-BBJ	M.S.880B Rallye Club	Weston Ltd.	
EI-BBK	A.109 Airedale	H. S. Igoe	
EI-BBL	R. Turbo 690A Commander	The Earl of Granard	
EI-BBM	Cameron O-65 balloon	Dublin Ballooning Club	
EI-BBN	Cessna F.150M	Sligo N.W. Aero Club	
EI-BBO	M.S.893E Rallye	J. G. Lacey & ptnrs.	
EI-BBV	Piper J-3C-65 Cub	Feidhlim Cronin	
EI-BBW	M.S.894A Rallye Minerva	J. J. Ladbrook	
EI-BCE	BN-2A-26 Islander	Aer Arann	

Notes	Reg.	Type	Owner or Operator
	EI-BCF	Bensen B.8M	P. E. & E. O'Loughlin
	EI-BCH	M.S.892A Rallye Commodore 150	The Condor Group
	EI-BCJ	F.8L Falco I Srs. 3	D. Kelly
	EI-BCK	Cessna F.172K	Iona National Airways
	EI-BCL	Cessna 182P	Iona National Airways
	EI-BCM	Piper J-3C-65 Cub	Kilmoon Flying Group
	EI-BCN	Piper J-3C-65 Cub	Snowflake Flying Group
	EI-BCO	Piper J-3C-65 Cub	J. Molloy
	EI-BCR	Boeing 737-281	Aer Lingus *Sir Oliver Plunkett*
	EI-BCS	M.S.880B Rallye Club	J. J. Martyn & J. O'Neill
	EI-BCT	Cessna 411A	Air Surveys International
	EI-BCU	M.S.880B Rallye Club	Weston Ltd.
	EI-BCV	Cessna F.150M	Iona National Airways
	EI-BCW	M.S.880B Rallye Club	B. E. Lalor & L. Prendergast
	EI-BCY	Beech 200 Super King Air (232)	Minister of Defence
	EI-BDD	PA-34-200 Seneca	D. Herlihy
	EI-BDE	Colt Ax7-77A balloon	Colting Balloons
	EI-BDH	M.S.880B Rallye Club	Munster Wings Ltd.
	EI-BDK	M.S.880B Rallye Club	M. F. Ledwith & L. McNamara
	EI-BDL	Evans VP-2	J. Duggan
	EI-BDM	PA-23 Aztec 250D	Executive Air Services
	EI-BDO	Cessna F.152	Iona National Airways
	EI-BDP	Cessna 182P	172 Flying Group
	EI-BDR	PA-28 Cherokee 180	Cork Flying Club
	EI-BDS	M.S.894A Rallye Minerva	Ireland West Airways
	EI-BDT	Dakota 4	Clyden Airways (*Stored*)
	EI-BDU	Dakota 4	Clyden Airways (*Stored*)
	EI-BDV	PA-32 Cherokee Six 260	Birr Flying Club
	EI-BEA	M.S.880B Rallye 100ST	Weston Ltd.
	EI-BEB	Boeing 737-248	Aer Lingus Teo *St. Eunan*
	EI-BEC	Boeing 737-248	Aer Lingus Teo *St. Fiacre*
	EI-BED	Boeing 747-130	Aer Linte Eireann Teo *St. Kieran*
	EI-BEE	Boeing 737-281	Air Tara Ltd.
	EI-BEF	Boeing 737-281	Air Tara Ltd.
	EI-BEG	—	Aer Lingus Teo
	EI-BEH	—	Aer Lingus Teo
	EI-BEI	—	Aer Lingus Teo
	EI-BEJ	—	Aer Lingus Teo
	EI-BEK	—	Aer Lingus Teo
	EI-BEL	—	Aer Lingus Teo
	EI-BEM	—	Aer Lingus Teo
	EI-BEN	Piper J-3C-65 Cub	Capt. J. J. Sullivan
	EI-BEO	Cessna 310Q	Iona National Airways
	EI-BEP	M.S.892A Rallye Commodore	H. Lynch & J. O'Leary
	EI-BET	Cessna F.337G	Messrs. Donegan, Leonard & Meade
	EI-BEY	Naval N3N-3	Huntley & Huntley Ltd.
	EI-BFB	M.S.880B Rallye 100ST	Weston Ltd.
	EI-BFC	Boeing 737-2H4	Air Tara Ltd.
	EI-BFD	Scheibe SF.28A	Dublin Gliding Club
	EI-BFE	Cessna F.150G	Joyce Aviation Ltd.
	EI-BFF	Beech A.23 Musketeer	M. R. Cody
	EI-BFH	Bell 212	Irish Helicopters Ltd.
	EI-BFI	M.S.880B Rallye 100ST	J. O'Neill
	EI-BFJ	Beech A.200 Super King Air	Minister of Defence
	EI-BFK	Bell 206B JetRanger 3	M. V. O'Brien
	EI-BFM	M.S.893E Rallye 235GT	M. Orr
	EI-BFO	Piper J-3C-90 Cub	J. Molloy
	EI-BFP	M.S.880B Rallye 100ST	Weston Ltd.
	EI-BFR	M.S.880B Rallye 100ST	Galway Flying Club
	EI-BFS	FRED Srs. 2	J. Farrant
	EI-BFT	Beech A200 Super King Air	Avair Ltd.
	EI-BFU	—	—
	EI-BFV	M.S.880B Rallye 100ST	Ormond Flying Club
	EI-BGA	M.S.880B Rallye 100ST	T. Daly
	EI-BGB	M.S.880B Rallye 100ST	M. Slattery & ptnrs.
	EI-BGC	M.S.880B Rallye Club	P. Moran
	EI-BGD	M.S.880B Rallye Club	S. O'Rourke & J. Lawlor
	EI-BGF	PA-28R Cherokee Arrow 180	Arrow Group
	EI-BGG	M.S.893E Rallye 180GT	D. A. Weldon
	EI-BGH	Cessna F.172N	Iona National Airways
	EI-BGI	Cessna F.152	Iona National Airways
	EI-BGJ	Cessna F.152	Kerry Aero Club
	EI-BGK	Cessna P206D	Shannon Executive Aviation
	EI-BGL	R. Turbo Commander 690B	Flightline Ltd.

Reg.	Type	Owner or Operator	Notes
EI–BGN	M.S.880B Rallye Club	T. Guckian & E. Browne	
EI–BGO	Canadair CL-44D-4J	Aer Turas Teo	
EI–BGP	Cessna 414A	Iona National Airways	
EI–BGS	M.S.893B Rallye 180GT	Mac Grids Ltd.	
EI–BGT	Colt 77A balloon	K. Haugh	
EI–BGU	M.S.880B Rallye Club	M. F. Neary	
EI–BGV	AA-5 Traveler	Crowe & Fitzgerald Ltd.	
EI–BHA	Beech A200 Super King Air	Sunbird Air Charter	
EI–BHB	M.S.887 Rallye 125	C. Burns	
EI–BHC	Cessna F.177RG	P. J. McGuire & B. Palfrey	
EI–BHD	M.S.893E Rallye 180GT	Arvan Homes Ltd.	
EI–BHF	M.S.892A Rallye Commodore 150	B. Mullen	
EI–BHH	—	—	
EI–BHI	Bell 206B JetRanger 2	International Helicopter Sales Ltd.	
EI–BHJ	M.65 Gemini 3C	R. E. Winn	
EI–BHK	M.S.880B Rallye Club	J. Lawlor & B. Lyons	
EI–BHL	Beech E90 King Air	Stewart Singlam Fabrics Ltd.	
EI–BHM	Cessna 337E	The Ross Flying Group	
EI–BHN	M.S.893A Rallye Commodore	K. O'Driscoll & ptnrs.	
EI–BHO	Sikorsky S-61N	Irish Helicopters Ltd.	
EI–BHP	M.S.893A Rallye Commodore	Wicklow Flying Group	
EI–BHT	Beech 77 Skipper	Hibernian Flying Club	
EI–BHV	Champion 7EC Traveler	M. MacDowell	
EI–BHW	Cessna F.150F	B. A. Carpenter	
EI–BHY	M.S.892E Rallye Commodore	D. Killian	
EI–BIB	Cessna F.152	Galway Aero Club	
EI–BIC	Cessna F.172N	Iona National Airways	
EI–BID	PA-18 Super Cub 95	J. A. McWilliams	
EI–BIE	Cessna FA.152	Wexford Aero Club	
EI–BIF	M.S.894 Rallye Minerva 235	Empire Entertainments Ltd.	
EI–BIG	Zlin 526	P. von Lonkhuyzen	
EI–BIJ	AB-206B JetRanger 2	Irish Helicopters Ltd.	
EI–BIK	PA-18-150 Super Cub	Dublin Gliding Club	
EI–BIL	Beech A35 Bonanza	W. J. Phelan	
EI–BIM	M.S.880B Rallye Club	D. Millar	
EI–BIN	Cessna F.172N	Iona National Airways Ltd.	
EI–BIO	Piper J-3C-65 Cub	Monasterevin Flying Club	
EI–BIP	Beech 200 Super King Air	Avair Ltd.	
EI–BIR	Cessna F.172M	M. M. Jurkoze & ptnrs.	
EI–BIS	Robin R.1180TD	Robin Aiglon Group	
EI–BIT	M.S.887 Rallye 125	P. Lacy & ptnrs.	
EI–BIU	Robin R.2112A	Bruton Aircraft Engineering Ltd.	
EI–BIV	Bellanca 8KCAB Citabria	Aerocrats Flying Group	
EI–BIW	M.S.880B Rallye Club	E. J. Barr	
EI–BJA	Cessna FRA.150L	Joyce Aviation Ltd.	
EI–BJB	Aeronca 7AC Champion	M. Rogers	
EI–BJC	Aeronca 7AC Champion	R. J. Bentley	
EI–BJD	Mooney M.20K	Elan Corporation Ltd.	
EI–BJE	Boeing 737-275	Air Tara Ltd. (leased to Nigeria Airways)	
EI–BJF	AA-5 Traveler	B. O'Shea	
EI–BJG	Robin R.1180	N. Hanley	
EI–BJH	Nipper T.66 Srs. 3	S. T. O'Rourke	
EI–BJI	Cessna FR.172E	Irish Parachute Club	
EI–BJJ	Aeronca 15AC Sedan	A. A. Alderdice & S. H. Boyd	
EI–BJK	M.S.880B Rallye 110ST	Barton Ltd.	
EI–BJL	Cessna 550 Citation II	Helicopter Maintenance Ltd.	
EI–BJM	Cessna A.152	Leinster Aero Club	
EI–BJN	Cessna 500 Citation	Tool & Mould Steel (Ireland) Ltd.	
EI–BJO	Cessna R.172K	M. Masterson	
EI–BJP	Boeing 737-275C	Air Tara Ltd.	
EI–BJS	AA-5B Tiger	A. Killian & C. Pearce	
EI–BJT	PA-38-112 Tomahawk	Westair	
EI–BJV	AB-206B JetRanger 3	J. Kelly	
EI–BJW	D.H.104 Dove 6	S. J. Filhol	
EI–BJY	Beech 200 Super King Air	Avair Ltd.	
EI–BKA	Pitts S-2A Special	A. Wignall	
EI–BKB	Aeronca 11AC	Huntley & Huntley Ltd.	
EI–BKC	Aeronca 15AC Sedan	G. Treacy	
EI–BKD	Mooney M.20J	Limerick Warehousing Ltd.	
EI–BKE	M.S.885 Super Rallye	C. Brady & G. Groom	
EI–BKF	Cessna F.172H	Kayfly Ltd.	
EI–BKG	Westland Bell 47G-3B1 (Soloy)	Irish Helicopters Ltd.	
EI–BKI	PA-31-350 Navajo Chieftain	Shannon Executive Aviation	

Notes	Reg.	Type	Owner or Operator
	EI–BKK	Taylor JT.1 Monoplane	F. J. Hoysted
	EI–BKL	Cessna FR.172F	Irish Parachute Club
	EI–BKM	Zenith CH.200	B. McGann
	EI–BKN	M.S.880B Rallye 100ST	Weston Ltd.
	EI–BKP	Zenith CH.200	L. McEnteggart
	EI–BKR	Cessna F.172N	B. Berry
	EI–BKS	Eipper Quicksilver	Irish Microlight Ltd.
	EI–BKT	AB-206B JetRanger	Irish Helicopters Ltd.
	EI–BKU	M.S.892A Rallye Commodore	Bruton Aircraft Engineering Ltd.
	EI–BKV	Beech 200 Super King Air	Avair Ltd.
	EI–BKW	Beech 76 Duchess	Avair Ltd.
	EI–BKY	Beech 99	Avair Ltd.
	EI–BLA	PA-23 Aztec 250	National Aluminium Ltd.
	EI–BLB	Stampe SV-4C	J. E. Hutchinson & R. A. Stafford
	EI–BLC	Boeing 707-347C	Aer Lingus
	EI–BLD	Bolkow Bo 105C	Irish Helicopters Ltd.
	EI–BLE	Eipper Microlight	B. Carpenter
	EI–BLF	Eipper Microlight	B. Carpenter
	EI–BLG	AB-206B JetRanger 3	Anglo Irish Meat Co. Ltd.
	EI–BLH	Cessna 421C	A. D. D. Rogers
	EI–BLI	Beech C90 King Air	Avair Ltd.
	EI–BLJ	—	—
	EI–BLK	PA-28-181 Archer II	—
	EI–BLL	—	—
	EI–BLM	—	—
	EI–BLN	—	—
	EI–BLO	—	—
	EI–BLP	Short SD3-30	Avair Ltd.

EP (Iran)

Notes	Reg.	Type	Owner or Operator
	EP–IAA	Boeing 747SP-86	Iran Air *Fars*
	EP–IAB	Boeing 747SP-86	Iran Air *Kurdistan*
	EP–IAC	Boeing 747SP-86	Iran Air *Khuzestan*
	EP–IAD	Boeing 747SP-86	Iran
	EP–IAG	Boeing 747-286B	Iran Air *Azarabadegan*
	EP–IAM	Boeing 747-186B	Iran Air
	EP–ICA	Boeing 747-2J9F	Iran Air
	EP–ICB	Boeing 747-2J9F	Iran Air
	EP–ICC	Boeing 747-2J9F	Iran Air
	EP–IRJ	Boeing 707-321B	Iran Air
	EP–IRK	Boeing 707-321C	Iran Air
	EP–IRL	Boeing 707-386C	Iran Air *Apadana*
	EP–IRM	Boeing 707-386C	Iran Air *Ekbatana*
	EP–IRN	Boeing 707-386C	Iran Air *Pasargad*

ET (Ethiopia)

Notes	Reg.	Type	Owner or Operator
	ET–AAH	Boeing 720-060B	Ethiopian Airlines *White Nile*
	ET–ABP	Boeing 720-060B	Ethiopian Airlines
	ET–ACQ	Boeing 707-379C	Ethiopian Airlines
	ET–AFA	Boeing 720-024B	Ethiopian Airlines
	ET–AFB	Boeing 720-024B	Ethiopian Airlines
	ET–AFK	Boeing 720-024B	Ethiopian Airlines

AIR FRANCE

F (France)

Reg.	*Type*	*Owner or Operator*	*Notes*
F-BCYX	Douglas DC-3	Trans-Europ Air	
F-BEIG	Douglas DC-3	Bretagne Air Services	
F-BHRQ	S.E.210 Caravelle III	Air Inter	
F-BHRR	S.E.210 Caravelle III	Air Inter	
F-BHRS	S.E.210 Caravelle III	Air Inter	
F-BHSV	Boeing 707-328B	Air France	
F-BHSX	Boeing 707-328B	Air France	
F-BHSY	Boeing 707-328B	Air France	
F-BIUK	F.27 Friendship Mk. 100	Uni-Air	
F-BJEN	S.E.210 Caravelle 10B	Europe Aero Service	
F-BJTE	S.E.210 Caravelle III	Air Charter International	
F-BJTG	S.E.210 Caravelle III	Air Charter International *Roussillon*	
F-BJTJ	S.E.210 Caravelle III	Air Charter International *Bourbonnais*	
F-BJTU	S.E.210 Caravelle 10B	Europe Aero Service	
F-BLCC	Boeing 707-328C (Cargo)	Air France	
F-BLCD	Boeing 707-328B	Air France	
F-BLCG	Boeing 707-328C (Cargo)	Air France *Chateau du Lude*	
F-BLCH	Boeing 707-328C	Air France	
F-BLCK	Boeing 707-328C	Air France	
F-BLCL	Boeing 707-328C	Air France	
F-BLLB	Boeing 707-328B	Air France	
F-BLOY	HPR-7 Herald 210	Europe Aero Service	
F-BMKS	S.E.210 Caravelle 10B	Europe Aero Service	
F-BNKA	S.E.210 Caravelle III	Air Inter	
F-BNKB	S.E.210 Caravelle III	Air Inter	
F-BNKC	S.E.210 Caravelle III	Air Inter	
F-BNKF	S.E.210 Caravelle III	Air Inter	
F-BNKG	S.E.210 Caravelle III	Air Inter	
F-BNKH	S.E.210 Caravelle III	Air Inter	
F-BNKJ	S.E.210 Caravelle III	Air Inter	
F-BNOG	S.E.210 Caravelle 12	Air Inter	
F-BNOH	S.E.210 Caravelle 12	Air Inter	
F-BOHA	S.E.210 Caravelle III	Air France *Comté de Nice*	
F-BOIZ	HPR-7 Herald 210	Europe Aero Service	
F-BOJA	Boeing 727-228	Air France	
F-BOJB	Boeing 727-228	Air France	
F-BOJC	Boeing 727-228	Air France	
F-BOJD	Boeing 727-228	Air France	
F-BOJE	Boeing 727-228	Air France	
F-BOJF	Boeing 727-228	Air France	
F-BPJG	Boeing 727-228	Air France	
F-BPJH	Boeing 727-228	Air France	
F-BPJI	Boeing 727-228	Air France	
F-BPJJ	Boeing 727-228	Air France	
F-BPJK	Boeing 727-228	Air France	
F-BPJL	Boeing 727-228	Air France	
F-BPJM	Boeing 727-228	Air France	
F-BPJN	Boeing 727-228	Air France	
F-BPJO	Boeing 727-228	Air France	
F-BPJP	Boeing 727-228	Air France	
F-BPJQ	Boeing 727-228	Air France	

Notes	Reg.	Type	Owner or Operator
	F–BPJR	Boeing 727-228	Air France
	F–BPJS	Boeing 727-228	Air France
	F–BPJT	Boeing 727-228	Air France
	F–BPJU	Boeing 727-214	Air Charter International
	F–BPJV	Boeing 727-214	Air Charter International
	F–BPNA	F.27 Friendship Mk. 500	Air Inter
	F–BPNB	F.27 Friendship Mk. 500	Air Inter
	F–BPNC	F.27 Friendship Mk. 500	Air Inter
	F–BPND	F.27 Friendship Mk. 500	Air Inter
	F–BPNE	F.27 Friendship Mk. 500	Air Inter
	F–BPNG	F.27 Friendship Mk. 500	Air Inter
	F–BPNH	F.27 Friendship Mk. 500	Air Inter
	F–BPNI	F.27 Friendship Mk. 500	Air Inter
	F–BPNJ	F.27 Friendship Mk. 500	Air Inter
	F–BPPA	Aero Spacelines Guppy-201	Aeromaritime
	F–BPUA	F.27 Friendship Mk. 500	Air France
	F–BPUB	F.27 Friendship Mk. 500	Air France
	F–BPUC	F.27 Friendship Mk. 500	Air France
	F–BPUD	F.27 Friendship Mk. 500	Air France
	F–BPUE	F.27 Friendship Mk. 500	Air France
	F–BPUF	F.27 Friendship Mk. 500	Air France
	F–BPUG	F.27 Friendship Mk. 500	Air France
	F–BPUH	F.27 Friendship Mk. 500	Air France
	F–BPUI	F.27 Friendship Mk. 500	Air France
	F–BPUJ	F.27 Friendship Mk. 500	Air France
	F–BPUK	F.27 Friendship Mk. 500	Air France
	F–BPUL	F.27 Friendship Mk. 500	Air France
	F–BPVA	Boeing 747-128	Air France
	F–BPVB	Boeing 747-128	Air France
	F–BPVC	Boeing 747-128	Air France
	F–BPVD	Boeing 747-128	Air France
	F–BPVE	Boeing 747-128	Air France
	F–BPVF	Boeing 747-128	Air France
	F–BPVG	Boeing 747-128	Air France
	F–BPVH	Boeing 747-128	Air France
	F–BPVL	Boeing 747-128	Air France
	F–BPVP	Boeing 747-128	Air France
	F–BPVR	Boeing 747-228F	Air France
	F–BPVS	Boeing 747-228B	Air France
	F–BPVT	Boeing 747-228B	Air France
	F–BPVV	Boeing 747-228F	Air France
	F–BPVX	Boeing 747-228B	Air France
	F–BPVY	Boeing 747-228B	Air France
	F–BPVZ	Boeing 747-228F	Air France
	F–BRGU	S.E.210 Caravelle VI-N	Minerve
	F–BRNI	Beech 70 Queen Air	Lucas Air Transport
	F–BSUM	F.27 Friendship Mk. 500	Air France
	F–BSUN	F.27 Friendship Mk. 500	Air France
	F–BSUO	F.27 Friendship Mk. 500	Air France
	F–BTGV	Aero Spacelines Guppy-201	Aeromaritime
	F–BTOA	S.E.210 Caravelle 12	Air Inter
	F–BTOB	S.E.210 Caravelle 12	Air Inter
	F–BTOC	S.E.210 Caravelle 12	Air Inter
	F–BTOD	S.E.210 Caravelle 12	Air Inter
	F–BTOE	S.E.210 Caravelle 12	Air Inter
	F–BTSC	Concorde 101	Air France
	F–BTSD	Concorde 101	Air France
	F–BTTA	Mercure 100	Air Inter
	F–BTTB	Mercure 100	Air Inter
	F–BTTC	Mercure 100	Air Inter
	F–BTTD	Mercure 100	Air Inter
	F–BTTE	Mercure 100	Air Inter
	F–BTTF	Mercure 100	Air Inter
	F–BTTG	Mercure 100	Air Inter
	F–BTTH	Mercure 100	Air Inter
	F–BTTI	Mercure 100	Air Inter
	F–BTTJ	Mercure 100	Air Inter
	F–BUAE	A.300B2 Airbus	Air Inter
	F–BUAF	A.300B2 Airbus	Air Inter
	F–BUAG	A.300B2 Airbus	Air Inter
	F–BUAH	A.300B2 Airbus	Air Inter
	F–BUAI	A.300B2 Airbus	Air Inter
	F–BUAJ	A.300B2 Airbus	Air Inter
	F–BUAK	A.300B2 Airbus	Air Inter
	F–BUAL	A.300B2 Airbus	Air Inter

Reg.	Type	Owner or Operator	Notes
F-BUTI	F-28 Fellowship 1000	Air Alpes (Air France)	
F-BUZC	S.E.210 Caravelle VI-R	Minerve	
F-BVFA	Concorde 101	Air France	
F-BVFB	Concorde 101	Air France	
F-BVFC	Concorde 101	Air France	
F-BVFD	Concorde 101	Air France	
F-BVFF	Concorde 101	Air France	
F-BVGA	A.300B2 Airbus	Air France	
F-BVGB	A.300B2 Airbus	Air France	
F-BVGC	A.300B2 Airbus	Air France	
F-BVGD	A.300B2 Airbus	Air France	
F-BVGE	A.300B2 Airbus	Air France	
F-BVGF	A.300B2 Airbus	Air France	
F-BVGG	A.300B4 Airbus	Air France	
F-BVGH	A.300B4 Airbus	Air France	
F-BVGI	A.300B4 Airbus	Air France	
F-BVGJ	A.300B4 Airbus	Air France	
F-BVGK	A.300B4 Airbus	Air France	
F-BVGL	A.300B4 Airbus	Air France	
F-BVGM	A.300B4 Airbus	Air France	
F-BVGN	A.300B4 Airbus	Air France	
F-BVGO	A.300B4 Airbus	Air France	
F-BVGP	A.300B4 Airbus	Air France	
F-BVGQ	A.300B4 Airbus	Air France	
F-BVGR	A.300B4 Airbus	Air France	
F-BVGS	A.300B4 Airbus	Air France	
F-BVGT	A.300B4 Airbus	Air France	
F-BVOX	Partenavia P.68B	Bretagne Air Services	
F-BVPL	SN-601 Corvette	Air Languedoc	
F-BVPZ	S.E.210 Caravelle VI-N	Corse Air	
F-BVSF	S.E.210 Caravelle VI-N	Europe Aero Service	
F-BXOO	S.E.210 Caravelle VI-N	Europe Aero Service *Languedoc*	
F-BYAA	F.27 Friendship Mk. 400	Air Alsace	
F-BYAB	F.27 Friendship Mk. 600	Air Alpes	
F-BYAO	F.27 Friendship Mk. 100	Uni-Air	
F-BYAP	F.27 Friendship Mk. 100	Uni-Air	
F-BYAR	F.27 Friendship Mk. 600	Air Alpes	
F-BYAT	S.E.210 Caravelle VI-N	Corse Air	
F-BYCA	S.E.210 Caravelle VI-N	Europe Aero Service *Alsace*	
F-BYCD	S.E.210 Caravelle VI-N	Europe Aero Service *Ile de France*	
F-BYCN	Boeing 707-321C (Cargo)	Air France	
F-BYCO	Boeing 707-321C (Cargo)	Air France	
F-BYCU	Douglas DC-3	Bretagne Air Services	
F-BYCY	S.E.210 Caravelle VI-N	Corse Air	
F-BYFM	Douglas DC-8-53	Minerve	
F-GAPA	S.E.210 Caravelle VI-R	Minerve	
F-GATP	S.E.210 Caravelle 10B	Minerve	
F-GATZ	S.E.210 Caravelle VI-N	Minerve	
F-GBBR	F.28 Fellowship 1000	T.A.T.	
F-GBBS	F.28 Fellowship 1000	T.A.T.	
F-GBBT	F.28 Fellowship 1000	T.A.T.	
F-GBBX	F.28 Fellowship 1000	Air Alsace	
F-GBDE	F.27 Friendship Mk. 400	Air Alpes	
F-GBEA	A.300B2 Airbus	Air France	
F-GBEB	A.300B2 Airbus	Air France	
F-GBEC	A.300B2 Airbus	Air France	
F-GBGA	EMB-110P2 Bandeirante	Brit. Air	
F-GBGI	F.27 Friendship Mk. 600	Air Alpes	
F-GBLE	EMB-110P2 Bandeirante	Brit. Air	
F-GBLY	Partenavia P.68B	Bretagne Air Services	
F-GBMG	EMB-110P2 Bandeirante	Brit. Air	
F-GBMJ	S.E.210 Caravelle VI-N	Europe Aero Service *Valentinois*	
F-GBMK	S.E.210 Caravelle VI-N	Europe Aero Service *Rousillon*	
F-GBRM	EMB-110P2 Bandeirante	Brit. Air	
F-GBRQ	FH.227B Friendship	T.A.T.	
F-GBRX	F.27 Friendship Mk. 200	Uni-Air	
F-GBRZ	FH.227B Friendship	T.A.T.	
F-GCBA	Boeing 747-228B	Air France	
F-GCBC	Boeing 747-228B	Air France	
F-GCDA	Boeing 727-228	Air France	
F-GCDB	Boeing 727-228	Air France	
F-GCDC	Boeing 727-228	Air France	
F-GCDD	Boeing 727-228	Air France	
F-GCDE	Boeing 727-228	Air France	
F-GCDF	Boeing 727-228	Air France	

Notes	Reg.	Type	Owner or Operator
	F-GCDG	Boeing 727-228	Air France
	F-GCDH	Boeing 727-228	Air France
	F-GCDI	Boeing 727-228	Air France
	F-GCDJ	Boeing 727-228	Air France
	F-GCFD	FH.227B Friendship	T.A.T.
	F-GCGH	FH.227B Friendship	T.A.T.
	F-GCGQ	Boeing 727-227	Europe Aero Service *Normandie*
	F-GCJL	Boeing 737-222	Euralair
	F-GCJO	FH.227B Friendship	T.A.T.
	F-GCJT	S.E.210 Caravelle 10B	Europe Aero Service
	F-GCLA	EMB-110P2 Bandeirante	Lucas Air Transport
	F-GCLL	Boeing 737-222	Euralair
	F-GCLM	FH.227B Friendship	T.A.T.
	F-GCLN	FH.227B Friendship	T.A.T.
	F-GCLO	FH.227B Friendship	T.A.T.
	F-GCLP	FH.227B Friendship	T.A.T.
	F-GCLQ	FH.227B Friendship	T.A.T.
	F-GCMA	F.27 Friendship Mk. 200	Air Alsace
	F-GCMR	F.27 Friendship Mk. 200	Air Alsace
	F-GCMV	Boeing 727-2X3	Air Charter International
	F-GCMX	Boeing 727-2X3	Air Charter International
	F-GCMY	Boeing 727-2X3	Air Charter International
	F-GCMZ	Boeing 727-2X3	Air Charter International
	F-GCPA	F.27 Friendship Mk. 100	Air Alsace
	F-GCPS	FH.227B Friendship	T.A.T.
	F-GCPT	FH.227B Friendship	T.A.T.
	F-GCPU	FH.227B Friendship	T.A.T.
	F-GCPX	FH.227B Friendship	T.A.T.
	F-GCPY	FH.227B Friendship	T.A.T.
	F-GCPZ	FH.227B Friendship	T.A.T.
	F-GCSL	Boeing 737-222	Euralair
	F-GCVI	S.E.210 Caravelle 12	Air Inter
	F-GCVJ	S.E.210 Caravelle 12	Air Inter
	F-GDAH	FH.227B Friendship	Uni-Air
	F-GDAQ	L-100-30 Hercules	Sfair
	F-GDFC	F.28 Fellowship 4000	Air France
	F-GDFD	F.28 Fellowship 4000	Air France

NOTE: Air France also operates ten more Boeing 747s which retain the US registrations N1252E, N1289E, N1305E, N4508E, N18815, N28366, N28899, N28903, N40116 and N63305.

HA (Hungary)

Notes	Reg.	Type	Owner or Operator
	HA-LBE	Tupolev Tu-134	Malev
	HA-LBF	Tupolev Tu-134	Malev
	HA-LBG	Tupolev Tu-134	Malev
	HA-LBH	Tupolev Tu-134	Malev
	HA-LBI	Tupolev Tu-134A	Malev
	HA-LBK	Tupolev Tu-134A	Malev
	HA-LBP	Tupolev Tu-134A	Malev
	HA-LBR	Tupolev Tu-134A	Malev
	HA-LCA	Tupolev Tu-154B	Malev
	HA-LCB	Tupolev Tu-154B	Malev
	HA-LCE	Tupolev Tu-154B	Malev
	HA-LCG	Tupolev Tu-154B	Malev
	HA-LCH	Tupolev Tu-154B	Malev
	HA-LCM	Tupolev Tu-154B	Malev
	HA-LCN	Tupolev Tu-154B	Malev
	HA-LCO	Tupolev Tu-154B	Malev
	HA-LCP	Tupolev Tu-154B	Malev

HB–INN (Top) Douglas DC-9-81 of Swissair

LN–BWG (Centre) Convair 580 of Nor-Fly

N623US (Bottom) Boeing 747-251B of Northwest Orient

HB (Switzerland)

BALAIR

Notes	*Reg.*	*Type*	*Owner or Operator*
	HB-ICI	S.E.210 Caravelle 10-R	CTA
	HB-ICN	S.E.210 Caravelle 10-R	CTA Ville de Genève
	HB-ICO	S.E.210 Caravelle 10-R	CTA
	HB-ICQ	S.E.210 Caravelle 10-R	CTA
	HB-IDF	Douglas DC-8-62	Swissair *Schwyz*
	HB-IDG	Douglas DC-8-62	Swissair *Neuchatel*
	HB-IDI	Douglas DC-8-62	Swissair *Solothurn*
	HB-IDK	Douglas DC-8-62CF	Swissair *Matterhorn*
	HB-IDL	Douglas DC-8-62	Swissair *Aargau*
	HB-IDO	Douglas DC-9-30	Swissair *Cointrin*
	HB-IDP	Douglas DC-9-30	Swissair *Basel-Land*
	HB-IDR	Douglas DC-9-30	Swissair *Baden*
	HB-IDT	Douglas DC-9-34	Balair
	HB-IDZ	Douglas DC-8-63PF	Balair
	HB-IFU	Douglas DC-9-30	Swissair *Chur*
	HB-IFV	Douglas DC-9-30	Swissair *Bülach*
	HB-IFW	Douglas DC-9-30F	Swissair *Payerne*
	HB-IFZ	Douglas DC-9-30	Balair
	HB-IGA	Boeing 747-257B	Swissair *Geneve*
	HB-IGB	Boeing 747-257B	Swissair *Zurich*
	HB-IHA	Douglas DC-10-30	Swissair *St. Gallen*
	HB-IHB	Douglas DC-10-30	Swissair *Schaffhausen*
	HB-IHC	Douglas DC-10-30	Swissair *Luzern*
	HB-IHD	Douglas DC-10-30	Swissair *Bern*
	HB-IHE	Douglas DC-10-30	Swissair *Vaud*
	HB-IHF	Douglas DC-10-30	Swissair *Nidwalden*
	HB-IHG	Douglas DC-10-30	Swissair *Grisons*
	HB-IHH	Douglas DC-10-30	Swissair *Basel-Stadt*
	HB-IHI	Douglas DC-10-30	Swissair *Fribourg*
	HB-IHK	Douglas DC-10-30	Balair
	HB-IHL	Douglas DC-10-30	Swissair *Ticino*
	HB-IHM	Douglas DC-10-30	Swissair *Valais/Wallis*
	HB-IHN	Douglas DC-10-30ER	Swissair
	HB-IHO	Douglas DC-10-30ER	Swissair
	HB-IKC	Douglas DC-9-32	Alisarda (Italy)
	HB-IKF	Douglas DC-9-51	Alisarda (Italy)
	HB-IKG	Douglas DC-9-51	Alisarda (Italy)
	HB-IKH	Douglas DC-9-51	Alisarda (Italy)
	HB-INA	Douglas DC-9-81	Swissair *Obwalden*
	HB-INB	Douglas DC-9-81	Balair
	HB-INC	Douglas DC-9-81	Swissair *Thurgau*
	HB-IND	Douglas DC-9-81	Swissair *Zug*
	HB-INE	Douglas DC-9-81	Swissair *Rümlang*
	HB-INF	Douglas DC-9-81	Swissair *Appenzell a.Rh.*
	HB-ING	Douglas DC-9-81	Swissair *Glarus*
	HB-INH	Douglas DC-9-81	Swissair *Winterthur*
	HB-INI	Douglas DC-9-81	Swissair *Kloten*
	HB-INK	Douglas DC-9-81	Swissair *Opfikon*
	HB-INL	Douglas DC-9-81	Swissair *Jura*
	HB-INM	Douglas DC-9-81	Swissair *Lausanne*
	HB-INN	Douglas DC-9-81	Swissair *Appenzell i.Rh.*
	HB-INO	Douglas DC-9-81	Swissair *Bellinzona*
	HB-INP	Douglas DC-9-81	Swissair *Oberglatt*
	HB-ISK	Douglas DC-9-51	Swissair *Höri*
	HB-ISL	Douglas DC-9-51	Swissair *Köniz*
	HB-ISM	Douglas DC-9-51	Swissair *Wettingen*
	HB-ISN	Douglas DC-9-51	Swissair *Sion*
	HB-ISO	Douglas DC-9-51	Swissair *Bienne*
	HB-ISP	Douglas DC-9-51	Swissair *Lugano*
	HB-ISR	Douglas DC-9-51	Swissair *Locarno*
	HB-ISS	Douglas DC-9-51	Swissair *Dietikon*
	HB-IST	Douglas DC-9-51	Swissair *Aarau*

Reg.	Type	Owner or Operator	Notes
HB–ISU	Douglas DC-9-51	Swissair *Bachenbülach*	
HB–ISV	Douglas DC-9-51	Swissair *Winkel*	
HB–ISW	Douglas DC-9-51	Swissair *Dubendorf*	

NOTE: Swissair is in the process of replacing its DC-9-30s with Series 81s.

HL (Korea)

HL7406	Boeing 707-385C	Korean Air Lines	
HL7425	Boeing 707-373C	Korean Air Lines	
HL7427	Boeing 707-321C	Korean Air Lines	
HL7431	Boeing 707-321C	Korean Air Lines	
HL7432	Boeing 707-338C	Korean Air Lines	
HL7433	Boeing 707-338C	Korean Air Lines	
HL7435	Boeing 707-321B	Korean Air Lines	

HS (Thailand)

HS–TGA	Boeing 747-2D7B	Thai Airways International *Visuthakasatriya*	
HS–TGB	Boeing 747-2D7B	Thai Airways International *Sirisobhakya*	
HS–TGC	Boeing 747-2D7B	Thai Airways International *Dararasmi*	
HS–TGD	Boeing 747-2D7B	Thai Airways International	
HS–TGF	Boeing 747-2D7B	Thai Airways International *Phimara*	
HS–TGG	Boeing 747-2D7B	Thai Airways International	

HZ (Saudi Arabia)

HZ–AHA	L.1011-385 TriStar 200	SAUDIA—Saudi Arabian Airlines	
HZ–AHB	L.1011-385 TriStar 200	SAUDIA—Saudi Arabian Airlines	
HZ–AHC	L.1011-385 TriStar 200	SAUDIA—Saudi Arabian Airlines	
HZ–AHD	L.1011-385 TriStar 200	SAUDIA—Saudi Arabian Airlines	
HZ–AHE	L.1011-385 TriStar 200	SAUDIA—Saudi Arabian Airlines	
HZ–AHF	L.1011-385 TriStar 200	SAUDIA—Saudi Arabian Airlines	
HZ–AHG	L.1011-385 TriStar 200	SAUDIA—Saudi Arabian Airlines	
HZ–AHH	L.1011-385 TriStar 200	SAUDIA—Saudi Arabian Airlines	
HZ–AHI	L.1011-385 TriStar 200	SAUDIA—Saudi Arabian Airlines	
HZ–AHJ	L.1011-385 TriStar 200	SAUDIA—Saudi Arabian Airlines	
HZ–AHL	L.1011-385 TriStar 200	SAUDIA—Saudi Arabian Airlines	
HZ–AHM	L.1011-385 TriStar 200	SAUDIA—Saudi Arabian Airlines	
HZ–AHN	L.1011-385 TriStar 200	SAUDIA—Saudi Arabian Airlines	
HZ–AHO	L.1011-385 TriStar 200	SAUDIA—Saudi Arabian Airlines	

Notes	Reg.	Type	Owner or Operator
	HZ–AHP	L.1011-385 TriStar 200	SAUDIA—Saudi Arabian Airlines
	HZ–AHQ	L.1011-385 TriStar 200	SAUDIA—Saudi Arabian Airlines
	HZ–AHR	L.1011-385 TriStar 200	SAUDIA—Saudi Arabian Airlines
	HZ–AJA	Boeing 747-168B	SAUDIA—Saudi Arabian Airlines
	HZ–AIB	Boeing 747-168B	SAUDIA—Saudi Arabian Airlines
	HZ–AIC	Boeing 747-168B	SAUDIA—Saudi Arabian Airlines
	HZ–AID	Boeing 747-168B	SAUDIA—Saudi Arabian Airlines
	HZ–AIE	Boeing 747-168B	SAUDIA—Saudi Arabian Airlines

NOTE: SAUDIA also operate DC-8s leased from various U.S. airlines.

I (Italy)

Alitalia

Notes	Reg.	Type	Owner or Operator
	I–ATIA	Douglas DC-9-32	Aero Trasporti Italiani (ATI)
	I–ATIE	Douglas DC-9-32	Aero Trasporti Italiani (ATI)
	I–ATIH	Douglas DC-9-32	Aero Trasporti Italiani (ATI)
	I–ATIJ	Douglas DC-9-32	Aero Trasporti Italiani (ATI)
	I–ATIK	Douglas DC-9-32	Aero Trasporti Italiani (ATI)
	I–ATIO	Douglas DC-9-32	Aero Trasporti Italiani (ATI)
	I–ATIQ	Douglas DC-9-32	Aero Trasporti Italiani (ATI)
	I–ATIU	Douglas DC-9-32	Aero Trasporti Italiani (ATI)
	I–ATIW	Douglas DC-9-32	Aero Trasporti Italiani (ATI)
	I–ATIX	Douglas DC-9-32	Aero Trasporti Italiani (ATI)
	I–ATIY	Douglas DC-9-32	Aero Trasporti Italiani (ATI)
	I–ATJA	Douglas DC-9-32	Aero Trasporti Italiani (ATI)
	I–ATJB	Douglas DC-9-32	Aero Trasporti Italiani (ATI)
	I–BUSB	A.300B4 Airbus	Alitalia *Tiziano*
	I–BUSC	A.300B4 Airbus	Alitalia *Botticelli*
	I–BUSD	A.300B4 Airbus	Alitalia *Caravaggio*
	I–BUSF	A.300B4 Airbus	Alitalia *Tintoretto*
	I–BUSG	A.300B4 Airbus	Alitalia *Canaletto*
	I–BUSH	A.300B4 Airbus	Alitalia *Mantegua*
	I–BUSJ	A.300B4 Airbus	Alitalia *Tiepolo*
	I–BUSK	A.300B4 Airbus	Alitalia *Pinturicchia*
	I–DEMA	Boeing 747-143A	Alitalia *Neil Alden Armstrong*
	I–DEMC	Boeing 747-243B	Alitalia *Taormina*
	I–DEMD	Boeing 747-243B	Alitalia
	I–DEMF	Boeing 747-243B	Alitalia *Portofino*
	I–DEMG	Boeing 747-243B	Alitalia
	I–DEML	Boeing 747-243B	Alitalia
	I–DEMN	Boeing 747-243B	Alitalia
	I–DEMO	Boeing 747-243B	Alitalia *Francesco de Pinedo*
	I–DEMP	Boeing 747-243B	Alitalia
	I–DEMU	Boeing 747-243B	Alitalia *Geo Chavez*
	I–DIBC	Douglas DC-9-32	Alitalia *Isola di Lampedusa*
	I–DIBD	Douglas DC-9-32	Alitalia *Isola di Montecristo*
	I–DIBJ	Douglas DC-9-32	Alitalia *Isola della Capraia*
	I–DIBN	Douglas DC-9-32	Alitalia *Isola della Palmaria*
	I–DIBO	Douglas DC-9-32	Aermediterranea *Conca D'Ora*
	I–DIBQ	Douglas DC-9-32	Alitalia *Isola di Pianosa*
	I–DIKA	Douglas DC-9-32	Alitalia *Isola di Capri*
	I–DIKC	Douglas DC-9-32	Alitalia *Isola di Ponza*
	I–DIKD	Douglas DC-9-32	Alitalia *Isola del Giglio*
	I–DIKE	Douglas DC-9-32	Alitalia *Isola d'Elba*
	I–DIKI	Douglas DC-9-32	Alitalia *Isola di Murano*
	I–DIKJ	Douglas DC-9-32	Alitalia *Isola di Lipari*
	I–DIKL	Douglas DC-9-32	Alitalia *Isola di Panarea*
	I–DIKM	Douglas DC-9-32	Alitalia *Isola di Tavolara*
	I–DIKN	Douglas DC-9-32	Alitalia *Isola di Nisida*
	I–DIKO	Douglas DC-9-32	Alitalia *Isola di Pantelleria*
	I–DIKP	Douglas DC-9-32	Aero Trasporti Italiani (ATI)
	I–DIKR	Douglas DC-9-32	Alitalia *Isola di Torcello*
	I–DIKS	Douglas DC-9-32	Alitalia *Isola di Filicudi*
	I–DIKT	Douglas DC-9-32	Alitalia *Isola di Ustica*

Reg.	Type	Owner or Operator	Notes
I-DIKU	Douglas DC-9-32	Alitalia *Isola d'Ischia*	
I-DIKV	Douglas DC-9-32	Alitalia *Isola di Vulcano*	
I-DIKW	Douglas DC-9-32	Alitalia *Isola di Giannutri*	
I-DIKY	Douglas DC-9-32	Aero Trasporti Italiani (ATI)	
I-DIKZ	Douglas DC-9-32	Alitalia *Isola di Linosa*	
I-DIRA	Boeing 727-243	Alitalia *Citta di Gubbio*	
I-DIRB	Boeing 727-243	Alitalia *Citta di Siracusa*	
I-DIRC	Boeing 727-243	Alitalia *Citta di Aosta*	
I-DIRD	Boeing 727-243	Alitalia *Citta di Bergamo*	
I-DIRF	Boeing 727-243	Alitalia *Citta di Lecce*	
I-DIRG	Boeing 727-243	Alitalia *Citta di Urbino*	
I-DIRI	Boeing 727-243	Alitalia *Citta di Siena*	
I-DIRJ	Boeing 727-243	Alitalia *Citta di Verona*	
I-DIRL	Boeing 727-243	Alitalia *Citta di Viterbo*	
I-DIRM	Boeing 727-243	Alitalia *Citta di Genova*	
I-DIRN	Boeing 727-243	Alitalia *Citta di Aquilea*	
I-DIRO	Boeing 727-243	Alitalia *Citta di Amalfi*	
I-DIRP	Boeing 727-243	Alitalia *Citta di Ivrea*	
I-DIRQ	Boeing 727-243	Alitalia *Citta di Sassari*	
I-DIRR	Boeing 727-243	Alitalia *Citta di Trento*	
I-DIRS	Boeing 727-243	Alitalia *Citta di Sulmona*	
I-DIRU	Boeing 727-243	Alitalia *Citta di Ravenna*	
I	Boeing 727-243	Alitalia	
I-DIZA	Douglas DC-9-32	Alitalia *Isola di Palmarola*	
I-DIZB	Douglas DC-9-32	Aero Trasporti Italiani (ATI)	
I-DIZC	Douglas DC-9-32	Aero Trasporti Italiani (ATI)	
I-DIZE	Douglas DC-9-32	Aero Trasporti Italiani (ATI)	
I-DIZF	Douglas DC-9-32	Aermediterranea *Dolomiti*	
I-DIZI	Douglas DC-9-32	Aero Trasporti Italiani (ATI)	
I-DIZO	Douglas DC-9-32	Alitalia *Isola di Tino*	
I-DIZU	Douglas DC-9-32	Aero Trasporti Italiani (ATI)	
I-DYNA	Douglas DC-10-30	Alitalia *Galileo Galilei*	
I-DYNB	Douglas DC-10-30	Alitalia *Giotto di Bondone*	
I-DYNC	Douglas DC-10-30	Alitalia *Luigi Pirandello*	
I-DYND	Douglas DC-10-30	Alitalia *Enrico Fermi*	
I-DYNE	Douglas DC-10-30	Alitalia *Dante Alghieri*	
I-DYNI	Douglas DC-10-30	Alitalia *Michelangelo Buonarrotti*	
I-DYNO	Douglas DC-10-30	Alitalia *Benvenuto Cellini*	
I-DYNU	Douglas DC-10-30	Alitalia *Guglielmo Marconi*	
I-GISA	S.E.210 Caravelle III	Altair	
I-SARW	Douglas DC-9-32	Alisarda	

NOTE: Alisarda also uses DC-9s which retain their Swiss registrations HB-IKC, HB-IKF, HB-IKG and HB-IKH.

JA (Japan)

Reg.	Type	Owner or Operator	Notes
JA8031	Douglas DC-8-62	Japan Air Lines	
JA8032	Douglas DC-8-62	Japan Air Lines	
JA8033	Douglas DC-8-62	Japan Air Lines	
JA8034	Douglas DC-8-62	Japan Air Lines	
JA8035	Douglas DC-8-62	Japan Air Lines	
JA8036	Douglas DC-8-62AF	Japan Air Lines	
JA8037	Douglas DC-8-62	Japan Air Lines	
JA8044	Douglas DC-8-62AF	Japan Air Lines	
JA8052	Douglas DC-8-62	Japan Air Lines	
JA8053	Douglas DC-8-62	Japan Air Lines	
JA8055	Douglas DC-8-62AF	Japan Air Lines	

Notes	Reg.	Type	Owner or Operator
	JA8056	Douglas DC-8-62AF	Japan Air Lines
	JA8101	Boeing 747-146	Japan Air Lines
	JA8102	Boeing 747-146	Japan Air Lines
	JA8103	Boeing 747-146	Japan Air Lines
	JA8104	Boeing 747-246B	Japan Air Lines
	JA8105	Boeing 747-246B	Japan Air Lines
	JA8106	Boeing 747-246B	Japan Air Lines
	JA8107	Boeing 747-146A	Japan Air Lines
	JA8108	Boeing 747-246B	Japan Air Lines
	JA8110	Boeing 747-246B	Japan Air Lines
	JA8111	Boeing 747-246B	Japan Air Lines
	JA8112	Boeing 747-146A	Japan Air Lines
	JA8113	Boeing 747-246B	Japan Air Lines
	JA8114	Boeing 747-246B	Japan Air Lines
	JA8115	Boeing 747-146A	Japan Air Lines
	JA8116	Boeing 747-146A	Japan Air Lines
	JA8122	Boeing 747-246B	Japan Air Lines
	JA8123	Boeing 747-246F	Japan Air Lines
	JA8125	Boeing 747-246B	Japan Air Lines
	JA8127	Boeing 747-246B	Japan Air Lines
	JA8128	Boeing 747-146A	Japan Air Lines
	JA8129	Boeing 747-246B	Japan Air Lines
	JA8130	Boeing 747-246B	Japan Air Lines
	JA8131	Boeing 747-246B	Japan Air Lines
	JA8132	Boeing 747-246F	Japan Air Lines
	JA8140	Boeing 747-246B	Japan Air Lines
	JA8141	Boeing 747-246B	Japan Air Lines
	JA8142	Boeing 747-146B	Japan Air Lines
	JA8143	Boeing 747-146B	Japan Air Lines
	JA8144	Boeing 747-246F	Japan Air Lines
	JA8149	Boeing 747-246B	Japan Air Lines
	JA8150	Boeing 747-246B	Japan Air Lines
	JA8151	Boeing 747-246F	Japan Air Lines
	JA	Boeing 747-246B	Japan Air Lines
	JA	Boeing 747-246B	Japan Air Lines

JY (Jordan)

Notes	Reg.	Type	Owner or Operator
	JY-ADP	Boeing 707-383C	Alia—The Royal Jordanian Airline *City of Amman*
	JY-ADT	Boeing 720-030B	Alia—The Royal Jordanian Airline *City of Madaba*
	JY-AEB	Boeing 707-384C	Alia—The Royal Jordanian Airline *City of Jerash*
	JY-AEC	Boeing 707-384C	Alia—The Royal Jordanian Airline *City of Umquais*
	JY-AED	Boeing 707-321C (Cargo)	Alia—The Royal Jordanian Airline *City of Al-Karameh*
	JY-AES	Boeing 707-321C	Alia—The Royal Jordanian Airline
	JY-AFA	Boeing 747-2D3B	Alia—The Royal Jordanian Airline *Prince Ali*
	JY-AFB	Boeing 747-2D3B	Alia—The Royal Jordanian Airline *Princess Haya*
	JY-AFQ	Boeing 707-344C	Alia—The Royal Jordanian Airline
	JY-AFR	Boeing 707-344C	Alia—The Royal Jordanian Airline
	JY-AFS	Boeing 747-2D3B	Alia—The Royal Jordanian Airline
	JY-AGA	L.1011-385 TriStar 500	Alia—The Royal Jordanian Airline
	JY-AGB	L.1011-385 TriStar 500	Alia—The Royal Jordanian Airline
	JY-AGC	L.1011-385 TriStar 500	Alia—The Royal Jordanian Airline
	JY-AGD	L.1011-385 TriStar 500	Alia—The Royal Jordanian Airline
	JY-AGE	L.1011-385 TriStar 500	Alia—The Royal Jordanian Airline

FRED OLSEN AIRTRANSPORT

LN (Norway)

Reg.	Type	Owner or Operator	Notes
LN–BWG	Convair 580	Nor-Fly	
LN–BWN	Convair 580	Nor-Fly	
LN–FOG	L-188AF Electra	Fred Olsen Airtransport	
LN–FOH	L-188AF Electra	Fred Olsen Airtransport	
LN–FOI	L-188CF Electra	Fred Olsen Airtransport	
LN–KLK	Convair 440	Nor-Fly	
LN–MAP	Convair 440	Nor-Fly	
LN–MOF	Douglas DC-8-63	Scanair *Tyra Viking*	
LN–MOG	Douglas DC-8-62	SAS *Odd Viking*	
LN–MOW	Douglas DC-8-62	SAS *Roald Viking*	
LN–NPB	Boeing 737-2R4C	Air Executive	
LN–NPH	F.27 Friendship MK. 300	Air Executive	
LN–NPI	F.27 Friendship Mk. 100	Air Executive	
LN–NPM	F.27 Friendship Mk. 100	Air Executive	
LN–RCA	A.300B2 Airbus	SAS *Snorry Viking*	
LN–RKA	Douglas DC-10-30	SAS *Olav Viking*	
LN–RKB	Douglas DC-10-30	SAS *Haakon Viking*	
LN–RLA	Douglas DC-9-41	SAS *Are Viking*	
LN–RLB	Douglas DC-9-41	SAS *Arne Viking*	
LN–RLC	Douglas DC-9-41	SAS *Gunnar Viking*	
LN–RLD	Douglas DC-9-41	SAS *Torleif Viking*	
LN–RLH	Douglas DC-9-41	SAS *Einar Viking*	
LN–RLJ	Douglas DC-9-41	SAS *Stein Viking*	
LN–RLK	Douglas DC-9-41	SAS *Erling Viking*	
LN–RLL	Douglas DC-9-21	SAS *Guttorm Viking*	
LN–RLN	Douglas DC-9-41	SAS *Halldor Viking*	
LN–RLO	Douglas DC-9-21	SAS *Gunder Viking*	
LN–RLP	Douglas DC-9-41	SAS *Froste Viking*	
LN–RLS	Douglas DC-9-41	SAS *Asmund Viking*	
LN–RLT	Douglas DC-9-41	SAS *Audun Viking*	
LN–RLU	Douglas DC-9-41	SAS *Eivind Viking*	
LN–RLW	Douglas DC-9-33AF	SAS *Rand Viking*	
LN–RLX	Douglas DC-9-41	SAS *Sote Viking*	
LN–RLZ	Douglas DC-9-41	SAS *Bodvar Viking*	
LN–RNA	Boeing 747-283B	SAS *Magnus Viking*	
LN–SUA	Boeing 737-205C	Braathens SAFE *Halvdan Svarte*	
LN–SUB	Boeing 737-205	Braathens SAFE *Magnus Den Gode*	
LN–SUC	F.28 Fellowship 1000	Braathens SAFE *Olav Kyrre*	
LN–SUD	Boeing 737-205	Braathens SAFE *Olav Tryggvason*	
LN–SUE	F.27 Friendship Mk. 100	Air Executive	
LN–SUF	F.27 Friendship Mk. 100	Air Executive	
LN–SUG	Boeing 737-205	Braathens SAFE *Harald Harfagre*	
LN–SUH	Boeing 737-205	Braathens SAFE *Sigurd Jorsalfar*	
LN–SUI	Boeing 737-205	Braathens SAFE *Haakon den Gode*	
LN–SUK	Boeing 737-205	Braathens SAFE *Olav Haraldsson*	
LN–SUL	F.27 Friendship Mk. 100	Air Executive	
LN–SUM	Boeing 737-205	Braathens SAFE *Magnus Lagabøter*	
LN–SUN	F.28 Fellowship 1000	Braathens SAFE *Haakon Sverresson*	
LN–SUO	F.28 Fellowship 1000	Braathens SAFE *Magnus Barfot*	
LN–SUP	Boeing 737-205	Braathens SAFE *Haakon IV*	
LN–SUS	Boeing 737-205	Braathens SAFE *Haakon V*	
LN–SUT	Boeing 737-205	Braathens SAFE *Oystein Magnusson*	
LN–SUX	F.28 Fellowship 1000	Braathens SAFE *Harald Hardrade*	

NOTE: SAS also operates two Boeing 747-283Bs which retain their U.S. registrations N4501Q and N4502R.

LV (Argentina)

Notes	Reg.	Type	Owner or Operator
	LV–LZD	Boeing 747-287B	Aerolineas Argentinas
	LV–MLO	Boeing 747-287B	Aerolineas Argentinas
	LV–MLP	Boeing 747-287B	Aerolineas Argentinas
	LV–MLR	Boeing 747-287B	Aerolineas Argentinas
	LV–MSG	Boeing 707-321C	Transporte Aereo Rioplatense
	LV–MZE	Boeing 707-338C	Transporte Aereo Rioplatense
	LV–OEP	Boeing 747-287B	Aerolineas Argentinas
	LV–OOZ	Boeing 747-287B	Aerolineas Argentinas
	LV–OPA	Boeing 747-287B	Aerolineas Argentinas

LX (Luxembourg)

Notes	Reg.	Type	Owner or Operator
	LX–BJV	Boeing 707-331C	Cargolux
	LX–DCV	Boeing 747-2R7F	Cargolux *City of Luxembourg*
	LX–ECV	Boeing 747-2R7F	Cargolux *City of Esch sur Alzette*
	LX–LGA	F.27 Friendship Mk. 100	Luxair *Prince Henri*
	LX–LGB	F.27 Friendship Mk. 100	Luxair *Prince Jean*
	LX–LGD	F.27 Friendship Mk. 400	Luxair *Princesse Margaretha*
	LX–LGH	Boeing 737-2C9	Luxair *Prince Guillaume*
	LX–LGI	Boeing 737-2C9	Luxair *Princesse Marie-Astrid*

LZ (Bulgaria)

Notes	Reg.	Type	Owner or Operator
	LZ–BTA	Tupolev Tu-154	Balkan Bulgarian Airlines
	LZ–BTC	Tupolev Tu-154	Balkan Bulgarian Airlines
	LZ–BTD	Tupolev Tu-154B	Balkan Bulgarian Airlines
	LZ–BTE	Tupolev Tu-154B	Balkan Bulgarian Airlines
	LZ–BTF	Tupolev Tu-154A	Balkan Bulgarian Airlines
	LZ–BTG	Tupolev Tu-154A	Balkan Bulgarian Airlines
	LZ–BTJ	Tupolev Tu-154B	Balkan Bulgarian Airlines
	LZ–BTK	Tupolev Tu-154B	Balkan Bulgarian Airlines
	LZ–BTL	Tupolev Tu-154B	Balkan Bulgarian Airlines
	LZ–BTM	Tupolev Tu-154B	Balkan Bulgarian Airlines
	LZ–BTO	Tupolev Tu-154B	Balkan Bulgarian Airlines
	LZ–BTP	Tupolev Tu-154B	Balkan Bulgarian Airlines
	LZ–BTR	Tupolev Tu-154B	Balkan Bulgarian Airlines
	LZ–BTS	Tupolev Tu-154B	Balkan Bulgarian Airlines
	LZ–BTT	Tupolev Tu-154B	Balkan Bulgarian Airlines
	LZ–BTU	Tupolev Tu-154B	Balkan Bulgarian Airlines
	LZ–TUA	Tupolev Tu-134	Balkan Bulgarian Airlines
	LZ–TUC	Tupolev Tu-134	Balkan Bulgarian Airlines
	LZ–TUD	Tupolev Tu-134	Balkan Bulgarian Airlines
	LZ–TUE	Tupolev Tu-134	Balkan Bulgarian Airlines
	LZ–TUF	Tupolev Tu-134	Balkan Bulgarian Airlines
	LZ–TUK	Tupolev Tu-134A	Balkan Bulgarian Airlines
	LZ–TUL	Tupolev Tu-134A	Balkan Bulgarian Airlines
	LZ–TUM	Tupolev Tu-134A	Balkan Bulgarian Airlines
	LZ–TUN	Tupolev Tu-134A	Balkan Bulgarian Airlines
	LZ–TUO	Tupolev Tu-134	Balkan Bulgarian Airlines
	LZ–TUP	Tupolev Tu-134A	Balkan Bulgarian Airlines
	LZ–TUR	Tupolev Tu-134A	Balkan Bulgarian Airlines
	LZ–TUS	Tupolev Tu-134A	Balkan Bulgarian Airlines
	LZ–TUZ	Tupolev Tu-134A	Balkan Bulgarian Airlines

N (USA)

Reg.	*Type*	*Owner or Operator*	*Notes*
N10ST	L-100-30 Hercules	Transamerica Airlines	
N11ST	L-100-30 Hercules	Transamerica Airlines	
N12ST	L-100-30 Hercules	Transamerica Airlines	
N15ST	L-100-30 Hercules	Transamerica Airlines	
N16ST	L-100-30 Hercules	Transamerica Airlines	
N17ST	L-100-30 Hercules	Transamerica Airlines	
N18ST	L-100-30 Hercules	Transamerica Airlines	
N19ST	L-100-30 Hercules	Transamerica Airlines	
N20ST	L-100-30 Hercules	Transamerica Airlines	
N21ST	L-100-30 Hercules	Transamerica Airlines	
N23ST	L-100-30 Hercules	Transamerica Airlines	
N24ST	L-100-30 Hercules	Transamerica Airlines	
N AF	Boeing 737-2T4	Air Florida/Air Europe	
N101AK	L-100-30 Hercules	Alaska International Air	
N101TV	Douglas DC-10-30CF	Air Florida	
N102TV	Douglas DC-10-30CF	Air Florida	
N103TV	Douglas DC-10-30CF	Transamerica Airlines	
N103WA	Douglas DC-10-30CF	World Airways	
N104AK	L-100-30 Hercules	Alaska International Air	
N104WA	Douglas DC-10-30CF	World Airways	
N105WA	Douglas DC-10-30CF	World Airways	
N106AK	L-100-30 Hercules	Alaska International Air	
N106WA	Douglas DC-10-30CF	World Airways	
N107AK	L-100-30 Hercules	Alaska International Air	
N107WA	Douglas DC-10-30CF	World Airways	
N108AK	L-100-30 Hercules	Alaska International Air	
N108WA	Douglas DC-10-30CF	World Airways	
N109WA	Douglas DC-10-30CF	World Airways	
N112WA	Douglas DC-10-30CF	World Airways	
N113WA	Douglas DC-10-30CF	World Airways	
N125TW	Boeing 747-136	Trans World Airlines	
N126TW	Boeing 747-136	Trans World Airlines	
N133TW	Boeing 747-146	Trans World Airlines	

Notes	Reg.	Type	Owner or Operator
	N134TW	Boeing 747-156	Trans World Airlines
	N323PA	Boeing 727-21	Pan Am *Clipper Natchez*
	N325PA	Boeing 727-21	Pan Am *Clipper Indiaman*
	N329PA	Boeing 727-21	Pan Am *Clipper Inca*
	N339PA	Boeing 727-21C	Pan Am *Clipper Flying Dutchman*
	N340PA	Boeing 727-21C	Pan Am *Clipper Goddess*
	N355PA	Boeing 727-21	Pan Am *Clipper Archer*
	N356PA	Boeing 727-21	Pan Am *Clipper Argonaut*
	N357PA	Boeing 727-21	Pan Am *Clipper Yankee*
	N358PA	Boeing 727-21	Pan Am *Clipper Driver*
	N361PA	Boeing 727-2D4	Pan Am *Clipper Friendship*
	N362PA	Boeing 727-2D4	Pan Am *Clipper Onward*
	N363PA	Boeing 727-221	Pan Am *Clipper Racer*
	N364PA	Boeing 727-221	Pan Am *Clipper Whistler*
	N365PA	Boeing 727-221	Pan Am *Clipper Peerless*
	N366PA	Boeing 727-221	Pan Am *Clipper Expounder*
	N367PA	Boeing 727-221	Pan Am *Clipper Matchless*
	N368PA	Boeing 727-221	Pan Am *Clipper Goodwill*
	N369PA	Boeing 727-221	Pan Am *Clipper Hotspur*
	N370PA	Boeing 727-221	Pan Am *Clipper Splendid*
	N447T	Conroy CL-44-O	*Skymonster*
	N501AK	L-100-30 Hercules	Alaska International Air
	N501PA	L-1011-385 TriStar 500	Pan Am *Clipper Eagle*
	N503PA	L-1011-385 TriStar 500	Pan Am *Clipper Flying Eagle*
	N504PA	L-1011-385 TriStar 500	Pan Am *Clipper National Eagle*
	N505PA	L-1011-385 TriStar 500	Pan Am *Clipper Eagle Wing*
	N507PA	L-1011-385 TriStar 500	Pan Am *Clipper Northern Eagle*
	N508PA	L-1011-385 TriStar 500	Pan Am *Clipper Bald Eagle*
	N509PA	L-1011-385 TriStar 500	Pan Am *Clipper Golden Falcon*
	N510PA	L-1011-385 TriStar 500	Pan Am *Clipper White Falcon*
	N511PA	L-1011-385 TriStar 500	Pan Am *Clipper Black Hawk*
	N512PA	L-1011-385 TriStar 500	Pan Am *Clipper Warhawk*
	N513PA	L-1011-385 TriStar 500	Pan Am *Clipper Wild Duck*
	N514PA	L-1011-385 TriStar 500	Pan Am *Clipper Game Cock*
	N530PA	Boeing 747SP-21	Pan Am *Clipper Mayflower*
	N531PA	Boeing 747SP-21	Pan Am *Clipper Freedom*
	N532PA	Boeing 747SP-21	Pan Am *Clipper Constitution*
	N533PA	Boeing 747SP-21	Pan Am *Clipper Young America*
	N534PA	Boeing 747SP-21	Pan Am *Clipper Great Republic*
	N536PA	Boeing 747SP-21	Pan Am *Clipper Lindbergh*
	N537PA	Boeing 747SP-21	Pan Am *Clipper Washington*
	N538PA	Boeing 747SP-21	Pan Am *Clipper Plymouth Rock*
	N539PA	Boeing 747SP-21	Pan Am *Clipper Liberty Bell*
	N540PA	Boeing 747SP-21	Pan Am *Clipper Star of the Union*
	N601BN	Boeing 747-127	Braniff International
	N601US	Boeing 747-151	Northwest Orient
	N602BN	Boeing 747-227B	Braniff International
	N602US	Boeing 747-151	Northwest Orient
	N603US	Boeing 747-151	Northwest Orient
	N604US	Boeing 747-151	Northwest Orient
	N605US	Boeing 747-151	Northwest Orient
	N606BN	Boeing 747SP-27	Braniff International
	N606US	Boeing 747-151	Northwest Orient
	N607US	Boeing 747-151	Northwest Orient
	N608US	Boeing 747-151	Northwest Orient
	N609US	Boeing 747-151	Northwest Orient
	N610BN	Boeing 747-130	Braniff International
	N610US	Boeing 747-151	Northwest Orient
	N611BN	Boeing 747-230B	Braniff International
	N611US	Boeing 747-251B	Northwest Orient
	N612US	Boeing 747-251B	Northwest Orient
	N613US	Boeing 747-251B	Northwest Orient
	N614US	Boeing 747-251B	Northwest Orient
	N615US	Boeing 747-251B	Northwest Orient
	N616US	Boeing 747-251F	Northwest Orient
	N617US	Boeing 747-251F	Northwest Orient
	N618US	Boeing 747-251F	Northwest Orient
	N619US	Boeing 747-251F	Northwest Orient
	N620US	Boeing 747-135	Northwest Orient
	N621US	Boeing 747-135	Northwest Orient
	N622US	Boeing 747-251B	Northwest Orient
	N623US	Boeing 747-251B	Northwest Orient
	N624US	Boeing 747-251B	Northwest Orient
	N625US	Boeing 747-251B	Northwest Orient
	N626US	Boeing 747-251B	Northwest Orient

Reg.	Type	Owner or Operator	Notes
N627US	Boeing 747-251B	Northwest Orient	
N628US	Boeing 747-251B	Northwest Orient	
N629US	Boeing 747-251F	Northwest Orient	
N652PA	Boeing 747-121	Pan Am *Clipper Mermaid*	
N653PA	Boeing 747-121	Pan Am *Clipper Pride of the Ocean*	
N655PA	Boeing 747-121	Pan Am *Clipper Sea Serpent*	
N656PA	Boeing 747-121	Pan Am *Clipper Empress of the Seas*	
N657PA	Boeing 747-121	Pan Am *Clipper Seven Seas*	
N658PA	Boeing 747-121F	Pan Am *Clipper Fortune*	
N659A	Boeing 747-121	Pan Am *Clipper Romance of the Seas*	
N707GB	Boeing 707-338C	Arrow Air	
N707ME	Boeing 707-327C	Arrow Air	
N731PA	Boeing 747-121	Pan Am *Clipper Ocean Express*	
N732PA	Boeing 747-121	Pan Am *Clipper Ocean Telegraph*	
N733PA	Boeing 747-121	Pan Am *Clipper Pride of the Sea*	
N734PA	Boeing 747-121	Pan Am *Clipper Champion of the Seas*	
N735PA	Boeing 747-121	Pan Am *Clipper Spark of the Ocean*	
N737PA	Boeing 747-121	Pan Am *Clipper Ocean Herald*	
N738PA	Boeing 747-121	Pan Am *Clipper Belle of the Sea*	
N739PA	Boeing 747-121	Pan Am *Clipper Maid of the Seas*	
N740PA	Boeing 747-121	Pan Am *Clipper Ocean Pearl*	
N741PA	Boeing 747-121	Pan Am *Clipper Sparkling Wave*	
N741PR	Boeing 747-2F6B	Philippine Airlines	
N741TV	Boeing 747-271C	Transamerica Airlines	
N742PA	Boeing 747-121	Pan Am *Clipper Neptune's Car*	
N742PR	Boeing 747-2F6B	Philippine Airlines	
N742TV	Boeing 747-271C	Transamerica Airlines *Uncle Sam*	
N743PA	Boeing 747-121	Pan Am *Clipper Black Sea*	
N743PR	Boeing 747-2F6B	Philippine Airlines	
N743TV	Boeing 747-271C	Transamerica Airlines	
N744PA	Boeing 747-121	Pan Am *Clipper Ocean Spray*	
N744PR	Boeing 747-2F6B	Philippine Airlines	
N744TV	Boeing 747-271C	Transamerica Airlines	
N745TV	Boeing 747-271C	Transamerica Airlines	
N747PA	Boeing 747-121	Pan Am *Clipper Sea Lark*	
N747TA	Boeing 747-212B	Metro International	
N748PA	Boeing 747-121	Pan Am *Clipper Crest of the Wave*	
N748TA	Boeing 747-212B	Metro International	
N748WA	Boeing 747-273C	World Airways	
N749PA	Boeing 747-121	Pan Am *Clipper Dashing Wave*	
N749TA	Boeing 747-212B	Metro International	
N749WA	Boeing 747-273C	World Airways	
N750PA	Boeing 747-121	Pan Am *Clipper Neptune's Favorite*	
N750WA	Boeing 747-124F	World Airways	
N751DA	L-1011-385 TriStar 500	Delta Air Lines	
N751PA	Boeing 747-121	Pan Am *Clipper Gem of the Sea*	
N752DA	L-1011-385 TriStar 500	Delta Air Lines	
N753DA	L-1011-385 TriStar 500	Delta Air Lines	
N753PA	Boeing 747-121	Pan Am *Clipper Queen of the Seas*	
N754PA	Boeing 747-121	Pan Am *Clipper Ocean Rover*	
N755PA	Boeing 747-121	Pan Am *Clipper Sovereign of the Seas*	
N760TW	Boeing 707-331B	Trans World Airlines	
N770PA	Boeing 747-121	Pan Am *Clipper Queen of the Pacific*	
N771PA	Boeing 747-121F	Pan Am *Clipper Messenger*	
N772FT	Douglas DC-8-63CF	Flying Tiger Line	
N773FT	Douglas DC-8-63CF	Air India Cargo	
N773TW	Boeing 707-331B	Trans World Airlines	
N774FT	Douglas DC-8-63CF	SAUDIA—Saudi Arabian Airlines	
N774TW	Boeing 707-331B	Trans World Airlines	
N775FT	Douglas DC-8-63CF	SAUDIA—Saudi Arabian Airlines	
N776FT	Douglas DC-8-63CF	Flying Tiger Line	
N779TW	Boeing 707-331B	Trans World Airlines	
N781FT	Douglas DC-8-63CF	Flying Tiger Line	
N783FT	Douglas DC-8-63AF	Flying Tiger Line	
N784FT	Douglas DC-8-63AF	Flying Tiger Line	
N786FT	Douglas DC-8-63AF	Flying Tiger Line	
N787FT	Douglas DC-8-63AF	Flying Tiger Line	
N788FT	Douglas DC-8-63AF	Flying Tiger Line	
N790FT	Douglas DC-8-63AF	Flying Tiger Line	
N791FT	Douglas DC-8-63CF	Flying Tiger Line	
N791TW	Boeing 707-331C	Global International Airways	
N792FT	Douglas DC-8-63CF	Flying Tiger Line	
N792TW	Boeing 707-331C	Global International Airways	
N793FT	Douglas DC-8-63CF	Flying Tiger Line	

Notes	Reg.	Type	Owner or Operator
	N795FT	Douglas DC-8-63CF	Flying Tiger Line
	N796FT	Douglas DC-8-63CF	Air India Cargo
	N797FT	Douglas DC-8-63CF	Flying Tiger Line
	N798FT	Douglas DC-8-63CF	Flying Tiger Line
	N800EV	Douglas DC-8-52	Evergreen International Airlines
	N801EV	Douglas DC-8-52	Evergreen International Airlines
	N801WA	Douglas DC-8-63CF	World Airways
	N802FT	Boeing 747-123F	Flying Tiger Line
	N803FT	Boeing 747-132F	Flying Tiger Line
	N804EV	Douglas DC-8-52	Evergreen International Airlines
	N804FT	Boeing 747-132F	Flying Tiger Line
	N804WA	Douglas DC-8-63CF	World Airways
	N805FT	Boeing 747-132F	Flying Tiger Line
	N805WA	Douglas DC-8-63CF	World Airways
	N806FT	Boeing 747-249F	Flying Tiger Line *Robert W. Prescott*
	N806WA	Douglas DC-8-63CF	World Airways
	N807FT	Boeing 747-249F	Flying Tiger Line *Thomas Haywood*
	N808FT	Boeing 747-249F	Flying Tiger Line *William E. Bartlett*
	N810EV	Douglas DC-8-61CF	SAUDIA—Saudi Arabian Airlines
	N810FT	Boeing 747-249F	Flying Tiger Line *Clifford G. Groh*
	N811EV	Douglas DC-8-63CF	Evergreen International Airlines
	N811FT	Boeing 747-245F	Flying Tiger Line
	N812FT	Boeing 747-245F	Flying Tiger Line
	N813FT	Boeing 747-245F	Flying Tiger Line
	N814FT	Boeing 747-245F	Flying Tiger Line
	N815FT	Boeing 747-245F	Flying Tiger Line *Henry L. Heguy*
	N821L	Douglas DC-10-30	Western Airlines
	N860FT	Douglas DC-8-61CF	Flying Tiger Line
	N861FT	Douglas DC-8-61CF	Flying Tiger Line
	N862FT	Douglas DC-8-61CF	Flying Tiger Line
	N863FT	Douglas DC-8-61CF	Flying Tiger Line
	N867FT	Douglas DC-8-61CF	Flying Tiger Line
	N868FT	Douglas DC-8-61CF	Flying Tiger Line
	N870TV	Douglas DC-8-63CF	Transamerica Airlines
	N871TV	Douglas DC-8-63CF	Transamerica Airlines
	N872TV	Douglas DC-8-63CF	Transamerica Airlines
	N901PA	Boeing 747-123F	Pan Am *Clipper Telegraph*
	N901WA	Douglas DC-10-10	Western Airlines
	N902PA	Boeing 747-132	Pan Am *Clipper Seaman's Bridge*
	N902WA	Douglas DC-10-10	Western Airlines
	N904PA	Boeing 747-221F	Pan Am *Clipper Industry*
	N904WA	Douglas DC-10-10	Capitol International Airways
	N905PA	Boeing 747-221F	Pan Am *Clipper Courier*
	N905WA	Douglas DC-10-10	Capitol International Airways
	N906WA	Douglas DC-10-10	Western Airlines
	N907CL	Douglas DC-8-63CF	Capitol International Airways
	N907WA	Douglas DC-10-10	Western Airlines
	N908WA	Douglas DC-10-10	Western Airlines
	N909WA	Douglas DC-10-10	Western Airlines
	N910CL	Douglas DC-8-63CF	Capitol International Airways
	N911CL	Douglas DC-8-61	Capitol International Airways
	N912R	Douglas DC-8-61	SAUDIA—Saudi Arabian Airlines
	N912WA	Douglas DC-10-10	Western Airlines
	N913WA	Douglas DC-10-10	Western Airlines
	N914WA	Douglas DC-10-10	Western Airlines
	N915WA	Douglas DC-10-10	Western Airlines
	N919CL	Douglas DC-8-63CF	Capitol International Airways
	N1035F	Douglas DC-10-30CF	Air Florida
	N1252E	Boeing 747-228B	Air France
	N1289E	Boeing 747-228B	Air France
	N1295E	Boeing 747-206B	KLM *Ganges*
	N1298E	Boeing 747-206B	KLM *Indus*
	N1304E	Boeing 747SP-J6	CAAC
	N1305E	Boeing 747-228B	Air France
	N1309E	Boeing 747-206B	KLM *Admiral Richard E. Byrd*
	N4225J	Boeing 707-338C	Arrow Air
	N4501Q	Boeing 747-283B	SAS *Dan Viking*
	N4502R	Boeing 747-283B	SAS *Huge Viking*
	N4508E	Boeing 747-228F	Air France
	N4864T	Douglas DC-8-63CF	SAUDIA—Saudi Arabian Airlines
	N4865T	Douglas DC-8-63CF	Transamerica Airlines
	N4866T	Douglas DC-8-63CF	Transamerica Airlines
	N4867T	Douglas DC-8-63CF	Transamerica Airlines
	N4868T	Douglas DC-8-63CF	Transamerica Airlines
	N4869T	Douglas DC-8-63CF	Transamerica Airlines

Reg.	*Type*	*Owner or Operator*	*Notes*
N6162A	Douglas DC-8-63CF	Capitol International Airways	
N6163A	Douglas DC-8-63CF	SAUDIA—Saudi Arabian Airlines	
N7036T	L.1011 TriStar 100	Trans World Airlines	
N7984S	L-100-20 Hercules	Southern Air Transport	
N8034T	L.1011 TriStar 100	Trans World Airlines	
N8417	Boeing 707-323C	Global International Airways	
N8705T	Boeing 707-331B	Trans World Airlines	
N8725T	Boeing 707-331B	Trans World Airlines	
N8733	Boeing 707-331B	Trans World Airlines	
N8735	Boeing 707-331B	Trans World Airlines	
N8736	Boeing 707-331B	Trans World Airlines	
N8737	Boeing 707-331B	Trans World Airlines	
N8762	Douglas DC-8-61	Capitol International Airways	
N8763	Douglas DC-8-61	Capitol International Airways	
N8764	Douglas DC-8-61	Capitol International Airways	
N8765	Douglas DC-8-61	Capitol International Airways	
N8766	Douglas DC-8-61	Capitol International Airways	
N9232R	L-100-30 Hercules	Southern Air Transport	
N9266R	L-100-20 Hercules	Southern Air Transport	
N15713	Boeing 707-331C	Global International Airways	
N17125	Boeing 747-136	Trans World Airlines	
N17126	Boeing 747-136	Trans World Airlines	
N18702	Boeing 707-331B	Trans World Airlines	
N18703	Boeing 707-331B	Trans World Airlines	
N18704	Boeing 707-331B	Trans World Airlines	
N18706	Boeing 707-331B	Trans World Airlines	
N18707	Boeing 707-331B	Trans World Airlines	
N18708	Boeing 707-331B	Trans World Airlines	
N18709	Boeing 707-331B	Trans World Airlines	
N18710	Boeing 707-331B	Trans World Airlines	
N18711	Boeing 707-331B	Trans World Airlines	
N18712	Boeing 707-331B	Trans World Airlines	
N18713	Boeing 707-331B	Trans World Airlines	
N18815	Boeing 747-228F	Air France	
N28366	Boeing 747-128	Air France	
N28724	Boeing 707-331B	Trans World Airlines	
N28726	Boeing 707-331B	Trans World Airlines	
N28727	Boeing 707-331B	Trans World Airlines	
N28899	Boeing 747-128	Air France	
N28903	Boeing 747-128	Air France	
N31029	L.1011-385 TriStar 100	Trans World Airlines	
N31030	L.1011-385 TriStar 100	Trans World Airlines	
N31031	L.1011-385 TriStar 100	Trans World Airlines	
N31032	L.1011-385 TriStar 100	Trans World Airlines	
N31033	L.1011-385 TriStar 100	Trans World Airlines	
N40116	Boeing 747-128	Air France	
N41020	L.1011-385 TriStar 100	Trans World Airlines	
N53110	Boeing 747-131	Trans World Airlines	
N53116	Boeing 747-131	Trans World Airlines	
N63305	Boeing 747-128	Air France	
N81025	L.1011-385 TriStar 100	Trans World Airlines	
N81026	L.1011-385 TriStar 100	Trans World Airlines	
N81027	L.1011-385 TriStar 100	Trans World Airlines	
N81028	L.1011-385 TriStar 100	Trans World Airlines	
N93104	Boeing 747-131	Trans World Airlines	
N93105	Boeing 747-131	Trans World Airlines	
N93106	Boeing 747-131	Trans World Airlines	
N93107	Boeing 747-131	Trans World Airlines	
N93108	Boeing 747-131	Trans World Airlines	
N93109	Boeing 747-131	Trans World Airlines	
N93115	Boeing 747-131	Trans World Airlines	
N93117	Boeing 747-131	Trans World Airlines	
N93119	Boeing 747-131	Trans World Airlines	

OD (Lebanon)

Notes	Reg.	Type	Owner or Operator
	OD–AFB	Boeing 707-3B4C	Middle East Airlines
	OD–AFD	Boeing 707-3B4C	Middle East Airlines
	OD–AFE	Boeing 707-3B4C	Middle East Airlines
	OD–AFL	Boeing 720-023B	Middle East Airlines
	OD–AFM	Boeing 720-023B	Middle East Airlines
	OD–AFN	Boeing 720-023B	Middle East Airlines
	OD–AFO	Boeing 720-023B	Middle East Airlines
	OD–AFP	Boeing 720-023B	Middle East Airlines
	OD–AFQ	Boeing 720-023B	Middle East Airlines
	OD–AFS	Boeing 720-023B	Middle East Airlines
	OD–AFU	Boeing 720-023B	Middle East Airlines
	OD–AFW	Boeing 720-023B	Middle East Airlines
	OD–AFY	Boeing 707-327C	Trans Mediterranean Airways
	OD–AFZ	Boeing 720-023B	Middle East Airlines
	OD–AGB	Boeing 720-023B	Middle East Airlines
	OD–AGD	Boeing 707-323C	Trans Mediterranean Airways
	OD–AGF	Boeing 720-047B	Middle East Airlines
	OD–AGG	Boeing 720-047B	Middle East Airlines
	OD–AGH	Boeing 747-2B4B	Middle East Airlines
	OD–AGI	Boeing 747-2B4B	Middle East Airlines
	OD–AGJ	Boeing 747-2B4B	Middle East Airlines
	OD–AGN	Boeing 707-323C	Trans Mediterranean Airways
	OD–AGO	Boeing 707-321C	Trans Mediterranean Airways
	OD–AGP	Boeing 707-321C	Trans Mediterranean Airways
	OD–AGQ	Boeing 720-047B	Middle East Airlines
	OD–AGR	Boeing 720-047B	Middle East Airlines
	OD–AGS	Boeing 707-331C	Trans Mediterranean Airways
	OD–AGT	Boeing 707-331C	Trans Mediterranean Airways
	OD–AGU	Boeing 707-347C	Middle East Airlines
	OD–AGV	Boeing 707-347C	Middle East Airlines
	OD–AGW	Boeing 707-327C	Trans Mediterranean Airways
	OD–AGX	Boeing 707-327C	Trans Mediterranean Airways
	OD–AGY	Boeing 707-327C	Trans Mediterranean Airways
	OD–AGZ	Boeing 707-327C	Trans Mediterranean Airways

OE (Austria)

AUSTRIAN AIRLINES

	OE–HLA	F.27 Friendship Mk.100	Lauda Air
	OE–ILB	F.27 Friendship Mk.600	Lauda Air
	OE–LDD	Douglas DC-9-32	Austrian Airlines *Steiermark*
	OE–LDE	Douglas DC-9-32	Austrian Airlines *Oberösterreich*
	OE–LDF	Douglas DC-9-32	Austrian Airlines *Salzburg*
	OE–LDG	Douglas DC-9-32	Austrian Airlines *Tirol*
	OE–LDH	Douglas DC-9-32	Austrian Airlines *Vorarlberg*
	OE–LDI	Douglas DC-9-32	Austrian Airlines *Bregenz*
	OE–LDK	Douglas DC-9-51	Austrian Airlines *Graz*
	OE–LDL	Douglas DC-9-51	Austrian Airlines *Linz*
	OE–LDM	Douglas DC-9-51	Austrian Airlines *Klagenfurt*
	OE–LDN	Douglas DC-9-51	Austrian Airlines *Innsbruck*
	OE–LDO	Douglas DC-9-51	Austrian Airlines *Eisenstädt*
	OE–LDP	Douglas DC-9-81	Austrian Airlines *Wien*
	OE–LDR	Douglas DC-9-81	Austrian Airlines *Niederösterreich*
	OE–LDS	Douglas DC-9-81	Austrian Airlines *Burgenland*
	OE–LDT	Douglas DC-9-81	Austrian Airlines *Kärnten*
	OE–LDU	Douglas DC-9-81	Austrian Airlines *Steirmark*
	OE–LDV	Douglas DC-9-81	Austrian Airlines *Oberösterreich*
	OE–LDW	Douglas DC-9-81	Austrian Airlines *Salzburg*
	OE–LDX	Douglas DC-9-81	Austrian Airlines *Tirol*
	OE–LDY	Douglas DC-9-81	Austrian Airlines *Vorarlberg*

NOTE: Austrian Airlines is in the process of replacing its DC-9-32s with Series 81s.

OH (Finland)

Reg.	*Type*	*Owner or Operator*	*Notes*
OH–KDM	Douglas DC-8-51	Kar Air	
OH–LFT	Douglas DC-8-62CF	Finnair *Paavo Nurmi*	
OH–LFZ	Douglas DC-8-62	Finnair *Jean Sibelius*	
OH–LHA	Douglas DC-10-30	Finnair *Iso Antti*	
OH–LHB	Douglas DC-10-30	Finnair	
OH–LHC	Douglas DC-10-30	Finnair	
OH–LNA	Douglas DC-9-41	Finnair	
OH–LNB	Douglas DC-9-41	Finnair	
OH–LNC	Douglas DC-9-41	Finnair	
OH–LSB	S.E.210 Caravelle 10-B	Finnair *Tampere*	
OH–LSD	S.E.210 Caravelle 10-B	Finnair *Oulu*	
OH–LSF	S.E.210 Caravelle 10-B	Finnair *Pori*	
OH–LSG	S.E.210 Caravelle 10-B	Finnair *Jyväskylä*	
OH–LSH	S.E.210 Caravelle 10-B	Finnair *Kuopio*	
OH–LYA	Douglas DC-9-14	Finnair	
OH–LYB	Douglas DC-9-14	Finnair	
OH–LYC	Douglas DC-9-14	Finnair	
OH–LYD	Douglas DC-9-14	Finnair	
OH–LYE	Douglas DC-9-14	Finnair	
OH–LYG	Douglas DC-9-14	Finnair	
OH–LYH	Douglas DC-9-15MC	Finnair	
OH–LYI	Douglas DC-9-15MC	Finnair	
OH–LYK	Douglas DC-9-15	Finnair	
OH–LYN	Douglas DC-9-51	Finnair	
OH–LYO	Douglas DC-9-51	Finnair	
OH–LYP	Douglas DC-9-51	Finnair	
OH–LYR	Douglas DC-9-51	Finnair	
OH–LYS	Douglas DC-9-51	Finnair	
OH–LYT	Douglas DC-9-51	Finnair	
OH–LYU	Douglas DC-9-51	Finnair	
OH–LYV	Douglas DC-9-51	Finnair	
OH–LYW	Douglas DC-9-51	Finnair	
OH–LYX	Douglas DC-9-51	Finnair	
OH–LYY	Douglas DC-9-51	Finnair	
OH–LYZ	Douglas DC-9-51	Finnair	

OK (Czechoslovakia)

OK–ABD	Ilyushin IL-62	Ceskoslovenske Aerolinie *Kosice*	
OK–AFA	Tupolev Tu-134A	Ceskoslovenske Aerolinie	
OK–AFB	Tupolev Tu-134A	Ceskoslovenske Aerolinie	
OK–CFC	Tupolev Tu-134A	Ceskoslovenske Aerolinie	
OK–CFE	Tupolev Tu-134A	Ceskoslovenske Aerolinie	
OK–CFF	Tupolev Tu-134A	Ceskoslovenske Aerolinie	
OK–CFG	Tupolev Tu-134A	Ceskoslovenske Aerolinie	
OK–CFH	Tupolev Tu-134A	Ceskoslovenske Aerolinie	
OK–DBE	Ilyushin IL-62	Ceskoslovenske Aerolinie *Brno*	
OK–DFI	Tupolev Tu-134A	Ceskoslovenske Aerolinie	

Notes	Reg.	Type	Owner or Operator
	OK–EBG	Ilyushin IL-62	Ceskoslovenske Aerolinie *Banska Bystrica*
	OK–EFJ	Tupolev Tu-134A	Ceskoslovenske Aerolinie
	OK–EFK	Tupolev Tu-134A	Ceskoslovenske Aerolinie
	OK–FBF	Ilyushin IL-62	Ceskoslovenske Aerolinie
	OK–GBH	Ilyushin IL-62	Ceskoslovenske Aerolinie *Usti Nad Labem*
	OK–HFL	Tupolev Tu-134A	Ceskoslovenske Aerolinie
	OK–HFM	Tupolev Tu-134A	Ceskoslovenske Aerolinie
	OK–IFN	Tupolev Tu-134A	Ceskoslovenske Aerolinie
	OK–JBI	Ilyushin IL-62M	Ceskoslovenske Aerolinie *Pizen*
	OK–JBJ	Ilyushin IL-62M	Ceskoslovenske Aerolinie *Hradec Kralové*
	OK–KBK	Ilyushin IL-62M	Ceskoslovenske Aerolinie *Ceske Budejovice*
	OK–YBA	Ilyushin IL-62	Ceskoslovenske Aerolinie *Praha*
	OK–YBB	Ilyushin IL-62	Ceskoslovenske Aerolinie *Bratislava*
	OK–ZBC	Ilyushin IL-62	Ceskoslovenske Aerolinie *Ostrava*

OO (Belgium)

Notes	Reg.	Type	Owner or Operator
	OO–ABB	Boeing 737-2P6	Air Belguim
	OO–ATJ	Boeing 727-30C	Transjet
	OO–DTA	FH-227B Friendship	Delta Air Transport
	OO–DTB	FH-227B Friendship	Delta Air Transport
	OO–DTC	FH-227B Friendship	Delta Air Transport
	OO–DTD	FH-227B Friendship	Delta Air Transport
	OO–DTE	FH-227B Friendship	Delta Air Transport
	OO–JPI	Swearingen SA226TC Metro II	Sabena
	OO–JPK	Swearingen SA226TC Metro II	Sabena
	OO–PLH	Boeing 737-247	Air Belguim
	OO–SBQ	Boeing 737-229	Sobelair
	OO–SBS	Boeing 737-229	Sobelair
	OO–SBT	Boeing 737-229	Sobelair
	OO–SBU	Boeing 707-373C	Sobelair
	OO–SDA	Boeing 737-229	Sabena
	OO–SDB	Boeing 737-229	Sabena
	OO–SDC	Boeing 737-229	Sabena
	OO–SDD	Boeing 737-229	Sabena
	OO–SDE	Boeing 737-229	Sabena
	OO–SDF	Boeing 737-229	Sabena
	OO–SDG	Boeing 737-229	Sabena
	OO–SDJ	Boeing 737-229C	Sabena
	OO–SDK	Boeing 737-229C	Sabena
	OO–SDL	Boeing 737-229	Sabena
	OO–SDM	Boeing 737-229	Sabena
	OO–SDN	Boeing 737-229	Sabena
	OO–SDO	Boeing 737-229	Sabena
	OO–SDP	Boeing 737-229C	Sabena
	OO–SDR	Boeing 737-229C	Sabena
	OO–SGA	Boeing 747-129	Sabena
	OO–SGB	Boeing 747-129	Sabena
	OO–SJJ	Boeing 707-329C	Sabena
	OO–SJL	Boeing 707-329C	Sabena
	OO–SJM	Boeing 707-329C	Sobelair
	OO–SJN	Boeing 707-329C	Sabena
	OO–SJO	Boeing 707-329C	Sabena
	OO–SLA	Douglas DC-10-30CF	Sabena
	OO–SLB	Douglas DC-10-30CF	Sabena
	OO–SLC	Douglas DC-10-30CF	Sabena

OO–SBT (Top) Boeing 737-229 of Sobelair

YV–138C (Centre) Douglas DC-10-30 of Viasa

5B–DAO (Bottom) Boeing 707-123B of Cyprus Airways

Notes	Reg.	Type	Owner or Operator
	OO-SLD	Douglas DC-10-30CF	Sabena
	OO-SLE	Douglas DC-10-30CF	Sabena
	OO-TEC	Boeing 707-131	Trans European Airways *Oostende*
	OO-TED	Boeing 707-131	Trans European Airways *Rena*
	OO-TEF	A.300B1 Airbus	Trans European Airways
	OO-TEH	Boeing 737-2M8	Trans European Airways *Marcus Johannes*
	OO-TEK	Boeing 737-2Q9	Trans European Airways *Liège*
	OO-TEL	Boeing 737-2M8	Trans European Airways *Antwerpen*
	OO-TEM	Boeing 737-2Q8	Trans European Airways
	OO-TEN	Boeing 737-2M8	Trans European Airways
	OO-TEO	Boeing 737-2M8	Trans European Airways *Jonathan*
	OO-TE	Boeing 737-2M8	Trans European Airways
	OO-WAY	Beech 99	Sabena
	OO-WAZ	Beech 99	Sabena

OY (Denmark)

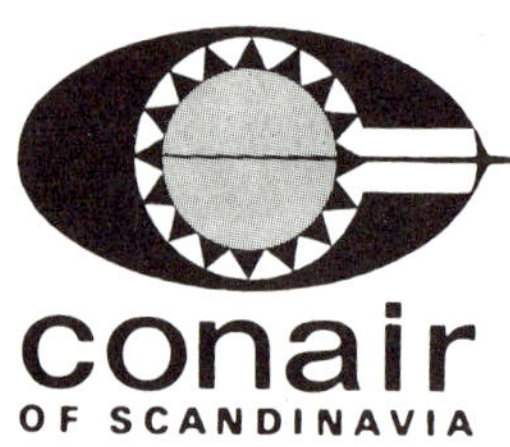

	OY-APJ	Boeing 737-2L9	Maersk Air
	OY-APK	Boeing 737-2L9	Maersk Air
	OY-APL	Boeing 737-2L9	Maersk Air
	OY-APO	Boeing 737-2L9	Maersk Air
	OY-APP	Boeing 737-2L9	Maersk Air
	OY-APR	Boeing 737-2L9	Maersk Air
	OY-APS	Boeing 737-2L9	Maersk Air
	OY-APU	Boeing 720-051B	Conair
	OY-APV	Boeing 720-051B	Conair
	OY-APW	Boeing 720-051B	Conair
	OY-APY	Boeing 720-051B	Conair
	OY-APZ	Boeing 720-051B	Conair
	OY-DSK	Boeing 720-025	Conair
	OY-DSP	Boeing 720-025	Conair
	OY-KAA	A.300B2 Airbus	SAS *Stig Viking*
	OY-KDA	Douglas DC-10-30	SAS *Gorm Viking*
	OY-KGA	Douglas DC-9-41	SAS *Heming Viking*
	OY-KGB	Douglas DC-9-41	SAS *Toste Viking*
	OY-KGC	Douglas DC-9-41	SAS *Helge Viking*
	OY-KGD	Douglas DC-9-21	SAS *Ubbe Viking*
	OY-KGE	Douglas DC-9-21	SAS *Orvar Viking*
	OY-KGF	Douglas DC-9-21	SAS *Rolf Viking*
	OY-KGG	Douglas DC-9-41	SAS *Sune Viking*
	OY-KGH	Douglas DC-9-41	SAS *Eiliv Viking*
	OY-KGI	Douglas DC-9-41	SAS *Bent Viking*
	OY-KGK	Douglas DC-9-41	SAS *Ebbe Viking*
	OY-KGL	Douglas DC-9-41	SAS *Angantyr Viking*
	OY-KGM	Douglas DC-9-41	SAS *Arnfinn Viking*
	OY-KGN	Douglas DC-9-41	SAS *Gram Viking*
	OY-KGO	Douglas DC-9-41	SAS *Holte Viking*
	OY-KGP	Douglas DC-9-41	SAS *Torbern Viking*
	OY-KGR	Douglas DC-9-41	SAS *Holger Viking*
	OY-KGS	Douglas DC-9-41	SAS *Hall Viking*
	OY-KHA	Boeing 747-283B	SAS *Ivar Viking*
	OY-KTE	Douglas DC-8-62CF	Scanair *Turid Viking*
	OY-KTF	Douglas DC-8-63	SAS *Frode Viking*
	OY-KTG	Douglas DC-8-63PF	Scanair
	OY-MBV	Boeing 737-2L9	Maersk Air
	OY-MBW	Boeing 737-2L9	Maersk Air
	OY-MBZ	Boeing 737-2L9	Maersk Air
	OY-SAA	S.E.210 Caravelle 12	Sterling Airways

Reg.	Type	Owner or Operator	Notes
OY-SAD	S.E.210 Caravelle 12	Sterling Airways	
OY-SAE	S.E.210 Caravelle 12	Sterling Airways	
OY-SAG	S.E.210 Caravelle 12	Sterling Airways	
OY-SAS	Boeing 727-2J4	Sterling Airways	
OY-SAT	Boeing 727-2J4	Sterling Airways	
OY-SAU	Boeing 727-2J4	Sterling Airways	
OY-SBE	Boeing 727-2J4	Sterling Airways	
OY-SBF	Boeing 727-2J4	Sterling Airways	
OY-SBG	Boeing 727-2J4	Sterling Airways	
OY-STF	S.E.210 Caravelle 10B	Sterling Airways	
OY-STH	S.E.210 Caravelle 10B	Sterling Airways	
OY-STI	S.E.210 Caravelle 10B	Sterling Airways	
OY-STM	S.E.210 Caravelle 10B	Sterling Airways	

NOTE: SAS also operates two Boeing 747-283Bs which retain their U.S. registrations N4501Q and N4502R.

PH (Netherlands)

Reg.	Type	Owner or Operator	Notes
PH-BUA	Boeing 747-206B	K.L.M. *The Mississippi*	
PH-BUB	Boeing 747-206B	K.L.M. *The Danube*	
PH-BUC	Boeing 747-206B	K.L.M. *The Amazon*	
PH-BUD	Boeing 747-206B	K.L.M. *The Nile*	
PH-BUE	Boeing 747-206B	K.L.M. *Rio de la Plata*	
PH-BUG	Boeing 747-206B	K.L.M. *The Orinoco*	
PH-BUH	Boeing 747-206B	K.L.M. *Dr Albert Plesman*	
PH-BUI	Boeing 747-206B	K.L.M. *Wilbur Wright*	
PH-BUK	Boeing 747-206B	K.L.M. *Louis Blériot*	
PH-BUL	Boeing 747-206B	K.L.M. *Charles Lindbergh*	
PH-BUM	Boeing 747-206B	K.L.M. *Sir Charles E. Kingsford-Smith*	
PH-BUN	Boeing 747-206B	K.L.M. *Anthony H. G. Fokker*	
PH-BUO	Boeing 747-206B	K.L.M. *Missouri*	
PH-CHB	F.28 Fellowship 4000	N.L.M. *Birmingham*	
PH-CHD	F.28 Fellowship 4000	N.L.M. *Maastricht*	
PH-CHF	F.28 Fellowship 4000	N.L.M. *Island of Guernsey*	
PH-DEA	Douglas DC-8-63	K.L.M. *Amerigo Vespucci*	
PH-DEB	Douglas DC-8-63	K.L.M. *Christopher Columbus*	
PH-DEC	Douglas DC-8-63	K.L.M. *Marco Polo*	
PH-DED	Douglas DC-8-63	K.L.M. *Leifur Eiriksson*	
PH-DEE	Douglas DC-8-63	K.L.M. *Abel Tasman*	
PH-DEF	Douglas DC-8-63	K.L.M. *Henry Hudson*	
PH-DEG	Douglas DC-8-63	K.L.M. *Jan van Reibeeck*	
PH-DEH	Douglas DC-8-63	K.L.M. *Vasco da Gama*	
PH-DEK	Douglas DC-8-63	K.L.M. *David Livingstone*	
PH-DEL	Douglas DC-8-63	K.L.M. *Fernao de Magalhaes*	
PH-DEM	Douglas DC-8-63	K.L.M. *James Cook* (on lease to Surinam Airways)	
PH-DNB	Douglas DC-9-15	K.L.M. *City of Brussels*	
PH-DNC	Douglas DC-9-15	K.L.M. *City of Luxembourg*	
PH-DNG	Douglas DC-9-32	K.L.M. *City of Rotterdam*	
PH-DNH	Douglas DC-9-32	K.L.M. *City of Zurich*	
PH-DNI	Douglas DC-9-32	K.L.M. *City of Istanbul*	
PH-DNK	Douglas DC-9-32	K.L.M. *City of Copenhagen*	
PH-DNL	Douglas DC-9-32	K.L.M. *City of London*	
PH-DNM	Douglas DC-9-33RC	K.L.M. *City of Madrid*	
PH-DNN	Douglas DC-9-33RC	K.L.M. *City of Vienna*	
PH-DNO	Douglas DC-9-33RC	K.L.M. *City of Oslo*	
PH-DNP	Douglas DC-9-33RC	K.L.M. *City of Athens*	
PH-DNR	Douglas DC-9-33RC	K.L.M. *City of Stockholm*	
PH-DNS	Douglas DC-9-32	K.L.M. *City of Arnhem*	

Notes	Reg.	Type	Owner or Operator
	PH–DNT	Douglas DC-9-32	K.L.M. *City of Lisbon*
	PH–DNV	Douglas DC-9-32	K.L.M. *City of Warsaw*
	PH–DNW	Douglas DC-9-32	K.L.M. *City of Moscow*
	PH–DNY	Douglas DC-9-33RC	K.L.M. *City of Paris*
	PH–DNZ	Douglas DC-9-33RC	K.L.M. *City of Rome*
	PH–DOA	Douglas DC-9-32	K.L.M. *City of Utrecht*
	PH–DOB	Douglas DC-9-32	K.L.M. *City of Santa Monica*
	PH–DTA	Douglas DC-10-30	K.L.M. *Johann Sebastian Bach*
	PH–DTB	Douglas DC-10-30	K.L.M. *Ludwig van Beethoven*
	PH–DTC	Douglas DC-10-30	K.L.M. *Frédéric François Chopin*
	PH–DTD	Douglas DC-10-30	K.L.M. *Maurice Ravel*
	PH–DTE	Douglas DC-10-30	K.L.M. *Wolfgang Amadeus Mozart*
	PH–DTI	Douglas DC-10-30	K.L.M. (leased to PAL)
	PH–DTK	Douglas DC-10-30	K.L.M. (leased to PAL)
	PH–DTL	Douglas DC-10-30	K.L.M. *Edvard Hagerup Grieg*
	PH–KFD	F.27 Friendship Mk. 200	N.L.M. *Jan Moll*
	PH–KFE	F.27 Friendship Mk. 600	N.L.M. *Jan Dellaert*
	PH–KFG	F.27 Friendship Mk. 200	N.L.M. *Koos Abspoel*
	PH–KFH	F.27 Friendship Mk. 200	British Midland Airways Ltd.
	PH–KFI	F.27 Friendship Mk. 500	N.L.M. *Bremen*
	PH–KFK	F.27 Friendship Mk. 500	N.L.M.
	PH–KFL	F.27 Friendship Mk. 500	N.L.M.
	PH–MAO	Douglas DC-9-33RC	Martinair *Desiderius Erasmus*
	PH–MAR	Douglas DC-9-33RC	Martinair *Jean Monnet*
	PH–MAX	Douglas DC-9-32	Martinair *Europe*
	PH–MBG	Douglas DC-10-30CF	Martinair *Kohoutek*
	PH–MBN	Douglas DC-10-30CF	Martinair *Anthony Ruys*
	PH–MBP	Douglas DC-10-30CF	Martinair *Hong Kong*
	PH–MBT	Douglas DC-10-30CF	Martinair
	PH–SAD	F.27 Friendship Mk. 200	N.L.M. *Evert van Dijk*
	PH–TVA	Boeing 707-123B	Transavia *Provincie Zeeland*
	PH–TVC	Boeing 737-2K2C	Transavia/Air Malta
	PH–TVE	Boeing 737-2K2C	Transavia/Air Malta
	PH–TVH	Boeing 737-222	Transavia *Neil Armstrong*
	PH–TVI	Boeing 737-222	Transavia *Michael Collins*
	PH–TVP	Boeing 737-2K2	Transavia/Air Malta
	PH–TVR	Boeing 737-2K2	Transavia
	PH–TVS	Boeing 737-2K2	Transavia
	PH–	Boeing 737-2K2	Transavia

NOTE: KLM also operate Boeing 747-206Bs N1295E, N1298E and N1309E.

PK (Indonesia)

Notes	Reg.	Type	Owner or Operator
	PK–GSA	Boeing 747-2U3B	Garuda Indonesian Airways *City of Jakarta*
	PK–GSB	Boeing 747-2U3B	Garuda Indonesian Airways *City of Bandung*
	PK–GSC	Boeing 747-2U3B	Garuda Indonesian Airways *City of Medan*
	PK–GSD	Boeing 747-2U3B	Garuda Indonesian Airways *City of Surabaya*

PP (Brazil)

Notes	Reg.	Type	Owner or Operator
	PP–VJH	Boeing 707-320C	VARIG
	PP–VJK	Boeing 707-379C	VARIG
	PP–VJS	Boeing 707-341C	VARIG
	PP–VJX	Boeing 707-345C	VARIG
	PP–VJY	Boeing 707-345C	VARIG
	PP–VLI	Boeing 707-385C	VARIG

Reg.	Type	Owner or Operator	Notes
PP–VLK	Boeing 707-324C	VARIG	
PP–VLL	Boeing 707-324C	VARIG	
PP–VLM	Boeing 707-324C	VARIG	
PP–VLN	Boeing 707-324C	VARIG	
PP–VLO	Boeing 707-324C	VARIG	
PP–VLP	Boeing 707-323C	VARIG	
PP–VMA	Douglas DC-10-30	VARIG	
PP–VMB	Douglas DC-10-30	VARIG	
PP–VMD	Douglas DC-10-30	VARIG	
PP–VMQ	Douglas DC-10-30	VARIG	
PP–VMS	Douglas DC-10-30	VARIG	
PP–VMT	Douglas DC-10-30	VARIG	
PP–VMU	Douglas DC-10-30	VARIG	
PP–VMV	Douglas DC-10-30	VARIG	
PP–VMW	Douglas DC-10-30	VARIG	
PP–VMX	Douglas DC-10-30	VARIG	
PP–VMY	Douglas DC-10-30	VARIG	
PP–VMZ	Douglas DC-10-30	VARIG	
PP–VNA	Boeing 747-2L5B	VARIG	
PP–VNB	Boeing 747-2L5B	VARIG	
PP–VNC	Boeing 747-2L5B	VARIG	

RP (Philippines)

NOTE: Philippine Airlines operate four Boeing 747s which retain their US registrations N741PR, N742PR, N743PR and N744PR.

S2 (Bangladesh)

বাংলাদেশ বিমান Bangladesh Biman

Reg.	Type	Owner or Operator	Notes
S2–ABN	Boeing 707-351C	Bangladesh Biman *City of Shah Jalal*	
S2–ACA	Boeing 707-351C	Bangladesh Biman *Khan Jahan Ali*	
S2–ACE	Boeing 707-351C	Bangladesh Biman *City of Tokyo*	
S2–ACF	Boeing 707-351C	Bangladesh Biman	
S2–	Boeing 707–369C	Bangladesh Biman	
S2–	Boeing 747–	Bangladesh Biman	

SE (Sweden)

Reg.	Type	Owner or Operator	Notes
SE–DAK	Douglas DC-9-41	SAS *Ragnvald Viking*	
SE–DAL	Douglas DC-9-41	SAS *Algot Viking*	
SE–DAM	Douglas DC-9-41	SAS *Starkad Viking*	
SE–DAN	Douglas DC-9-41	SAS *Alf Viking*	
SE–DAO	Douglas DC-9-41	SAS *Asgaut Viking*	
SE–DAP	Douglas DC-9-41	SAS *Torgils Viking*	
SE–DAR	Douglas DC-9-41	SAS *Agnar Viking*	
SE–DAS	Douglas DC-9-41	SAS *Garder Viking*	
SE–DAT	Douglas DC-9-41	SAS *Gissur Viking*	
SE–DAU	Douglas DC-9-41	SAS *Hadding Viking*	
SE–DAW	Douglas DC-9-41	SAS *Gotrik Viking*	
SE–DAX	Douglas DC-9-41	SAS *Helsing Viking*	
SE–DBG	Douglas DC-8-62	Scanair *Jorund Viking*	
SE–DBI	Douglas DC-8-62CF	Scanair *Ylva Viking*	
SE–DBK	Douglas DC-8-63	Scanair *Sigyn Viking*	
SE–DBL	Douglas DC-8-63	SAS *Tord Viking*	
SE–DBM	Douglas DC-9-41	SAS *Ossur Viking*	
SE–DBN	Douglas DC-9-33AF	SAS *Sigtrygge Viking*	
SE–DBO	Douglas DC-9-21	SAS *Siger Viking*	
SE–DBP	Douglas DC-9-21	SAS *Rane Viking*	
SE–DBR	Douglas DC-9-21	SAS *Skate Viking*	
SE–DBS	Douglas DC-9-21	SAS *Svipdag Viking*	
SE–DBT	Douglas DC-9-41	SAS *Agne Viking*	

Notes	Reg.	Type	Owner or Operator
	SE–DBU	Douglas DC-9-41	SAS *Hjalmar Viking*
	SE–DBW	Douglas DC-9-41	SAS *Adils Viking*
	SE–DBX	Douglas DC-9-41	SAS *Arnljot Viking*
	SE–DDL	Boeing 747-283B	SAS *Huge Viking*
	SE–DDP	Douglas DC-9-41	SAS *Brun Viking*
	SE–DDR	Douglas DC-9-41	SAS *Atle Viking*
	SE–DDS	Douglas DC-9-41	SAS *Alrik Viking*
	SE–DDT	Douglas DC-9-41	SAS *Amund Viking*
	SE–DDU	Douglas DC-8-62	SAS *Knud Viking*
	SE–DFD	Douglas DC-10-30	SAS *Dag Viking*
	SE–DFE	Douglas DC-10-30	SAS *Sverker Viking*
	SE–DFK	A.300B2 Airbus	SAS *Sven Viking*
	SE–DFL	A.300B2 Airbus	SAS *Ingemar Viking*
	SE–DFZ	Boeing 747-283B	SAS *Knut Viking*
	SE–DGA	F.28 Fellowship 1000	Linjeflyg
	SE–DGB	F.28 Fellowship 1000	Linjeflyg
	SE–DGC	F.28 Fellowship 1000	Linjeflyg
	SE–DGD	F.28 Fellowship 4000	Linjeflyg
	SE–DGE	F.28 Fellowship 4000	Linjeflyg
	SE–DGF	F.28 Fellowship 4000	Linjeflyg
	SE–DGG	F.28 Fellowship 4000	Linjeflyg
	SE–DGH	F.28 Fellowship 4000	Linjeflyg
	SE–DGI	F.28 Fellowship 4000	Linjeflyg
	SE–DGK	F.28 Fellowship 4000	Linjeflyg
	SE–DGL	F.28 Fellowship 4000	Linjeflyg
	SE–DGM	F.28 Fellowship 4000	Linjeflyg
	SE–DGN	F.28 Fellowship 4000	Linjeflyg
	SE–	F.28 Fellowship 4000	Linjeflyg
	SE–	F.28 Fellowship 4000	Linjeflyg
	SE–IEG	F.27 Friendship	Aerocenter Trafikflyg

NOTE: SAS also operates two Boeing 747-283Bs, which retain their U.S. registrations N4501Q and N4502R.

SP (Poland)

Notes	Reg.	Type	Owner or Operator
	SP–LAB	Ilyushin IL-62	Polskie Linie Lotnicze (LOT) *Tadeusz Kosciuszko*
	SP–LAC	Ilyushin IL-62	Polskie Linie Lotnicze (LOT) *Fryderyk Chopin*
	SP–LAD	Ilyushin IL-62	Polskie Linie Lotnicze (LOT) *Kazimierz Pulaski*
	SP–LAE	Ilyushin IL-62	Polskie Linie Lotnicze (LOT) *Henryk Sienkiewicz*
	SP–LAF	Ilyushin IL-62	Polskie Linie Lotnicze (LOT) *Adam Michiewicz*
	SP–LAG	Ilyushin IL-62	Polskie Linie Lotnicze (LOT) *Maria Sklodowska-Curie*
	SP–LBA	Ilyushin IL-62M	Polskie Linie Lotnicze (LOT) *Juliusz Sowacki*
	SP–LBB	Ilyushin IL-62M	Polskie Linie Lotnicze (LOT) *Jgnacy Paderewski*
	SP–LBC	Ilyushin IL-62M	Polskie Linie Lotnicze (LOT) *Joseph Conrad-Korzeniowski*
	SP–LBD	Ilyushin IL-62M	Polskie Linie Lotnicze (LOT)
	SP–LBE	Ilyushin IL-62MK	Polskie Linie Lotnicze (LOT)
	SP–LBG	Ilyushin IL-62M	Polskie Linie Lotnicze (LOT)
	SP–LGA	Tupolev Tu-134	Polskie Linie Lotnicze (LOT)
	SP–LGC	Tupolev Tu-134	Polskie Linie Lotnicze (LOT)
	SP–LGD	Tupolev Tu-134	Polskie Linie Lotnicze (LOT)
	SP–LGE	Tupolev Tu-134	Polskie Linie Lotnicze (LOT)
	SP–LHA	Tupolev Tu-134A	Polskie Linie Lotnicze (LOT)
	SP–LHB	Tupolev Tu-134A	Polskie Linie Lotnicze (LOT)
	SP–LHC	Tupolev Tu-134A	Polskie Linie Lotnicze (LOT)
	SP–LHD	Tupolev Tu-134A	Polskie Linie Lotnicze (LOT)
	SP–LHE	Tupolev Tu-134A	Polskie Linie Lotnicze (LOT)
	SP–LHF	Tupolev Tu-134A	Polskie Linie Lotnicze (LOT)
	SP–LHG	Tupolev Tu-134A	Polskie Linie Lotnicze (LOT)

Reg.	*Type*	*Owner or Operator*	*Notes*
SP–LSA	Ilyushin IL-18V (Cargo)	Polskie Linie Lotnicze (LOT)	
SP–LSB	Ilyushin IL-18V	Polskie Linie Lotnicze (LOT)	
SP–LSC	Ilyushin IL-18V (Cargo)	Polskie Linie Lotnicze (LOT)	
SP–LSD	Ilyushin IL-18V	Polskie Linie Lotnicze (LOT)	
SP–LSE	Ilyushin IL-18V	Polskie Linie Lotnicze (LOT)	
SP–LSF	Ilyushin IL-18E	Polskie Linie Lotnicze (LOT)	
SP–LSG	Ilyushin IL-18E (Cargo)	Polskie Linie Lotnicze (LOT)	
SP–LSH	Ilyushin IL-18V	Polskie Linie Lotnicze (LOT)	
SP–LSI	Ilyushin IL-18D	Polskie Linie Lotnicze (LOT)	

ST (Sudan)

Reg.	Type	Owner or Operator	Notes
ST–AFA	Boeing 707-3J8C	Sudan Airways	
ST–AFB	Boeing 707-3J8C	Sudan Airways	

SU (Egypt)

Reg.	Type	Owner or Operator	Notes
SU–AOU	Boeing 707-366C	EgyptAir *Khopho*	
SU–APD	Boeing 707-366C	EgyptAir *Khafrah*	
SU–APE	Boeing 707-366C	EgyptAir *Mankara*	
SU–AVX	Boeing 707-366C	EgyptAir *Tutankhamun*	
SU–AVY	Boeing 707-366C	EgyptAir *Akhenaton*	
SU–AVZ	Boeing 707-366C	EgyptAir *Mena*	
SU–AXK	Boeing 707-366C	EgyptAir *Seti I*	
SU–BBA	Boeing 707-338C	Air Cargo Egypt	

SX (Greece)

Reg.	Type	Owner or Operator	Notes
SX–BEB	A.300B4 Airbus	Olympic Airways *Odysseus*	
SX–BEC	A.300B4 Airbus	Olympic Airways *Achilles*	
SX–BED	A.300B4 Airbus	Olympic Airways *Telemachos*	
SX–BEE	A.300B4 Airbus	Olympic Airways *Nestor*	
SX–BEF	A.300B4 Airbus	Olympic Airways *Ajax*	
SX–BEG	A.300B4 Airbus	Olympic Airways *Diomides*	
SX–DBC	Boeing 707-384C	Olympic Airways *City of Knossos*	
SX–DBD	Boeing 707-384C	Olympic Airways *City of Sparta*	
SX–DBE	Boeing 707-384B	Olympic Airways *City of Pella*	
SX–DBF	Boeing 707-384B	Olympic Airways *City of Mycenae*	
SX–DBO	Boeing 707-351C	Olympic Airways *City of Lindos*	
SX–DBP	Boeing 707-351C	Olympic Airways *City of Thebes*	
SX–OAA	Boeing 747-284B	Olympic Airways *Olympic Zeus*	
SX–OAB	Boeing 747-284B	Olympic Airways *Olympic Eagle*	

TC (Turkey)

Notes	Reg.	Type	Owner or Operator
	TC–JAU	Douglas DC-10-10	Türk Hava Yollari (THY) *Istanbul*
	TC–JAY	Douglas DC-10-10	Türk Hava Yollari (THY) *Izmir*
	TC–JBF	Boeing 727-2F2	Türk Hava Yollari (THY) *Adana*
	TC–JBG	Boeing 727-2F2	Türk Hava Yollari (THY) *Ankara*
	TC–JBJ	Boeing 727-2F2	Türk Hava Yollari (THY) *Diyarbakir*
	TC–JBM	Boeing 727-2F2	Türk Hava Yollari (THY) *Menderes*
	TC–JBR	Boeing 727-2F2	Türk Hava Yollari (THY) *Afyon*
	TC–	Boeing 727-2F2	Turk Hava Yollari (THY)
	TC–	Boeing 727-2F2	Turk Hava Yollari (THY)
	TC–	Boeing 727-2F2	Turk Hava Yollari (THY)
	TC–JBS	Boeing 707-321B	Türk Hava Yollari (THY) *Basak*
	TC–JBT	Boeing 707-321B	Türk Hava Yollari (THY) *Baris*
	TC–JBU	Boeing 707-321B	Türk Hava Yollari (THY) *Yurdum*
	TC–JCC	Boeing 707-321C	Türk Hava Yollari (THY) *Kervanl*

ICELANDAIR

TF (Iceland)

Notes	Reg.	Type	Owner or Operator
	TF–BCV	Douglas DC-8-63CF	Cargolux
	TF–CCV	Douglas DC-8-63CF	Cargolux *City of Echternach*
	TF–FLB	Douglas DC-8-63CF	Icelandair
	TF–FLC	Douglas DC-8-63CF	Icelandair
	TF–FLE	Douglas DC-8-63CF	SAUDIA—Saudi Arabian Airlines
	TF–FLG	Boeing 727-185C	Icelandair
	TF–FLH	Boeing 727-108C	Icelandair
	TF–FLI	Boeing 727-208	Icelandair
	TF–IUB	Douglas DC-6A	Iscargo
	TF–ISC	L-188 Electra	Iscargo
	TF–VLB	Boeing 720-047B	Eagle Air
	TF–VLJ	Boeing 707-324C	Libyan Arab Airlines
	TF–VLK	Boeing 737-2Q8	Eagle Air
	TF–VLL	Boeing 707-321C	British Midland/Eagle Air

TJ (Cameroon)

Notes	Reg.	Type	Owner or Operator
	TJ–CAA	Boeing 707-3H7C	Cameroon Airlines *La Sanaga*

TS (Tunisia)

Notes	Reg.	Type	Owner or Operator
	TS–IOC	Boeing 737-2H3	Tunis-Air *Salammbo*
	TS–IOD	Boeing 737-2H3C	Tunis-Air *Bulla Regia*
	TS–IOE	Boeing 737-2H3	Tunis-Air
	TS–IOF	Boeing 737-2H3	Tunis-Air

Reg.	Type	Owner or Operator	Notes
TS–JHN	Boeing 727-2H3	Tunis-Air *Carthago*	
TS–JHO	Boeing 727-2H3	Tunis-Air *Jerba*	
TS–JHP	Boeing 727-2H3	Tunis-Air *Monastir*	
TS–JHQ	Boeing 727-2H3	Tunis-Air *Tozeur-Nefta*	
TS–JHR	Boeing 727-2H3	Tunis-Air *Bizerte*	
TS–JHS	Boeing 727-2H3	Tunis-Air *Kairouan*	
TS–JHT	Boeing 727-2H3	Tunis-Air *Sidi Bousaid*	
TS–JHU	Boeing 727-2H3	Tunis-Air *Hannibal*	
TS–JHV	Boeing 727-2H3	Tunis-Air *Jugurtha*	
TS–JHW	Boeing 727-2H3	Tunis-Air *Ibn Khaldoun*	

VH (Australia)

Reg.	Type	Owner or Operator	Notes
VH–EBA	Boeing 747-238B	Qantas Airways *City of Canberra*	
VH–EBC	Boeing 747-238B	Qantas Airways *City of Sydney*	
VH–EBD	Boeing 747-238B	Qantas Airways *City of Perth*	
VH–EBE	Boeing 747-238B	Qantas Airways *City of Brisbane*	
VH–EBF	Boeing 747-238B	Qantas Airways *City of Adelaide*	
VH–EBG	Boeing 747-238B	Qantas Airways *City of Hobart*	
VH–EBH	Boeing 747-238B	Qantas Airways *City of Newcastle*	
VH–EBI	Boeing 747-238B	Qantas Airways *City of Darwin*	
VH–EBJ	Boeing 747-238B	Qantas Airways *City of Geelong*	
VH–EBK	Boeing 747-238B	Qantas Airways *City of Wollongong*	
VH–EBL	Boeing 747-238B	Qantas Airways *City of Townsville*	
VH–EBM	Boeing 747-238B	Qantas Airways *City of Parramatta*	
VH–EBN	Boeing 747-238B	Qantas Airways *City of Albury*	
VH–EBO	Boeing 747-238B	Qantas Airways *City of Elizabeth*	
VH–EBP	Boeing 747-238B	Qantas Airways *City of Freemantle*	
VH–EBQ	Boeing 747-238B	Qantas Airways *City of Bunbury*	
VH–EBR	Boeing 747-238B	Qantas Airways *City of Dubbo*	
VH–EBS	Boeing 747-238B	Qantas Airways	
VH–ECA	Boeing 747-238B	Qantas Airways *City of Sale*	
VH–ECB	Boeing 747-238B	Qantas Airways *City of Swan Hill*	
VH–ECC	Boeing 747-238B	Qantas Airways *City of Shepparton*	

VP–W (Zimbabwe)

Reg.	Type	Owner or Operator	Notes
VP–WKR	Boeing 707-330B	Air Zimbabwe	
VP–WKS	Boeing 707-330B	Air Zimbabwe	
VP–WKT	Boeing 707-330B	Air Zimbabwe	

VR–H (Hong Kong)

Reg.	Type	Owner or Operator	Notes
VR–HIA	Boeing 747-267B	Cathay Pacific Airways	
VR–HIB	Boeing 747-267B	Cathay Pacific Airways	
VR–HIC	Boeing 747-267B	Cathay Pacific Airways	
VR–HID	Boeing 747-267B	Cathay Pacific Airways	
VR–HIE	Boeing 747-267B	Cathay Pacific Airways	
VR–HKG	Boeing 747-267B	Cathay Pacific Airways	

VT (India)

Notes	*Reg.*	*Type*	*Owner or Operator*
	VT–EBE	Boeing 747-237B	Air-India *Emperor Shahjehan*
	VT–EBN	Boeing 747-237B	Air-India *Emperor Rajendra Chola*
	VT–EBO	Boeing 747-237B	Air-India *Emperor Nikramaditya*
	VT–EDU	Boeing 747-237B	Air-India *Emperor Akbar*
	VT–EFJ	Boeing 747-237B	Air-India *Emperor Chandragupta*
	VT–EFO	Boeing 747-237B	Air-India *Emperor Kanishka*
	VT–EFU	Boeing 747-237B	Air-India *Emperor Krishna Deva*
	VT–EGA	Boeing 747-237B	Air-India *Samudra Gupto*
	VT–EGB	Boeing 747-237B	Air-India *Mahendra Varman*
	VT–EGC	Boeing 747-237B	Air-India *Harsha Vardhuma*

NOTE: Air India Cargo operates two Douglas DC-8-63CFs which retain their U.S. registrations N773FT and N796FT.

YA (Afghanistan)

	YA–LAS	Douglas DC-10-30	Ariana

YI (Iraq)

	YI–AGE	Boeing 707-370C	Iraqi Airways
	YI–AGF	Boeing 707-370C	Iraqi Airways
	YI–AGG	Boeing 707-370C	Iraqi Airways
	YI–AGN	Boeing 747-270C	Iraqi Airways
	YI–AGO	Boeing 747-270C	Iraqi Airways

YK (Syria)

شركة الطيران العربية السورية

SYRIAN ARAB AIRLINES

	YK–AHA	Boeing 747SP-94	Syrian Arab Airlines *16 Novembre*
	YK–AHB	Boeing 747SP-94	Syrian Arab Airlines *Arab Solidarity*

YR (Romania)

Reg.	Type	Owner or Operator	Notes
YR–ABA	Boeing 707-3K1C	Tarom	
YR–ABC	Boeing 707-3K1C	Tarom	
YR–ABM	Boeing 707-321C	Tarom	
YR–ABN	Boeing 707-321C	Tarom	
YR–BCB	BAC One-Eleven 424	Tarom	
YR–BCE	BAC One-Eleven 424	Tarom	
YR–BCG	BAC One-Eleven 401	Tarom	
YR–BCH	BAC One-Eleven 402	Tarom	
YR–BCI	BAC One-Eleven 525FT	Tarom	
YR–BCJ	BAC One-Eleven 525FT	Tarom	
YR–BCK	BAC One-Eleven 525FT	Tarom	
YR–BCL	BAC One-Eleven 525FT	Tarom	
YR–BCM	BAC One-Eleven 525FT	Tarom	
YR–BCN	BAC One-Eleven 525FT	Tarom	
YR–BCQ	BAC One-Eleven 525RC	Tarom	
YR–BCR	BAC One-Eleven 487GK	Tarom	
YR–IRA	Ilyushin IL-62	Tarom	
YR–IRB	Ilyushin IL-62	Tarom	
YR–IRC	Ilyushin IL-62	Tarom	
YR–IRD	Ilyushin IL-62M	Tarom	
YR–IRE	Ilyushin IL-62M	Tarom	
YR–TPA	Tupolev Tu-154B	Tarom	
YR–TPB	Tupolev Tu-154B	Tarom	
YR–TPC	Tupolev Tu-154B	Tarom	
YR–TPD	Tupolev Tu-154B	Tarom	
YR–TPE	Tupolev Tu-154B	Tarom	
YR–TPF	Tupolev Tu-154B	Tarom	
YR–TPG	Tupolev Tu-154B	Tarom	
YR–TPI	Tupolev Tu-154B	Tarom	
YR–TPJ	Tupolev Tu-154B	Tarom	
YR–TPK	Tupolev Tu-154B	Tarom	
YR–TPL	Tupolev Tu-154B	Tarom	

YU (Yugoslavia)

Reg.	Type	Owner or Operator	Notes
YU–AGE	Boeing 707-340C	Jugoslovenski Aerotransport	
YU–AGG	Boeing 707-340C	Jugoslovenski Aerotransport	
YU–AGI	Boeing 707-351C	Jugoslovenski Aerotransport	
YU–AGJ	Boeing 707-351C	Jugoslovenski Aerotransport	
YU–AHJ	Douglas DC-9-32	Inex Adria Airways *Ljubljana*	
YU–AHL	Douglas DC-9-32	Jugoslovenski Aerotransport	
YU–AHM	Douglas DC-9-32	Jugoslovenski Aerotransport *Tivat*	
YU–AHN	Douglas DC-9-32	Jugoslovenski Aerotransport	
YU–AHO	Douglas DC-9-32	Jugoslovenski Aerotransport	
YU–AHP	Douglas DC-9-32	Jugoslovenski Aerotransport	

Notes	Reg.	Type	Owner or Operator
	YU–AHU	Douglas DC-9-32	Jugoslovenski Aerotransport
	YU–AHV	Douglas DC-9-32	Jugoslovenski Aerotransport
	YU–AHW	Douglas DC-9-33CF	Inex Adria Airways *Sarajevo*
	YU–AHX	Tupolev Tu-134A	Aviogenex *Beograd*
	YU–AHY	Tupolev Tu-134A	Aviogenex *Zagreb*
	YU–AJA	Tupolev Tu-134A	Aviogenex *Titograd*
	YU–AJB	Douglas DC-9-32	Inex Adria Airways
	YU–AJD	Tupolev Tu-134A	Aviogenex *Skopje*
	YU–AJF	Douglas DC-9-32	Inex Adria Airways
	YU–AJH	Douglas DC-9-32	Jugoslovenski Aerotransport
	YU–AJI	Douglas DC-9-32	Jugoslovenski Aerotransport
	YU–AJJ	Douglas DC-9-32	Jugoslovenski Aerotransport
	YU–AJK	Douglas DC-9-32	Jugoslovenski Aerotransport
	YU–AJL	Douglas DC-9-32	Jugoslovenski Aerotransport
	YU–AJM	Douglas DC-9-32	Jugoslovenski Aerotransport
	YU–AJP	Douglas DC-9-33CF	Inex Adria Airways
	YU–AJT	Douglas DC-9-51	Inex Adria Airways
	YU–AJU	Douglas DC-9-51	Inex Adria Airways *Maribor*
	YU–AJV	Tupolev Tu-134A	Aviogenex *Mostar*
	YU–AJW	Tupolev Tu-134A	Aviogenex *Pristina*
	YU–AJZ	Douglas DC-9-81	Inex Adria Airways
	YU–AKA	Boeing 727-2H9	Jugoslovenski Aerotransport
	YU–AKB	Boeing 727-2H9	Jugoslovenski Aerotransport
	YU–AKD	Boeing 727-2L8	Yugoslav Government/J.A.T.
	YU–AKE	Boeing 727-2H9	Jugoslovenski Aerotransport
	YU–AKF	Boeing 727-2H9	Jugoslovenski Aerotransport
	YU–AKG	Boeing 727-2H9	Jugoslovenski Aerotransport
	YU–AKI	Boeing 727-2H9	Jugoslovenski Aerotransport
	YU–AKJ	Boeing 727-2H9	Jugoslovenski Aerotransport
	YU–AKK	Boeing 727-2H9	Jugoslovenski Aerotransport
	YU–AKL	Boeing 727-2H9	Jugoslovenski Aerotransport
	YU–AMA	Douglas DC-10-30	Jugoslovenski Aerotransport *Nikola Tesla*
	YU–AMB	Douglas DC-10-30	Jugoslovenski Aerotransport *Edvard Rusijan*
	YU–ANA	Douglas DC-9-82	Inex Adria Airways
	YU–ANB	Douglas DC-9-82	Inex Adria Airways
	YU–ANE	Tupulev Tu-134A	Aviogenex *Novi Sad*

YV (Venezuela)

Notes	Reg.	Type	Owner or Operator
	YV–133C	Douglas DC-10-30	Viasa
	YV–134C	Douglas DC-10-30	Viasa
	YV–135C	Douglas DC-10-30	Viasa
	YV–136C	Douglas DC-10-30	Viasa
	YV–137C	Douglas DC-10-30	Viasa
	YV–138C	Douglas DC-10-30	Viasa

ZK (New Zealand)

Notes	Reg.	Type	Owner or Operator
	ZK–NZV	Boeing 747-219B	Air New Zealand
	ZK–NZW	Boeing 747-219B	Air New Zealand
	ZK–NZX	Boeing 747-219B	Air New Zealand
	ZK–NZY	Boeing 747-219B	Air New Zealand
	ZK–NZZ	Boeing 747-219B	Air New Zealand

ZS (South Africa)

Reg.	Type	Owner or Operator	Notes
ZS–SAL	Boeing 747-244B	South African Airways *Tafelberg*	
ZS–SAM	Boeing 747-244B	South African Airways *Drakensberg*	
ZS–SAN	Boeing 747-244B	South African Airways *Lebombo*	
ZS–SAO	Boeing 747-244B	South African Airways *Magaliesberg*	
ZS–SAP	Boeing 747-244B	South African Airways *Swartberg*	
ZS–SAR	Boeing 747-244B	South African Airways *Waterberg*	
ZS–SAS	Boeing 747-244B	South African Airways *Helderberg*	
ZS–SPA	Boeing 747SP-44	South African Airways *Matroosberg*	
ZS–SPB	Boeing 747SP-44	South African Airways *Outeniqua*	
ZS–SPC	Boeing 747SP-44	South African Airways *Maluti*	
ZS–SPD	Boeing 747SP-44	South African Airways *Majuba*	
ZS–SPE	Boeing 747SP-44	South African Airways *Hantam*	
ZS–SPF	Boeing 747SP-44	South African Airways *Soutpansberg*	

3B (Mauritius)

Reg.	Type	Owner or Operator	Notes
3B–NAE	Boeing 707-344B	Air Mauritius	

3X (Guinea)

Reg.	Type	Owner or Operator	Notes
3X–GAZ	Boeing 707-351C	Air Guinee	
3X–	Boeing 707-328B	Air Guinee	
3X–	Boeing 707-328B	Air Guinee	

4R (Sri Lanka)

Reg.	Type	Owner or Operator	Notes
4R–ALE	L.1011-385 TriStar 1	Air Lanka	
4R–ALF	L.1011-385 TriStar 1	Air Lanka	
4R–ALG	L.1011-385 TriStar 1	Air Lanka	
4R–ULA	L.1011-385 TriStar 500	Air Lanka	
4R–ULB	L.1011-385 TriStar 500	Air Lanka	

4W (Yemen)

Reg.	Type	Owner or Operator	Notes
4W–ACF	Boeing 727-2N8	Yemen Airways	
4W–ACG	Boeing 727-2N8	Yemen Airways	
4W–ACH	Boeing 727-2N8	Yemen Airways	
4W–ACI	Boeing 727-2N8	Yemen Airways	
4W–ACJ	Boeing 727-2N8	Yemen Airways	

4X (Israel)

Reg.	Type	Owner or Operator	Notes
4X–ATA	Boeing 707-458	El Al	
4X–ATB	Boeing 707-458	El Al	
4X–ATR	Boeing 707-358B	El Al	
4X–ATS	Boeing 707-358B	El Al	

Notes	Reg.	Type	Owner or Operator
	4X–ATT	Boeing 707-358B	El Al
	4X–ATX	Boeing 707-358C	El Al
	4X–ATY	Boeing 707-358C	El Al
	4X–AXA	Boeing 747-258B	El Al
	4X–AXB	Boeing 747-258B	El Al
	4X–AXC	Boeing 747-258B	El Al
	4X–AXD	Boeing 747-258C	El Al
	4X–AXF	Boeing 747-258C	El Al
	4X–AXG	Boeing 747-258F	El Al
	4X–AXH	Boeing 747-258B	El Al
	4X–AXZ	Boeing 747-124F	El Al
	4X–BMA	Boeing 720-023B	Maof Airlines
	4X–BMB	Boeing 720-023B	Maof Airlines

5A (Libya)

Notes	Reg.	Type	Owner or Operator
	5A–DAI	Boeing 727-224	Libyan Arab Airlines
	5A–DAK	Boeing 707-3L5C	Libyan Arab Airlines
	5A–DGJ	Canadair CL-44J	United African Airlines
	5A–DHJ	Canadair CL-44D4	United African Airlines
	5A–DHL	Boeing 707-321C	United African Airlines
	5A–DHM	Boeing 707-123B	United African Airlines
	5A–DIA	Boeing 727-2L5	Libyan Arab Airlines
	5A–DIB	Boeing 727-2L5	Libyan Arab Airlines
	5A–DIC	Boeing 727-2L5	Libyan Arab Airlines
	5A–DID	Boeing 727-2L5	Libyan Arab Airlines
	5A–DIE	Boeing 727-2L5	Libyan Arab Airlines
	5A–DIF	Boeing 727-2L5	Libyan Arab Airlines
	5A–DIG	Boeing 727-2L5	Libyan Arab Airlines
	5A–DIH	Boeing 727-2L5	Libyan Arab Airlines
	5A–DII	Boeing 727-2L5	Libyan Arab Airlines
	5A–DIX	Boeing 707-348C	United African Airlines
	5A–DIZ	Boeing 707-351C	United African Airlines
	5A–DJM	Boeing 707-321B	United African Airlines
	5A–DJO	Boeing 707-338C	United African Airlines

5B (Cyprus)

Notes	Reg.	Type	Owner or Operator
	5B–DAG	BAC One-Eleven 537GF	Cyprus Airways
	5B–DAH	BAC One-Eleven 537GF	Cyprus Airways
	5B–DAJ	BAC One-Eleven 537GF	Cyprus Airways
	5B–DAK	Boeing 707-123B	Cyprus Airways
	5B–DAL	Boeing 707-123B	Cyprus Airways
	5B–DAO	Boeing 707-123B	Cyprus Airways
	5B–DAP	Boeing 707-123B	Cyprus Airways

5N (Nigeria)

Notes	Reg.	Type	Owner or Operator
	5N–ABJ	Boeing 707-3F9C	Nigeria Airways
	5N–ABK	Boeing 707-3F9C	Nigeria Airways
	5N–ANN	Douglas DC-10-30	Nigeria Airways
	5N–ANO	Boeing 707-3F9C	Nigeria Airways
	5N–ANR	Douglas DC-10-30	Nigeria Airways

5X (Uganda)

Reg.	Type	Owner or Operator	Notes
5X–UAC	Boeing 707-351C	Uganda Airlines	
5X–UBC	Boeing 707-338C	Uganda Airlines	
5X–UCF	Lockheed L382G Hercules	Uganda Airlines	

5Y (Kenya)

Reg.	Type	Owner or Operator	Notes
5Y–BBI	Boeing 707-351B	Kenya Airways	
5Y–BBJ	Boeing 707-351B	Kenya Airways	
5Y–BBK	Boeing 707-351B	Kenya Airways	
5Y–BBX	Boeing 720-047B	Kenya Airways	

7T (Algeria)

Reg.	Type	Owner or Operator	Notes
7T–VEA	Boeing 727-2D6	Air Algerie *Tassili*	
7T–VEB	Boeing 727-2D6	Air Algerie *Hoggar*	
7T–VEC	Boeing 737-2D6	Air Algerie *Aures Nemencha*	
7T–VED	Boeing 737-2D6C	Air Algerie *Atlas Saharien*	
7T–VEE	Boeing 737-2D6C	Air Algerie *Oasis*	
7T–VEF	Boeing 737-2D6	Air Algerie *Saoura*	
7T–VEG	Boeing 737-2D6	Air Algerie *Monts des Ouleds Neils*	
7T–VEH	Boeing 727-2D6	Air Algerie *Lalla Khadidja*	
7T–VEI	Boeing 727-2D6	Air Algerie *Djebel Amour*	
7T–VEJ	Boeing 737-2D6	Air Algerie *Chrea*	
7T–VEK	Boeing 737-2D6	Air Algerie *Edough*	
7T–VEL	Boeing 737-2D6	Air Algerie *Akfadou*	
7T–VEM	Boeing 727-2D6	Air Algerie *Mont du Ksall*	
7T–VEN	Boeing 737-2D6	Air Algerie *La Soummam*	
7T–VEO	Boeing 737-2D6	Air Algerie *Le Titteri*	
7T–VEP	Boeing 727-2D6	Air Algerie *Mont du Tessala*	
7T–VEQ	Boeing 737-2D6	Air Algerie *Le Zaccar*	
7T–VER	Boeing 737-2D6	Air Algerie *Le Souf*	
7T–VES	Boeing 737-2D6C	Air Algerie *Le Tadmaït*	
7T–VET	Boeing 727-2D6	Air Algerie	
7T–VEU	Boeing 727-2D6	Air Algerie	
7T–VEV	Boeing 727-2D6	Air Algerie	
7T–VEW	Boeing 727-2D6	Air Algerie	
7T–	Boeing 727-2D6	Air Algerie	
7T–	Boeing 737-2D6	Air Algerie	

9G (Ghana)

Notes	Reg.	Type	Owner or Operator
	9G–ACE	Britannia 253F	Geminair
	9G–	Douglas DC-10-30	Ghana Airways

9H (Malta)

Notes	Reg.	Type	Owner or Operator
	9H–AAK	Boeing 720-047B	Air Malta
	9H–AAL	Boeing 720-047B	Air Malta
	9H–AAM	Boeing 720-040B	Air Malta
	9H–AAN	Boeing 720-040B	Air Malta
	9H–AAO	Boeing 720-047B	Air Malta

NOTE: Air Malta also operates Boeing 737s PH–TVC, PH–TVE and PH–TVP on lease from Transavia Holland.

9J (Zambia)

Notes	Reg.	Type	Owner or Operator
	9J–ADY	Boeing 707–349C	Zambia Airways
	9J–AEB	Boeing 707–351C	Zambia Airways
	9J–AEL	Boeing 707–338C	Zambia Airways
	9J–AEQ	Boeing 707-321C (Cargo)	Zambia Airways

9K (Kuwait)

Notes	Reg.	Type	Owner or Operator
	9K–ACJ	Boeing 707-369C	Kuwait Airways *Wara*
	9K–ACK	Boeing 707-369C	Kuwait Airways *Kadhma*
	9K–ACL	Boeing 707-369C	Kuwait Airways *Al-Jahra*
	9K–ACM	Boeing 707-369C	Kuwait Airways *Failaka*
	9K–ACN	Boeing 707-369C	Kuwait Airways *Burghan*
	9K–ACS	Boeing 707-321C	Kuwait Airways *Gharnada*
	9K–ACU	Boeing 707-321C	Kuwait Airways *Qurtoba*
	9K–ACX	Boeing 707-311C	Kuwait Airways
	9K–ADA	Boeing 747-269B	Kuwait Airways *Al Sabahiya*
	9K–ADB	Boeing 747-269B	Kuwait Airways *Al Jaberyia*
	9K–ADC	Boeing 747-269B	Kuwait Airways *Al Murbarakiya*
	9K–ADD	Boeing 747-269B	Kuwait Airways
	9K–	Boeing 747-269B	Kuwait Airways

9L (Sierra Leone)

Sierra Leone Airways' services between Freetown and London are operated on their behalf by British Caledonian.

9M (Malaysia)

Reg.	Type	Owner or Operator	Notes
9M–MAS	Douglas DC-10-30	Malaysian Airline System	
9M–MAT	Douglas DC-10-30	Malaysian Airline System	
9M–MAV	Douglas DC-10-30	Malaysian Airline System	

9Q (Zaïre)

Reg.	Type	Owner or Operator	Notes
9Q–CDA	Boeing 707-338C	Zaïre Cargo	
9Q–CLI	Douglas DC-10-30	Air Zaïre *Mont Ngaliema*	
9Q–CLT	Douglas DC-10-30	Air Zaïre *Mont Ngafula*	

9V (Singapore)

Reg.	Type	Owner or Operator	Notes
9V–SQD	Boeing 747-212B	Singapore Airlines	
9V–SQE	Boeing 747-212B	Singapore Airlines	
9V–SQF	Boeing 747-212B	Singapore Airlines	
9V–SQG	Boeing 747-212B	Singapore Airlines	
9V–SQH	Boeing 747-212B	Singapore Airlines	
9V–SQI	Boeing 747-212B	Singapore Airlines	
9V–SQJ	Boeing 747-212B	Singapore Airlines	
9V–SQK	Boeing 747-212B	Singapore Airlines	
9V–SQL	Boeing 747-212B	Singapore Airlines	
9V–SQM	Boeing 747-212B	Singapore Airlines	
9V–SQN	Boeing 747-212B	Singapore Airlines	
9V–SQO	Boeing 747-212B	Singapore Airlines	
9V–SQP	Boeing 747-212B	Singapore Airlines	
9V–SQQ	Boeing 747-212B	Singapore Airlines	
9V–SQR	Boeing 747-212B	Singapore Airlines	
9V–SQS	Boeing 747-212B	Singapore Airlines	

9Y (Trinidad and Tobago)

Reg.	Type	Owner or Operator	Notes
9Y–TED	Boeing 707-351C	B.W.I.A. *Scarlet Ibis*	
9Y–TEJ	Boeing 707-351C	B.W.I.A. *Bird of Paradise*	
9Y–TEK	Boeing 707-351C	B.W.I.A. *Toucan*	
9Y–TEX	Boeing 707-321B	B.W.I.A.	
9Y–TEZ	Boeing 707-321B	B.W.I.A.	
9Y–TGJ	L.1011 TriStar 500	B.W.I.A. *Flamingo*	
9Y–TGN	L.1011 TriStar 500	B.W.I.A.	
9Y–THA	L.1011 TriStar 500	B.W.I.A.	
9Y–THB	L.1011 TriStar 500	B.W.I.A.	

Radio Frequencies

The frequencies used by the larger airfields/airports are listed below. Abbreviations used: TWR – Tower, APP – Approach, A/G – Air-ground advisory. It is possible for changes to be made from time to time with the frequencies allocated which are all quoted in Megahertz (MHz).

Airfield	*TWR*	*APP*	*A/G*
Aberdeen	118·1	120·4	
Aldergrove	118·3	120·0	
Alderney	123·6		
Andrewsfield			130·55
Barton			122·7
Bembridge			122·2
Biggin Hill	129·4		
Birmingham	118·3	120·5	
Bitteswell	123·25	130·15	
Blackbushe			122·3
Blackpool	118·4	118·4	
Bodmin			122·7
Booker			121·15
Bournemouth	125·6	118·65	
Cambridge	122·2	123·6	
Cardiff	121·12	125·85	
Carlisle			123·6
Compton Abbas			122·7
Coventry	119·25	122·0	
Cranfield	123·2	122·85	
Denham			123·5
Doncaster			122·9
Dundee	122·9		
Dunkeswell			123·5
Dunsfold	130·0	122·55	
Duxford			123·5
East Midlands	124·0	119·65	
Edinburgh	118·7	121·2	
Elstree			122.4
Exeter	119·8	128·15	
Fairoaks			130·42
Felthorpe			123·5
Fenland			123·05
Filton	124·95	130·85	
Gatwick	119·45	119·6	
Glasgow	118·8	119·1	
Goodwood	119·7	122·45	
Guernsey	119·95	128·65	
Halfpenny Green			123·0
Hamble	120·65	125·0	
Hatfield	130·8	123·35	
Haverfordwest			122·2
Hawarden	124·95	123·35	
Headcorn			122·0
Heathrow	118·7	119·2	
	121·0	119·5	
Hethal			122·35
Hucknall			130·8
Humberside	118·55	123·15	
Inverness	122·6	122·6	
Ipswich	123·25		
Jersey	119·45	120·3	
Kidlington	119·8	130·3	
Land's End			122·3
Leavesden	122·15		
Leeds	120·3	123·75	
Leicester E.			122·25
Little Snoring			122·4
Liverpool	118·1	119·85	
Luton	120·2	129·55	
Lydd	126·7	120·7	
Manchester	118·7	119·4	
Netherthorpe			123·5
Newcastle	126·35		
North Denes			120·45
Norwich	118·9	119·35	
Perth	119·8	122·3	
Plymouth	122·6	123·2	
Prestwick	118·1	120·55	
Redhill			123·45
Rochester			122·25
Ronaldsway	118·9	120·85	
Sandown			123·5
Seething			122·6
Sherburn			122·6
Shobdon			123·5
Shoreham	125·4	123·15	
Sibson			122·3
Sleap			122·45
Southampton	118·2	128·85	
Southend	119·7	128·95	
Stansted	118·15	126·95	
Stapleford			122·8
Sumburgh	118·25	123·15	
Sunderland	122·7	122·2	
Swansea	122·6	119·7	
Swanton Morley	123·5		
Sywell			122·7
Tees-side	119·8	118·85	
Thruxton			125·35
Tollerton			122·8
Wellesbourne			130·45
Weston	122·5		
White Waltham	122.6		
Wick	119·7		
Wickenby			122·45
Yeovil	125·4	130·8	

Airline Codes

Two letter codes are used by airlines to prefix flight numbers in timetables, airport movement boards, etc. Those listed below identify both UK and overseas airlines appearing in the book.

Code	Airline	Registration
AC	Air Canada	C
AE	Air Europe	G
AF	Air France	F
AH	Air Algerie	7T
AI	Air India	VT
AK	Air Bridge	G
AO	Aviaco	EC
AR	Aerolineas Argentinas	LV
AT	Royal Air Maroc	CN
AY	Finnair	OH
AZ	Alitalia	I
BA	British Airways	G
BB	Balair	HB
BC	Brymon Aviation	G
BD	British Midland	G
BF	Alaska Intl. Air	N
BG	Bangladesh Biman	S2
BM	ATI	I
BN	Braniff	N
BR	British Caledonian	G
BS	Air Executive	LN
BU	Braathens	LN
BW	B.W.I.A.	9Y
BX	Spantax	EC
BY	Britannia	G
BZ	Brit. Air	F
CA	CAAC	B
CC	Air Freight Egypt	SU
CD	Air Commuter	G
CL	Capitol Intl. A/W	N
CP	CP Air	C
CV	Cargolux	LX
CW	Air Continental	G
CX	Cathay Pacific	VR-H
CY	Cyprus A/W	5B
DA	Dan-Air	G
DE	Delta Air Transport	OO
DF	Condor	D
DG	Air Atlantique	G
DK	Scanair	SE
DL	Delta A/L	N
DM	Maersk	OY
EI	Aer Lingus	EI
EL	Euralair	F
EN	Eastern A/W	G
EO	Cabair	G
ET	Ethiopian A/L	ET
EX	Executive Express	G
EY	Europe Aero Service	F
EZ	Euroair	G
EZ	Evergreen Intl.	N
FC	Fairflight	G
FD	Ford	G
FI	Icelandair	TF
FO	Fred Olsen	LN
FQ	Minerve	F
FT	Flying Tiger	N
GA	Garuda	PK
GE	Guernsey A/L	G
GE	German Cargo	D
GF	Gulf Air	A40
GG	Air London	G
GH	Ghana A/W	9G
GK	Laker A/W	G
GP	Geminair	9G
GR	Aurigny A/S	G
GX	Global Intl.	N
HE	Trans European A/W	OO
HF	Hapag-Lloyd	G
HG	Centreline A/S	G
HN	NLM	PH
HR	TAR	LV
HV	Transavia	PH
HZ	Thurston Aviation	G
IA	Iraq A/W	YI
IB	Iberia	EC
IF	Interflug	DDR
IG	Alisarda	I
IJ	TAT	F
IR	Iran Air	EP
IT	Air Inter	F
IY	Yemen A/W	4W
JA	Scimitar A/L	G
JJ	Aviogenex	YU
JK	TAE	EC
JL	Japan A/L	JA
JP	Inex Adria	
JU	JAT	YU
JY	Jersey European	G
KG	Orion A/W	G
KJ	Iscargo	TF
KL	KLM	PH
KM	Air Malta	9H
KQ	Kenya A/W	5Y
KR	Kar-Air	OH
KT	British Airtours	G
KU	Kuwait A/W	9K
KZ	Trans Europe Air	F
LC	Loganair	G
LF	Linjeflyg	SE
LG	Luxair	LX
LH	Lufthansa	D
LJ	Sierra Leone A/W	9L
LK	Lucas A/T	F
LL	Aero Lloyd	D
LN	Libyan Arab A/L	5A
LO	Polish A/L (LOT)	SP
LP	Air Alpes	F
LS	Express A/S	G
LT	LTU	D
LW	Lauda Air	OE
LY	El Al	4X
LZ	Bulgarian A/L	LZ
MA	Malev	HA
ME	Middle East A/L	OD
MF	Maof A/L	4X
MH	Malaysian A/L	9M
MK	Mauritius	3B
MP	Martinair	PH
MS	Egyptair	SU
NB	Sterling A/W	OY
NP	Heavy Lift	G
NQ	NW Territorial A/W	C
NW	Northwest Orient	N
OA	Olympic A/W	SX
OK	Czech A/L	OK
OM	Monarch A/L	G
OO	Sobelair	OO
OS	Austrian A/L	OE
OY	Conair	OY
PA	Pan Am	N
PJ	Peregrine A/S	G
PK	Pakistan Intl.	AP
PR	Philippine A/L	RP
PW	Pacific Western	C
QA	Alidair	G
QC	Air Zaïre	9Q
QF	Qantas	VH
QH	Air Florida	N
QK	Aeromaritime	F
QT	Inter City A/L	G
QU	Uganda A/L	5X
QZ	Zambia A/W	9J
RB	Syrian Arab	YK
RD	Airlift Intl.	N
RD	Alderney A/F	G
RG	Varig	PP
RH	Air Zimbabwe	VP-W
RJ	Alia	JY
RM	McAlpine	G
RO	Tarom	YR
RU	CTA	HB
RY	Redcoat	G
SA	South African A/W	ZS
SD	Sudan A/W	ST
SF	Air Charter Intl.	F
SG	Spacegrand	G
SJ	Southern A/T	N
SK	SAS	SE OY LN
SN	Sabena	OO
SQ	Singapore A/L	9Q
SR	Swissair	HB
SU	Aeroflot	CCCP
SY	Air Alsace	F
TC	Thai Intl.	HS
TK	Turkish A/L	TC
TL	Trans Mediterranean	OD
TP	Air Portugal	CS
TR	Transeuropa	EC
TU	Tunis Air	TS
TV	Transamerica	N
TW	TWA	N
UH	Bristow	G
UJ	Air Lanka	4R
UK	Air UK	G
UP	Air Foyle	G
UY	Cameroon A/L	TJ
VA	Viasa	YV
VF	British Air Ferries	G
VL	Eagle Air	TF
VS	Tradewinds	G
VY	Air Belgium	OO
WA	Western A/L	N
WD	Wardair	C
WE	WDL Flugdienst	D
WG	Air Ecosse	G
WN	Norfly	LN
WO	World A/W	N
WT	Nigeria A/W	5N
ZA	Genair	G
ZR	Casair	G

BRITISH AIRCRAFT PRESERVATION COUNCIL REGISTER

The British Aircraft Preservation Council was formed in 1967 to co-ordinate the works of all bodies involved in the preservation, restoration and display of historical aircraft. Membership covers the whole spectrum of national, Service, commercial and voluntary groups, and meetings are held regularly at the bases of member organisations. The Council is able to provide a means of communication, helping to resolve any misunderstandings or duplication of effort. Every effort is taken to encourage the raising of standards of both organisation and technical capacity amongst the member groups to the benefit of everyone interested in aviation. To assist historians, the B.A.P.C. register has been set up and provides an identity for those aircraft which do not qualify for a Service serial or inclusion in the UK Civil Register.

Aircraft on the current BAPC Register are as follows:

Notes	*Reg.*	*Type*	*Owner or Operator*
	1	Roe Triplane Type IV (replica)	The Shuttleworth Trust, Old Warden, Beds.
	2	Bristol Boxkite (replica)	The Shuttleworth Trust
	3	Blériot XI	The Shuttleworth Trust
	4	Deperdussin monoplane	The Shuttleworth Trust
	5	Blackburn monoplane	The Shuttleworth Trust
	6	Roe Triplane Type IV (replica)	Historic Aircraft Museum, Southend
	7	Southampton University MPA	The Shuttleworth Trust
	8	Dixon ornithopter	The Shuttleworth Trust
	9	Humber Monoplane (replica)	Midland Air Museum, Coventry
	10	Hafner R.II Revoplane	British Rotorcraft Museum
	11	English Electric Wren	The Shuttleworth Trust
	12	Mignet HM.14 Pou-du-Ciel	Museum of Flight, E. Fortune
	13	Mignet HM.14 Pou-du-Ciel	The Aeroplane Collection Ltd.
	14	Addyman standard training glider	N. H. Ponsford
	15	Addyman standard training glider	The Aeroplane Collection Ltd.
	16	Addyman ultra-light aircraft	N. H. Ponsford
	17	Woodhams Sprite	The Aeroplane Collection Ltd.
	18	Killick MP Gyroplane	The Aeroplane Collection Ltd.
	19	Bristol F.2b	Anne Lindsay
	20	Lee-Richards annular biplane (replica)	Newark Air Museum
	22	Mignet HM.14 Pou-du-Ciel (G–AEOE)	Newark Air Museum
	25	Nyborg TGN-III glider	Midland Air Museum
	26	Auster AOP.9	S. Wales Aircraft Preservation Soc.
	27	Mignet HM.14 Pou-du-Ciel	M. J. Abbey
	28	Wright Flyer (replica)	RAF Museum, Cardington
	29	Mignet HM.14 Pou-du-Ciel (G–ADRY)	J. J. Penney
	31	Slingsby T.7 Tutor	S. Wales Aircraft Preservation Soc.
	32	Crossley Tom Thumb	Midland Air Museum
	33	DFS.108-49 Grunau Baby 116	Russavia Collection, Duxford
	34	DFS.108-49 Grunau Baby 116	D. Elsdon
	35	EoN primary glider	The Shuttleworth Trust
	36	FZG-76 (V.I) (replica)	The Shuttleworth Trust
	37	Blake Bluetit	The Shuttleworth Trust
	38	Bristol Scout replica (A1742)	RAF St. Athan
	40	Bristol Boxkite (replica)	Bristol City Museum
	41	B.E.2C (replica) (6232)	RAF St. Athan
	42	Avro 504 (replica)	RAF St. Athan
	43	Mignet HM.14 Pou-du-Ciel	Lincolnshire Aviation Museum
	44	Miles Magister (L6906)	G. H. R. Johnson
	45	Pilcher Hawk (replica)	Lord Braye
	46	Mignet HM.14 Pou-du-Ciel	Alan McKechnie Racing Ltd.
	47	Watkins monoplane	C. M. Watkins
	48	Pilcher Hawk (replica)	Glasgow Museum of Transport
	49	Pilcher Hawk	Royal Scottish Museum
	50	Roe Triplane Type I	Science Museum
	51	Vickers Vimy IV	Science Museum
	52	Lilienthal glider	Science Museum
	53	Wright Flyer (replica)	Science Museum
	54	JAP-Harding monoplane	Science Museum
	55	Levavasseur Antoinette VII	Science Museum
	56	Fokker E.III	Science Museum
	57	Pilcher Hawk (replica)	Science Museum
	58	Yokosuka MXY–7 Ohka II	Science Museum

Reg.	Type	Owner or Operator	Notes
59	Sopwith Camel (replica) (D3419)	RAF St. Athan	
60	Murray M.1 helicopter	The Aeroplane Collection Ltd.	
61	Stewart man-powered ornithopter	Lincolnshire Aviation Museum	
62	Cody Biplane	Science Museum	
63	Hurricane (replica) (L1592)	Torbay Aircraft Museum	
64	Hurricane (replica)	Historic Aircraft Museum, Southend	
65	Spitfire (replica)	Torbay Aircraft Museum	
66	Bf 109 (replica)	Historic Aircraft Museum, Southend	
67	Bf 109 (replica)	Midland Air Museum	
68	Hurricane (replica)	Midland Air Museum	
69	Spitfire (replica)	Midland Air Museum	
70	Auster AOP.5 (TJ472)	Aircraft Preservation Soc. of Scotland	
71	Spitfire (replica)	Norfolk & Suffolk Aviation Museum	
72	Hurricane (replica)	Midland Air Museum	
73	Hurricane (replica)	J. P. Berkeley	
74	Bf 109 (replica)	Torbay Aircraft Museum	
75	Mignet HM.14 Pou-du-Ciel	Nigel Ponsford	
76	Mignet HM.14 Pou-du-Ciel (G–AFFI)	Nostell Aviation Museum	
77	Mignet HM.14 Pou-du-Ciel	Skyfame Collection	
78	Hawker Hind (Afghan) (K5457)	The Shuttleworth Trust	
79	Fiat G.46–4	Historic Aircraft Museum, Southend	
80	Airspeed Horsa	Museum of Army Flying	
81	Hawkridge Dagling	Russavia Collection, Duxford	
82	Hawker Hind (Afghan)	RAF Museum	
83	Kawasaki Ki-100-IB	Aerospace Museum, Cosford	
84	Nakajima Ki-46 (Dinah III)	RAF St. Athan	
85	Weir W-2 autogyro	Museum of Flight, E. Fortune	
86	de Havilland Tiger Moth (replica)	Yorkshire Aircraft Preservation Soc.	
87	Bristol Babe (replica)	Yorkshire Aircraft Preservation Soc.	
88	Fokker Dr I (replica)	Fleet Air Arm Museum	
89	Cayley glider (replica)	RAF Museum	
90	Colditz Cock (replica)	Torbay Aircraft Museum	
91	Fieseler Fi 103/FZG.76 (V.1)	Lashenden Air Warfare Museum	
92	Fieseler Fi 103/FZG.76 (V.1)	RAF Museum	
93	Fieseler Fi 103/FZG.76 (V.1)	RAF St. Athan	
94	Fieseler Fi 103/FZG.76 (V.1)	Cosford Aerospace Museum	
95	Gizmer autogyro	N.E. Aircraft Museum	
96	Brown helicopter	N.E. Aircraft Museum	
97	Luton L.A.4a Minor	N.E. Aircraft Museum	
98	Yokosuka MXY-7 Ohka II	RAF Museum, Henlow	
99	Yokosuka MXY-7 Ohka II	Aerospace Museum, Cosford	
100	Clarke glider	Science Museum	
101	Mignet HM.14 Pou-du-Ciel	Lincolnshire Aviation Museum	
102	Mignet HM.14 Pou-du-Ciel	W. Sneesby	
103	Pilcher glider (replica)	Personal Plane Services Ltd.	
104	Blériot XI (replica)	RAF St. Athan	
105	Blériot XI (replica)	Mosquito Aircraft Museum	
106	Blériot XI	RAF Museum	
107	Blériot XXVII	RAF Museum, Cardington	
108	Fairey Swordfish	RAF Museum, Henlow	
109	Slingsby Kirby Cadet	RAF Museum, Henlow	
110	Fokker D.VII replica (static) (5125)	Leisure Sport Ltd.	
111	Sopwith Triplane replica (static) (N5492)	Leisure Sport Ltd.	
112	D.H.2 replica (static) (5964)	Leisure Sport Ltd.	
113	S.E.5A replica (static) (B4863)	Leisure Sport Ltd.	
114	Vickers Type 60 Viking (static)	Leisure Sport Ltd.	
115	Mignet HM.14 Pou-du-Ciel	Essex Aviation Group	
116	Santos-Dumont Demoiselle (replica)	Cornwall Aero Park, Helston	
117	B.E.2C (replica)	British Broadcasting Corp.	
118	Albatross D.V. (replica)	British Broadcasting Corp.	
119	Benson B.7	N.E. Aircraft Museum	
120	Mignet HM.14 Pou-du-Ciel	1324 (Brough) Squadron ATC	
121	Mignet HM.14 Pou-du-Ciel (G–AEKR)	Nostell Aviation Museum	
122	Avro 504 (replica)	British Broadcasting Corp.	
123	Vickers FB.5 Gunbus (replica)	British Broadcasting Corp.	
124	Lilienthal Glider Type XI (replica)	Science Museum	

Notes	Reg.	Type	Owner or Operator
	125	Clay Cherub	Midland Air Museum
	126	D.31 Turbulent (static)	Midland Air Museum
	127	Halton Jupiter	Shuttleworth Trust
	128	Watkinson Cyclogyroplane Mk. IV	British Rotorcraft Museum
	129	Blackburn 1911 Monoplane (replica)	Cornwall Aero Park, Helston
	130	Blackburn 1912 Monoplane (replica)	Cornwall Aero Park, Helston
	131	Pilcher Hawk (replica)	C. Paton
	132	Blériot XI	Aerospace Museum, Cosford
	133	Fokker Dr I (replica)	Torbay Aircraft Museum
	134	Pitts S-2A static (G–RKSF)	Rothmans of Pall Mall
	135	Bristol M.IC (replica) (C4912)	Leisure Sport Ltd.
	136	Deperdussin Seaplane (replica)	Leisure Sport Ltd.
	137	Sopwith Baby Floatplane (replica)	Leisure Sport Ltd.
	138	Hansa Brandenburg W.29 Floatplane (replica) (2292)	Leisure Sport Ltd.
	139	Fokker Dr I (replica) (DR 1/17)	Leisure Sport Ltd.
	140	Curtiss R3C-2 Floatplane (replica)	Leisure Sport Ltd.
	141	Macchi M.39 Floatplane (replica)	Leisure Sport Ltd.
	142	SE-5A (replica)	Cornwall Aero Park, Helston
	143	Paxton MPA	R. A. Paxton
	144	Weybridge Mercury	Weybridge MPAG
	145	Oliver MPA	D. Oliver
	146	Pedal Aeronauts Toucan MPA	Shuttleworth Trust
	147	Benson B.7	Norfolk & Suffolk Aviation Museum
	148	Hawker Fury II (replica) (K7271)	Aerospace Museum, Cosford
	149	Short S.27 (replica)	Fleet Air Arm Museum
	150	SEPECAT Jaguar GR.1 (replica) (XX732)	RAF Exhibition Flight
	151	SEPECAT Jaguar GR.1 (replica) (XX824)	RAF Exhibition Flight
	152	BAe Hawk T.1 (replica) (XX163)	RAF Exhibition Flight
	153	Westland WG.33	Westland Helicopters Ltd.
	154	D.31 Turbulent	Lincolnshire Aviation Museum
	155	Panavia Tornado GR.1 (replica) (ZA322)	RAF Exhibition Flight
	156	Supermarine S-6B (replica) (S5195)	Leisure Sport Ltd.
	157	Waco CG-4A	Pennine Aviation Museum
	158	Fieseler Fi 103/FZG.76 (V.1)	Joint Bomb Disposal School
	159	Fuji Oka	Joint Bomb Disposal School
	160	Chargus 108 hang glider	Museum of Flight, E. Fortune
	161	Stewart Ornithopter Cappela	Bomber County Museum, Cleethorpes
	162	Goodhart Newbury Manflier MPA	Bomber County Museum, Cleethorpes
	163	AFEE 10/42 Rotabuggy (replica)	Wessex Aviation Soc. Wimbourne
	164	Wight Quadraplane Type I (replica)	Wessex Aviation Soc. Wimbourne
	165	Bristol F.2b	RAF Museum, Cardington
	166	Bristol F.2b	Shuttleworth Trust

NOTE: Registrations/Serials carried are mostly false identities. MPA = Man Powered Aircraft.

ADDENDA

New in-sequence registrations

Reg.	Type	Owner or Operator	Notes
G-AAOK	Curtis Wright Travel Air 120	Shipping & Airlines Ltd.	
G-AEOF	Rearwin 8500	P. Mann	
G-AFGM	Piper J-4A Cup Coupe	J. A. Thomas	
G-AFWH	Piper J-4A Cub Coupe	C. C. Lovell	
G-AMEN	PA-19 Super Cub 95	A. Lovejoy & ptnrs.	
G-AMSG	SIPA 903	S. W. Markham	
G-AYSD	Slingsby T.61A	J. Conolly	
G-BIVW	Z.326 Trener Master	G. C. Masterson	
G-BJGT	Mooney M.20J	Express Aviation Services Ltd.	
G-BJHB	Mooney M.20K	Express Aviation Services Ltd.	
G-BJRH	Rango NA.36 balloon	Rango Kite Co.	
G-BJSA	BN-2A Islander	Harvest Air Ltd.	
G-BJSP	Guido 1A Srs. 61 balloon	G. A. Newsome	
G-BJST	AT-16 Harvard IV	V. Norman & M. Lawrence	
G-BJSV	PA-28-161 Warrior II	A.F. Aviation Ltd.	
G-BJSY	Beech C90 King Air	Airmore Aviation Ltd.	
G-BJSZ	Piper J-3C-65 Cub	H. Gilbert	
Out-of-sequence registrations			
G-BZZZ	Enstrom F-28A	Chiltern Handbags Ltd. (G-BBBZ)	
G-GENE	Cessna 501 Citation	Falmer Aircraft Ltd.	
G-GOLF	SOCATA TB.10 Tobago	Pipe Chemical Holdings Ltd.	
G-HIRE	GA-7 Cougar	Cabair Ltd.	
G-JENN	AA-5B Tiger	Cabair Ltd.	
G-JLBI	Bell 206L-1 LongRanger	Tower Transport Ltd.	
G-KOOL	D.H.104 Devon C.2	J. D. Rees	
G-MBDN	Hornet Atlas	K. R. Wilson	
G-MBDP	Flexiform Sealander Skytrike	A. N. Baumber	
G-MBDX	Electra Eagle	Ardenco Ltd.	
G-MBEO	Flexiform Sealander	H. Williams	
G-MBFE	Eagle Rainbow Microlight	P. Cole	
G-MBFT	Sigma 12 Meter	D. P. Watts	
G-MBFU	Ultra Sports Tripacer	T. Prowse	
G-MBFV	Comet Skytrike	R. Willis	
G-MBFW	Hiway Skytrike	B. Bayes	
G-MBFX	Hiway Skytrike II	Noel Whittall Ltd.	
G-MBFY	Mirage Microlight II	J. P. Metcalf	
G-MBFZ	MSS Goldwing	I. T. Barr	
G-MBGA	Typhoon Tripacer 250	S. E. Bond	
G-MBGB	American Eagle Microlight	J. C. Miles	
G-MBGD	Pterodactyl 430C Replica	C. Wilkinson	
G-MBGE	Hiway Scorpion Trike	J. A. Rudd	
G-MBGF	Twamley Trike	R. W. Twamley	
G-MBGG	Chargus Titan 38	A. G. Doubtfire	
G-MBGH	Chargus T.250	A. G. Doubtfire	
G-MBGI	Chargus Titan 38	A. G. Doubtfire	
G-MBGJ	Hiway Skytrike Mk. 2	Peninsula Flight Ltd.	
G-MBGK	Electra Flyer Eagle	R. J. Osbourne	
G-MBGL	Flexiform Sealander Skytrike	H. Field	
G-MBGM	Eipper Quicksilver MX	R. Gill	
G-MBGN	Weedhopper Model A	D. Roberts	
G-MBGO	American Eagle Microlight	I. Willsher	
G-MBGP	Typhoon Skytrike	M. Collett	
G-MBGR	Goldwing Canard Microlight	I. D. Stokes	
G-MBGS	Rotec Rally 2B	P. C. Bell	
G-MBGT	Eagle Microlight	D. C. Lloyd	
G-MBGU	Chargus Microlight	I. D. B. Hayes	
G-MBGV	Skyhook Cutlass	D. M. Parsons	
G-MBGW	Hiway Skytrike	G. W. R. Cooke	
G-MBGX	Lightning Microlight	R. B. D. Baker	
G-MBGY	Hiway Demon Skytrike	W. Hopkins	
G-MBGZ	American Eagle Microlight	J. K. Davies	
G-MBHA	Trident Trike	P. Jackson	
G-MBHB	Moto Delta G.11	R. Leech & J. Earley	
G-MBHC	Lightning T.250	R. E. Worth	
G-MBHD	Vulcan Trike	D. Kiddy	
G-MBHE	Eagle Microlight	D. K. W. Paterson	
G-MBHF	Pterodactyl Ptraveller	D. B. Gurry	
G-MBHG	Solar Storm Buggy	A. J. Sharpe	

Notes	Reg.	Type	Owner or Operator
	G–MBHH	Flexiform Sealander Skytrike	G. J. Norris
	G–MBHI	Tripacer 250 Kamouette Atlas 18	P. T. Anstey
	G–MBHJ	Hornet Microlight	G. Haigh
	G–MBHK	Flexiform Skytrike	G. Barfoot
	G–MBHL	Skyhook Skytrike	C. R. Brewitt
	G–MBHM	Weedhopper Microlight	J. Hopkinson
	G–MBHN	Weedhopper Microlight	S. Hopkinson
	G–MBHP	Eagle Microlight II	P. V. Trollope & H. Caldwell
	G–MBHR	Flexiform Skytrike	Y. P. Ogborne
	G–MBHS	Flexiform Skysails	M. V. Rainford & J. Hardy
	G–MBHT	Chargus T.250	S. F. Dawe
	G–MBHU	Flexiform Hilander Skytrike	R. M. Strange
	G–MBHV	Pterodactyl Ptraveller	H. Partridge
	G–MBHW	Eagle Microlight	P. D. Lloyd-Davies
	G–MBHX	Pterodactyl Ptraveller	W. F. Tremayne
	G–MBHY	Pterodactyl Ptraveller	T. C. N. Carroll
	G–MBHZ	Pterodactyl Ptraveller	T. Deeming
	G–MBRB	Eagle Mk. I Microlight	R. C. Bott
	G–MJST	Pterodactyl Ptraveller	G. C. Sutton
	G–NAVY	D.H.104 Sea Devon C.20	J. S. Flavell & K. E. Fehrenbach
	G–NEIL	Thunder Ax3 balloon	Thunder Balloons Ltd.
	G–NICE	Short SD3-30	Eastern Airways Ltd.
	G–OLIN	PA-30 Twin Comanche 160B	Skyhawk Ltd. (G–AWMB)
	G–OMAV	AS.355F Twin Squirrel	Massallaz Helicopters Ltd.
	G–PINT	Cameron 65 barrel balloon	Charles Wells Ltd.
	G–REID	Rotorway Scorpion 133	J. Reid (G–BGAW)
	G–RENO	SOCATA TB.10 Tobago	East Bilney Garage Ltd.
	G–RORG	Bell 206A JetRanger	R. G. Woodward (G–AXMM)
	G–SKSA	Skyship 500 airship	Airship Industries Ltd.
	G–SOOO	PA-30 Twin Comanche 160	Leisair Avionics Ltd. (G–AXMY)
	G–TALL	SA. Jetstream 3001	British Aerospace Public Ltd. Co.
	G–TJET	Lockheed T-33A	Aces High Ltd.
	G–WELD	Hughes 369HS	Uni-Weld Ltd. (G–FROG)

Overseas addenda

Reg.	Type	Owner or Operator
A40–TT	L.1011-385 Tristar 209	Gulf Air
D–AIAF	A.300B2 Airbus	Lufthansa
EI–BHH	Cessna FRA.150L	H. Harold
EI–BLJ	Cessna T.210H	N. V. Holt Ltd.
EI–BLL	Cessna F.172P	J. J. Spollen
EI–BLM	Hiway Skytrike II	R. Hudson
EI–BLN	Eipper Quicksilver MX	B. A. Carpenter
EI–BLO	Microlight CP.16	R. W. Hall
EI–BLR	PA-34-200T Seneca	R. Paris
EI–BLS	Cessna 150M	Phoenix Flying
EI–BLT	PA-38-112 Tomahawk	B. A. Carpenter
EI–BLU	Evans VP-1	A. Bailey
F–GBRV	FH.227B Friendship	T.A.T.
F–GCFC	FH.227B Friendship	T.A.T.
HA–LBO	Tupolev Tu-134	Malev
HZ–AIF	Boeing 747SP-68	SAUDIA- Saudi Arabian Airlines
HZ–AIG	Boeing 747-168B	SAUDIA- Saudi Arabian Airlines
HZ–AIH	Boeing 747-168B	SAUDIA- Saudi Arabian Airlines
HZ–AII	Boeing 747-168B	SAUDIA- Saudi Arabian Airlines
N724DA	L.1011-385 TriStar 200	Delta Air Lines
00–RVM	Boeing 737-2Q8	Air Belgium (ex TF-VLK)
PH–MOL	F.28 Fellowship 1000	Air Alsace
RP–C2003	Douglas DC-10-30	Nigeria Airways
S2–ACK	Boeing 707-321C	Bangladesh Biman (ex 9K-ACU)
S2–ACM	Boeing 707-369C	Bangladesh Biman (ex 9K-ACK)
4R–TNK	L.1011-385 TriStar 100	Air Lanka (ex C–FTNK)
7T–VHG	L.100-30 Hercules	Air Algerie
7T–VHK	L.100–30 Hercules	Air Algerie
7T–VHL	L.100-30 Hercules	Air Algerie
9Q–CLY	Boeing 707-336C	G.K.N. Sankey Ltd. (ex G-Aylt)

Future Allocations Log (In-Sequence)

The grid provides the facility to record future in-sequence registrations as they are issued or seen. To trace a particular code, refer to the left hand column which contains the three letters following the G prefix. The final letter can be found by reading across the columns headed A to Z. For example, the box for G–BKAD is located five rows down (BKA) and then four across to the D column.

G-	A	B	C	D	E	F	G	H	I	J	K	L	M	N	O	P	R	S	T	U	V	W	X	Y	Z
BJW																									
BJX																									
BJY																									
BJZ																									
BKA																									
BKB																									
BKC																									
BKD																									
BKE																									
BKF																									
BKG																									
BKH																									
BKI																									
BKJ																									
BKK																									
BKL																									
BKM																									
BKN																									
BKO																									
BKP																									
BKR																									
BKS																									
BKT																									
BKU																									
BKV																									
BKW																									
BKX																									
BKY																									
BKZ																									
BLA																									
BLB																									
BLC																									
BLD																									
	A	B	C	D	E	F	G	H	I	J	K	L	M	N	O	P	R	S	T	U	V	W	X	Y	Z

Credit: *Wal Gandy*

Future Allocations Log (Out-of-Sequence)

This grid can be used to record out-of-sequence registrations as they are issued or seen. The first two columns are provided for the ranges prefixed with G–B, ie. from G–BLxx to G–BZxx. The remaining columns cover the sequences from G–Cxxx to G–Zxxx and in this case it is necessary to insert the last three letters in the appropriate section.

G-B	G-B	G-C	G-E	G-G	G-I	G-L	G-N	G-P	G-S	G-U
L	R									
M	S									
										G-V
					G-J					
										G-W
	T									
N							G-O			
		G-D	G-F	G-H		G-M		G-R		
	U									
									G-T	
	V									G-X
O										
	W									
					G-K					
										G-Y
	X									
P	Y									
										G-Z
	Z									

Credit: *Wal Gandy*